sanitary landfill design handbook

The Science and Art of Site Selection, Investigation and Design

by George Noble P.E.

TECHNOMIC PUBLISHING

a TECHNOMIC® publication

TECHNOMIC Publishing Co., Inc.

265 Post Road West, Westport, CT 06880

SANITARY LANDFILL DESIGN HANDBOOK

265 Post Road West, Westport, CT 06880

a TECHNOMIC® publication

Printed in U.S.A.
Library of Congress Card No. 76-1134
I.S.B.N. 0-87762-100-4

To

Richard W. Eldredge, P.E.

ABOUT THE AUTHOR

George Noble, P.E., Assoc. A.I.P., Diplomate A.A.E.E.

Mr. Noble has a degree in civil engineering from the University of London and graduate qualifications in town planning and land use planning from University College, London. He has been engaged in all aspects of solid waste management work from the preliminary design stage to on-site construction management. He has planned and designed landfills in several states for both private industry and municipal government. He has conducted statewide and countywide solid waste management plans in several states and conducted a number of solid waste inventories for industry in Iowa, Pennsylvania and Wisconsin.

Mr. Noble is a registered professional engineer in the State of Pennsylvania and also in the United Kingdom and Europe. He is presently Manager of the econENVIRONomics Department of Roy F. Weston Inc. in the Midwest Regional Office in Chicago, Illinois.

ACKNOWLEDGMENTS

I wish to acknowledge the assistance of a number of professional colleagues at Roy F. Weston, Inc. Many friends both in the West Chester, Philadelphia Office and in the Chicago Office have assisted in reviewing sections of this handbook and making invaluable suggestions for improvement. Whilst, I would not have them implicated in any of this book's shortcomings, I willingly acknowledge their contribution in all other areas. In particular, I wish to thank Walter Satterthwaite and Walter Beers of the Earth Sciences Department in West Chester for their assistance on the geology sections of the book. I owe special thanks to Ron Stephenson of the econENVIRONomics Department in Chicago for his invaluable assistance with the landscaping section and to Dr. David Arndorfer of the same Department for his assistance with some of the climatological discussion. I am also indebted to my good friends Dr. Derek Stubbs and Dr. Joe Bruno for their assistance with the first aid section. I am much indebted to Richard W. Eldredge, P.E. (now of Eldredge Engineering Associates, Inc.) for all I learned from him while we worked together at Roy F. Weston, Inc. and for his painstakingly (and sometimes painfully) thorough review of the draft and for the many suggestions he made for improving the text.

Finally, I wish to thank my wife, Sally, and my secretary, Liz Crosby, for the many hours of their own time spent typing the draft. Without that help and the many small improvements that went with it, there would be no final product.

PREFACE

For many years, there has been a surprising lack of good practical information on the subject of landfill design. Only the Federal EPA and several manufacturers of earth moving equipment have sought to fill the gap in any organized fashion. Many excellent research papers have also been published. In the early days of the Bureau of Solid Waste Management, there were several excellent training programs available on the subject of landfill design. Now instruction is confined chiefly to written material of a more general nature and in particular the excellent "Sanitary Landfill Design and Operation" by Brunner and Keller which is a good source of general information on the subject of landfill design.

In recent years attention has turned to more exotic forms of solid waste management and a parallel lack of respect for the old methods now seems to be in order. This is unfortunate from the perspective that no alternative method has yet proved itself to be both technically and economically feasible.

It is my conviction that sanitary landfill remains a safe and acceptable method of ultimate disposal. It is also my contention that it is time we gave it a serious try since the great majority of landfills presently in operation in this country cannot be considered sanitary landfills by any stretching of the definition.

Unfortunately, it often seems that the only thing to change in the last few years has been employment for sign painters to add the word sanitary before the word landfill. While this kind of action shows a thoughtful regard for public display, there is clearly more to a good landfill than a readiness to change the name.

I have tried in this book to "pull together" all the various pieces of information necessary to complete a good design. In addition, I have tried to examine the "sticky" question of "minimum depth to water table." At present, those few states in the forefront of attempts to regulate sanitary landfill design appear content to apply rules of thumb which vary from 2 feet of separation to 20 feet or even greater. I have always felt we could do better than that. Recent work suggests that the renovative action of the soil interface between water table and refuse base can be approximated reasonably well. I have proposed a methodology in the design chapter which can be used to "design" the depth to water table. I have no pretensions about the method. There is considerable room for improvement, but it does enable the designer to obtain a "feel" for the problems involved and the order of magnitude of separation required.

Finally, I have tried to offer useful and practical advice on many other aspects of design. All the nomographs, curves and design criteria necessary to "do a good job" are contained in the appendix including information which is now out of print or difficult to obtain. A simple method of estimating evapo-transpiration is also included in addition to many useful hints on landscaping methods.

In short, this is a working and practical handbook.

I hope it may make a modest contribution to the technology and help in the design of many good sanitary landfills.

The designer who searches these pages for information on the use of impermeable membranes to contain leachate will search in vain. The value of membranes in landfill design remains controversial. I do not see them as significant. It is my own feeling that membranes should be the last resort when all efforts at finding a suitable site have failed.

Many of the States contacted for the recent Waste Age survey of solid waste practices indicated that "any site selected, which is submitted with a liner design is an example of a poorly selected site. Liners are a last resort for areas where urbanization or geological configurations mandate the use of a site regardless of soils, geology or hydrogeology." Those responding in this manner indicated that site rejection because of liner requirements has more often than not resulted in a new site more suitably located.

I am uneasy about a technique which depends upon total "containment" since total collection and treatment of leachate must automatically follow. This arrangement would suggest that future landfill operators will be responsible for leachate treatment "in perpetuity" and I find that prospect not only in-appropriate, but inconceivable. I am not even sure that "containment" in perpetuity is a reasonable possibility since membrane rupture is an ever present risk. For this very reason, some states require that sites should be *designed* for membrane failure at some stage in their life and in that event, the site selection and design criteria are identical to those presently required for non-membrane landfills. Thus, we have the peculiar spectacle of landfills being located with all the advantages of natural renovation, but with an expensive membrane reducing the value of renovative capacity to zero. Except, of course, in an emergency when presumably the slow and measured process of leachate renovation is expected to act with the surety of a safety valve against a niagara of leachate flow from one point source. It is time we returned to a more rational approach.

Properly sited, correctly engineered and responsibly operated sanitary landfills are fully capable of effective disposal and "treatment" of residential waste. I hope this book may contribute to their continued and successful use.

Chicago, Illinois

George Noble, P.E.

TABLE OF CONTENTS

CHAPTER

TABLES

FIGURES

APPENDICES

Chapter 1

PUBLIC INVOLVEMENT

Public involvement in the selection and implementation of a proposed sanitary landfill is critical to the success of the enterprise. No matter how carefully selected and prepared the site or how carefully planned the proposed operation, unless the public is involved "before the event," the whole enterprise is at risk. Bite the bullet and get the public involved.

The days when a site could be chosen and the disposal operation put into effect without public consultation are numbered. In any event, State regulations which call for submission of detailed plans and public notice have essentially eliminated this practice. Even in states where no regulations currently exist, it is unlikely that our presently environmentally conscious society will allow preparations for such a controversial event as the opening of a landfill to go unannounced. Thus the invitation to get the public involved may be an invitation to the inevitable, but it is perhaps better to do it willingly rather than grudgingly.

Once the step is taken to present the case to the public, there is no turning back. The acceptance or rejection of the case for sanitary landfill depends upon the strength of the argument and from the moment the step is taken, the standard of public presentation had better be good. It can be confidently expected that the case will be subjected to public damnation by many who would prefer the problem of solid waste disposal to just go away. However, as the following pages show, it is not at all impossible to make a good case for sanitary landfill, particularly in the present state of the art of alternative technology. Also, as much as many would prefer the problem to leave, there is a ready acceptance among the public that something must be done to get rid of solid waste and whatever that something is, it should be efficient and preferably not too expensive.

To summarize the situation so far, anyone contemplating establishing a new landfill who wishes to reduce the risk of failure, should involve the public from the beginning, regardless of whether or not it is a requirement. Every effort should be made that the case be a good one. Properly involved and completely informed, the public will naturally be more inclined to accept a good plan than a bad one.

Let us assume (with all the benefit of the knowledge and experience contained in this handbook) that the plan is a good one. What happens next? How is the public to become involved? What are the pitfalls and where is the path to success?

The simple answer is that ideally you do not develop a good plan first and get the public involved afterwards. In fact, that may be a prime example of a good plan lost by lack of forethought. Difficult though it may seem, the best insurance against rejection is *early* public involvement. This should be the first principle of public participation.

1. Get the Public Involved Early

For many proposed landfills, this may not be too difficult. In many states in the nation it is now essential that some form of solid waste management plan should be developed in order to plan for the orderly phase-out of old dumps and the phase-in of new landfills. There is scarcely one of these solid waste management plans which does not require a sanitary landfill at some stage in the disposal plan.

Very often a statewide or a county wide plan requires a number of stategically located landfills. These are often fairly large undertakings designed to make maximum benefit of any savings in scale involved. If the plan is a good one, it will present a number of carefully evaluated alternatives and present logical reasons for coming to the final decision.

This is an ideal beginning for public involvement. Often the consultant responsible for developing the plan will take the first step towards public participation by calling the first public meeting. Indeed many state planning regulations require a public hearing at the conclusion of the solid waste management study. Landfill operators should take the lead from these plans. They frequently present valuable cost information in support of the final conclusions. If the plan is fair and impartial the final conclusions will be hard to contradict. Landfills in the general areas proposed by State, Regional or County plans therefore have a head start on those for which the case must be developed from scratch.

If the consultant is not required to hold a public meeting, then a meeting should be arranged to bring local groups together to discuss the subject of solid waste management. The purpose of such a meeting should be to "take the pulse" of local feeling on the subject of sanitary landfills. It will no doubt bring up some astonishing gaps in knowledge on the subject of sanitary landfill and if landfill is the right way to go for the communities involved, this simply defines the amount of public education which will be required to get started when the time comes.

It may be thought that a general public meeting this early in the game is a waste of time and effort. Under certain circumstances this may be so, particularly for the private operator who knows his territory very well. In many instances he may already know who the opposition will be, where the problem areas are and who might be the best allies. This is all one can hope to achieve from a general public meeting, except that very often the mere fact of holding a meeting is a small plus in the public mind. For the municipal operation an initial public meeting is a must if the municipal government has any hope of successfully establishing a new landfill.

In these days of municipal encouragement of volunteer endeavour, especially in the solid waste business, it pays to keep the public informed.

Private or Public, the format for the initial public meeting might be as follows:

> Public Meeting
> Our Solid Waste Problems
>
> 1. Introductions
> 2. The Solid Waste Problem, National and Local
> 3. Where we are now — Present methods of disposal
>   How long will they last?
>   Present problems
> 4. Where are we going — Possible alternatives
>   Costs to you, the user
> 5. General discussion

If possible, the local press should be present and others who may be expected to feel strongly about the subject should be personally invited to attend. Obviously care should be taken that adequate notification of the meeting should be given, (a minimum of three weeks). The venue should be as local as possible and the data should not clash with major local or national events.

The addition of guest speakers or films is probably not appropriate at this point since both tend to have a case to make. The whole tone of this initial meeting should be neutral. The meeting is intended to acquaint members of the public with the problems and generally measure reaction to alternative solutions. Remember the public has a right to know.

For dyed-in-the-wool landfill operators the meeting may not always be a one hundred percent success. The public has been lately nurtured by a diet of resource recovery and hazardous waste. There is very properly much concern about the way landfills have been operated in the past and naturally this has led many to look in the direction of viable alternatives. For large municipalities or regions with a collection area representing about 200,000 persons or more, there may be some benefits in resource recovery and it is important that they be examined. However, for the great majority of systems, sanitary landfill represents the most reliable method of ultimate disposal currently available. In areas large enough to support a growing volunteer demand for resource recovery, there are several areas where active volunteer groups can make valuable contributions and reduce the ultimate volume going to landfill. This type of effort can substantially prolong the life of a landfill site and also, incidentally, reduce the number of items going to the landfill which are subject to wind loss and consequent litter problems.

At this point we will assume that all that can be done in the way of reducing the total load going to the landfill has been done. This should be stressed at the public meeting, since there remains a tendency in the public mind to consider that recycle is somehow a total alternative to landfill.

The second principle of public participation is as follows:

2. Have Good Graphics

In general the public is only faintly aware of the problems involved in the business of solid waste disposal. The initial public meeting is not only a trial balloon to obtain a measure of the opposition, but also an exercise in public education. Aim for the best visual aids which can be produced to illustrate the story. Wherever possible, graphic material should be prepared for the occasion and not "store bought." Somehow graphics which have been used for something else never quite have the same freshness and local appeal needed in critical situations like this. It is worthwhile investing some time and effort in putting together some originals including, if possible, some pictures of successful landfill operations.

To some extent the results of the meeting may be very predictable, but nevertheless, the fact of having a meeting is invaluable in itself. The most important feature of this exposure is that it is clearly in support of the basic principle of getting the public involved early. You obtain a feel for specific sensitivities in the area and you can start to plan your specific program for public participation immediately after the meeting.

Before that it might be wise to make sure that news of the event has been transmitted to as many local newspapers as possible. It is important that the process of public participation should be seen to be done. To the extent that you have been successful in inviting other members of the media, you may also acquire some further publicity from radio or television.

Also, if the invitation campaign was successful, the meeting should have been attended by a number of local interest groups such as the League of Women Voters and general environmental interests. After the meeting, while the issues are still fresh and new, some mention should be made to these interest groups of possible future meetings to talk about specific sites.

Veterans of public meetings will realize at this stage that the audience at this preliminary meeting, where philosophies are being discussed rather than issues, will be predominantly thoughtful public citizens, rather than locals aroused by specific issues. It could hardly be otherwise. General areas where suitable sites might be found may be discussed at this first meeting, but with nothing more specific, local battle is not ready to be joined. However, thoughtful public citizens are good allies and every effort should be made to stay close to them in the months ahead.

After the meeting and after a certain amount of reappraisal of general response, the next step is to develop a strategy for landfill siting. Chapter 3 deals with both technical and non-technical factors which are important in the selection of landfill sites. In general, as Chapter 3 illustrates, the complete fulfullment of the technical requirements is a worthy aim, but this counts for nothing if the site is politically in the wrong place. With the guidance of Chapter 3 we will assume everything has been done to choose the best spot for the new landfill. What happens next?

Slowly, quietly and with the minimum of fuss, prospects should be discussed with any potential neighbors of the site. It is still an option, and it is important to find out how they feel about it. Public or private, senior officials ought to do this as evidence of interest in the site and more particularly to make a commitment as to the way the site could be run, the time frame (vital information) and the final use of the site. Outlined drawings of typical final uses are very important at this stage, but they must be realistic and even a little conservative in terms of what can be achieved. Artists renderings of play areas which look as though they were designed for Buckingham Palace are as likely to arouse hostility as to make friends. Above all, the main aim of these meetings will be to show willingness to listen. After all this, you go away and check whether your site decision was the best one under the circumstances. If you think it was, speed is essential. Now get your supporters lined up and ready for the announcement. Political representatives in the community and other important figures should be notified ahead of time that an announcement will be made. Then make it. The local newspaper is probably best, but radio and television may be only too ready to climb aboard at this stage. "As a result of extensive investigation. . .etc. the site selected for sanitary landfill is Murky Hollow" (or wherever).

By the time of the announcement, all those to whom it might come as a shock will have been notified. Those who were most concerned about the prospect during previous visits should now be contacted again. Invite them to the next meeting and start preparations for the most important presentation of the whole campaign.

At this point the options must be considered dead issues. Forget them. There exists a kind of "green pastures" law when it comes to landfill, which states that for the purposes of solid waste disposal, the land in the neighbors' field is always more suited than the field selected. The burden of proof is yours. Your decision must now be justified on the grounds that:

- it can solve the problems of the community.
- it can be engineered and operated in such a way that it will not constitute a nuisance.
- the final use for the site will be a real value to the community.

Above all, stay close to the neighbors. Very properly it is they who have most concern about the impact upon their own community and its property values. As the chapter on site design illustrates, there are a great many ways in which the site can be designed to literally put your best face forward. You can even phase-out in stages so that the community does not have to wait forever for you to come through on promises to provide a playground. With care and attention, to some simple rules, it is possible to operate a model landfill. You could not purchase a better insurance against finding yourself without a site in the future. Everything done on the site should indicate to the public that you kept the faith that it would be a good operation.

Speed is of the essence in the next few months. No doubt the press announcements will have generated some concerns not covered by selective meetings. In as much as those who are concerned can be identified, efforts should be made to discuss problems with them. For preference, always meet face to face, rather than at meetings. However, sometimes meetings must be held almost as punctuation marks in a sentence, to show that the process is moving forward.

The next meeting, for example should mark the turning point from general to specific. The same techniques for public notification should be used as previously described. Now the agenda for the meeting should be somewhat as follows:

Public Meeting
The Proposed Murky Hollow Sanitary Landfill

1. Introductions
2. Recap of the last meeting
3. Results of detailed investigation
4. The proposed plan
   a. Plan of operation
   b. Proposed controls
   c. Phased public use
   d. Final use
5. General discussion

If there is massive public opposition, even after your carefully laid plans to meet with all interested parties, this is going to be a tough meeting. Prepare for it with care. Graphics to illustrate the plans should be clear, simple and unambiguous. Speakers should be well rehearsed and capable of picking up the general tenor and style of the meeting. There are a number of small details to keep in mind about meetings of this type and when the outcome is as important as it is in this case, it is worth planning ahead a little. The following pointers may help:

1. Avoid adversary proceedings. The local court room is often chosen for such meetings, with disastrous results. The public is physically isolated from the speakers by the bar and the whole place adds an unwanted air of formality to the proceedings. Usually you have to be up front with a table for water and an easel for making presentations but have others speak from the floor to add a general air of informality to the meeting.
2. Do not be totally identified as an outsider by your dress. Consultants commonly

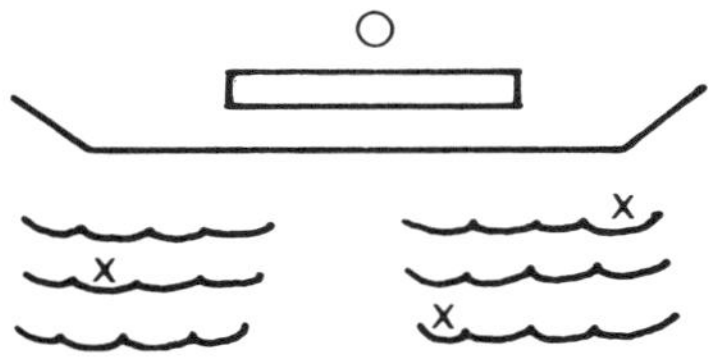

Have others speak from the floor to add informality to the proceedings.

Figure 1. Courtroom layout for effective public meetings.

make the mistake of overdressing for these occasions. There is no need to overdo the camaraderie to the extent of being conspicuous, but if informality is the key at least dispense with jackets for a period.

3. Take a tip out of the politicians handbook and specifically "plant" questions. There is nothing wrong with this approach. It is a very effective way of emphasizing important aspects of the plan. It also helps to generate questions when there is some hesitation to take the plunge.

By the end of the meeting you should know whether or not you have a fight on your hands. If the all important preparatory work behind the scenes was done right, there should be few problems at this stage. If opposition turns out to be tough and well organized, even careful groundwork may not do the trick. What happens next depends upon whether this is a public or a private proposal.

Paradoxically the public operation has the disadvantage in situations of this kind. Rights to eminent domain may be exercised by public bodies and as we will see later in considering Legal Aspects this often leads to exclusion from zoning regulations. However, eminent domain is rarely a popular means of establishing a site since the exercise of the right is often considered a moral defeat. So far in this discussion, we have not discussed the zoning issue since the assumption is made that regardless of zoning, the public still impacts the final decision. Much of the foregoing public involvement may superficially center around a dispute over zoning. However, the real issue is landfill siting and the real concern is public education.

In the public arena, the risk of controversy becoming "political" is very great. For this reason, election periods should be carefully avoided when planning the foregoing series of public meetings. However, the "open season" for political controversy is sometimes very extensive and notwithstanding all carefully laid plans the landfill issue may become political and sanitary landfill is hardly ever a "popular" issue.

Obviously, for a proposed public landfill plan to become a political football is a serious embarrassment particularly to the well intentioned staff who put the plan together. If the plan has been the subject of careful lobbying and forewarning in the manner previously discussed and the issue still becomes political, there is very little that the well-intentioned landfill planner can do. No amount of earnest discussion of the needs and the lack of alternatives seems to make much impression upon politicians in full pursuit of a "popular" issue. I can only suggest that the would be operator accept defeat gratefully and await a better political climate.

For the private operator who faces the same controversy, the only solution is to go underground. This may seem extraordinary advice under the circumstances since the last several pages have been bent on putting the case for public disclosure. However, at this point, public disclosure has failed (at least temporarily) and a discrete exit to gather support for a counter-attack seems appropriate.

Outside the political arena (and the private operator can usually opt out very easily by simply taking the issue out of public view), it may become clear what the next move should be. Almost certainly a delay of several months may be necessary. For the new operator with new customers waiting an alternative site might be the best solution. For the seasoned operator who can "hold out" for a period, a few months wait is generally the best first step in the offensive.

In some instances, where an election is involved, the political problems may solve themselves. After a discrete wait in memory of departed politicians, it may then be appropriate to begin again the round of public counseling and consultation. If the political forces simply become more entrenched, even the seasoned operator may feel it politic to move on to another site.

Personally, I feel that whenever the resurrected plan puts in an appearance, it ought to be different in many important respects. If the earlier plan ran into trouble, it will still carry a few battle scars and there is no sense in suffering the same wounds twice. If there were concerns about litter, I would think it important to take another look at the litter control measures. Perhaps the moveable litter rack could be bigger. Maybe is should be redesigned. Perhaps it should be a continuous flexible net across the critical end of the proposed landfill.

If noise was a problem, perhaps the site could be replanned to "back away" from the sensitive area. A landscape mound makes a good dense noise barrier. The noise reducing capability of trees is largely illusory, but if trees cut out the *view* of the noise makers, it works.

Whatever the public concerns, there are generally concessions which can be made to address those issues. If the site is a re-emerging issue, I suggest it is only politic to make those concessions.

Chapter 2

FUNDAMENTALS

The fundamental principle of the sanitary landfill method of solid waste disposal is to achieve efficient disposal with minimum environmental impact. It is the aim of this book to show that this is feasible and that the generally poor publicity which the method has suffered is largely due to bad practice.

There are many important features of good landfill practice. In this chapter, it is not the intention to review them all. The aim here is to "nail down" the fundamentals so that we know what we are trying to achieve in discussing finding a site and designing the landfill in later chapters.

The often quoted basic definition of sanitary landfill comes from a Sanitary Landfill Manual of Practice, prepared by the American Society of Civil Engineers (ASCE) in 1959. It is still today an excellent working definition: "Sanitary Landfill is a method of disposing of refuse on land without creating nuisances or hazards to public health or safety, by utilizing the principles of engineering to confine the refuse to the smallest practical area, to reduce it to the smallest practical volume, and to cover it with a layer of earth at the conclusion of each day's operation or at such more frequent intervals as may be necessary."[1]

The two basic principles which are embodied in this definition are:

1. Compaction
2. Cover

Any method of landfill must seek to achieve these two principles at a minimum. Compaction achieves an efficient use of land by reducing the volume of the solid waste to be buried. Cover achieves the necessary "burial" of the waste for aesthetic and health reasons.

Let us assume for the moment that this is done. What happens to the solid waste compacted, buried and essentially forgotten?

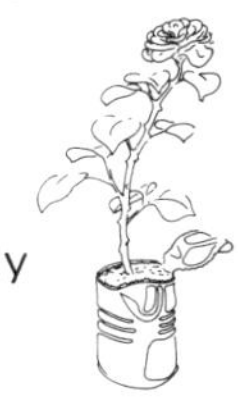

Decomposition

Generally, the landfill contains a heterogeneous mix of wastes; probably chiefly

[1] Sanitary Landfill Manual of Practice. American Society of Civil Engineers 1959.

of residential origin. Food wastes will be broken down by micro-organisms present in the soil. Metal wastes will be oxidized and by this process, some metals such as aluminum and copper will be protected from further reduction by the development of protective oxides. Ferrous metals will ultimately be completely oxidized. Organic wastes of all kinds will be broken down by microbial action. Inorganic wastes may remain inert or be subject to breakdown by chemical action. The following table summarizes these activities.

TABLE 1

Processes of First Stage Solids Decomposition in a Sanitary Landfill

Waste	Agent	Product
Food	Micro organisms	
Metals	Oxidation	Metal Compounds
Organic	Micro organisms	
Inorganic	Chemical Action	Complex Compounds
Inert Materials	All	Small Changes

Clearly few of these reactions occur in isolation. In practice, the mixing of the waste ensures that these activities take place in continual interaction. In fact, efficiency of the processes is significantly improved by this mixing and interaction.

It is also important to realize that because these activities take place in the body of the landfill, they are essentially controlled by that environment. The major controlling factors are:

- oxygen
- temperature
- water

Oxygen

The availability of oxygen clearly controls the period of activity of oxygen-using micro-organisms (aerobic bacteria). The amount of oxygen is quite limited in a landfill, since it is only available from air spaces, dissolved in water, and as a bi-product of other chemical reactions. During the transitional period, when the oxygen supply is falling off, bacteria thrive which can adapt to conditions in which there is very little oxygen (facultative anaerobic bacteria). Finally, when all the oxygen has been used up, micro organisms which thrive only in conditions of no oxygen, begin to develop and continue the process of breakdown (anaerobic bacteria).

It is important to realize that these groups of bacteria are only broad general classifications. In fact, in each group there are thousands of different kinds of bacteria. It is merely convenient in this discussion to classify them according to oxygen utilization. For the purposes of summary, the classifications are repeated in Table 2.

TABLE 2

Broad Classification of Bacteria Present in a Sanitary Landfill

Bacteria Type	Classification (According to Oxygen Use)
Aerobic	Requires oxygen
Faculative anaerobic	Adaptable
Anaerobic	Feeble in presence of oxygen.

The way in which bacteria breakdown organic solid waste is not unlike the way in which we digest food. The bacteria possess the capacity to manufacture enzymes which are used to convert insoluble material into a soluble form which can be readily absorbed through the membrane of the cell wall.

The specific types of bacteria which thrive in the landfill at any one time are entirely dependent upon the environment which exists in the landfill. Different types of bacteria will develop at different temperatures, moisture content and waste characteristics. Clearly all types of bacteria will probably exist in the landfill at the same time since the small environments (micro-environments) in the landfill are many and varied.

It may be helpful to consider here the final products of the breakdown effected by the bacteria. This depends upon the waste itself. Basically, the organic material may be classified as foodstuffs or materials continuing cellulose, such as paper and rags. The foodstuffs are, of course, either protein, carbohydrates, fats or mineral salts. Most of the proteins are initially decomposed by aerobes into organic acids and other complex organic compounds. In turn, these are further broken down by anaerobic bacteria. Considerable heat is liberated in these reactions. Carbohydrates are typically decomposed by aerobic bacteria into glucose which in turn may be further broken down, either by aerobic or anaerobic activity. It is in the bacterial reduction of fats that methane is released as a bi-product. In a sewage treatment works, this reaction is commonly controlled in a sludge digestion tank and methane may be collected and used as an energy resource. In a sanitary landfill, it is difficult to control, though some efforts have been made to collect the gas[2]. Carbon di-

[2] "Garbage Routes to Methane", Larry J. Ricci, Chemical Engineering, May 27, 1974.

oxide, water, hydrogen sulfide and heat are common byproducts of the breakdown of protein, carbohydrates and fats and their release into the complex landfill environment furthers the chemical change of much of the inorganic material in the fill.

Cellulose materials are decomposed by the action of the chemical environment and also by the action of bacterial enzymes. More detailed discussion of the specific chemical and bacteriological activity in a landfill can be obtained from an excellent publication of the Institute of Public Cleansing in England, called "Notes on the Science and Practice of Controlled Tipping of Refuse." A series of charts from that publication[3] showing the processes of decomposition of domestic solid waste in a sanitary landfill, are reproduced here as a summary of the foregoing.

Temperature

Temperature also controls the type of bacteria which thrive and grow in the landfill. During the period of aerobic decomposition only oxygen using bacteria are present and the heat bi-product of their activity further elevates the initial temperature. Thus a rise in temperature in the body of the landfill is an early indication of

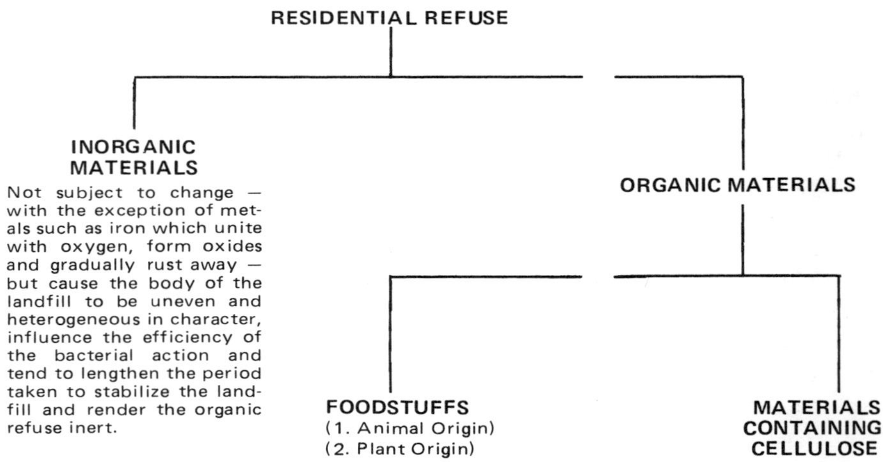

Figure 2. Chart showing the Processes of Decomposition of Residential Refuse into Simpler Elements (Gases and Water) caused by the Action of Bacteria and Bacterial Enzymes in a Sanitary Landfill.

Source: Figures 2, 3 and 4 are based on information contained in "Notes on the Science and Practice of Controlled Tipping of Refuse", by R. E. Bevan, Published by the Institute of Public Cleansing, U.K. 1967.

---

[3] With minor modifications to account for the differences in technical jargon.

FOODSTUFFS
(1. Of Animal Origin – Examples: Bacon, Beef, Fish)
(2. Of Plant Origin – Examples: Apples, Beans, Cabbage)

- **PROTEIN-ALBUMEN**
  Consists of Complex Nitrogenous Compounds in form of Amino-Acids
  - **LEUCINE**
    Aerobic bacteria, proteolytic enzymes and oxygen act on this amino-acid, which is split up into carbon dioxide gas, water, and ammonia gas. Considerable heat is given out (755 Calories) during this process.
  - **GLYCOCOLL**
    Is decomposed by aerobic bacteria much in the same way as leucine and into similar gases and water. Again heat is liberated (152 Calories).
  - **SULPHUR COMPOUNDS**
    Sulphur compounds are frequently present in protein. Bacteria and their enzymes reduce sulphur compounds first to sulphuretted hydrogen gas, then secondly to sulphuric acid or to sulphur and water.
- **CARBOHYDRATES**
  Sugar, Starch, etc.
  Aerobic bacteria and bacterial enzyme – diastase – change the carbohydrate into the gum, dextrin, and glucose.
  Glucose is decomposed in one of two ways:
  1. Oxidation by aerobic bacteria into carbon dioxide gas and water, with the liberation of heat.

  or

  2. An aerobic lactic acid bacteria will rearrange the glucose molecule into the molecule of lactic acid.

  Lactic acid is reduced to acetic acid, carbon dioxide, and water.

  Finally acetic acid is oxidized to carbon dioxide and water.
- **FATS**
  Bacterial action through the bacterial enzyme, lipase, splits up the fat into acids and Glycerol (glycerine).
  - **FATTY ACIDS**
    Generally are oxidized into carbon dioxide gas and methane gas by a biological oxidation involving the addition of water. An alternative method of reduction, by the enzyme lipase, produces alcohols and thence hydrogen gas.
  - **GLYCEROL**
    Is first oxidized by B. xylinum to dihydroxyacetone, water being split off.

    Dihydroxyacetone is changed to one of three forms:
    1. By the loss of a molecule of water methyl glyoxal is formed;

    or

    2. A rearrangement of the molecule produces lactic acid;

    or

    3. The dihydroxyacetone is split up into ethyl alcohol and carbon dioxide.

    The product of the reduction of dihydroxyacetone, whether 1, 2, or 3 is oxidized into acetic acid and carbon dioxide.

    Acetic acid is broken down into carbon dioxide and water or into carbon dioxide and methane.
- **MINERAL SALTS**
  Exist in the foodstuffs in very small quantities only. Have a value as manurial constituents.
- **MOISTURE**
  Is essential to all processes of decomposition. In practically every one of the final steps of reduction of foodstuffs some water is released and in many of the earlier processes also. The surplus water helps to start the reduction of cellulose matter, later being again released.

Figure 3. Decomposition of foodstuffs.

MATERIALS CONTAINING CELLULOSE

Paper, Rags, Fruit Skins, etc.

ALPHA (PURE) CELLULOSE

The bacterial enzyme cellulase adds on the elements of water, transforming the cellulose into the disaccharide (sugar) cellobiose.

By the bacterial enzyme, cellobiase, a further molecule of water is added to produce glucose.

Glucose is decomposed in one of two ways:

1. Oxidation by aerobic bacteria into carbon dioxide gas and water with the liberation of heat (674 Calories).

or

2. An aerobic lactic acid bacteria will rearrange the glucose molecule into the molecule of lactic acid.

Lactic acid is reduced to acetic acid, carbon dioxide and water.

Finally acetic acid is oxidised to carbon dioxide and water.

COMPLEX CELLULOSES

(Hemicellulose and Ligno-Cellulose)

Are reduced in a similar manner to alpha cellulose but, in addition to carbon dioxide gas, the gases methane and hydrogen are evolved.

PECTIN

Two actions are possible in the decomposition of pectin:

1. The aerobic organism pectinobacter amilophilum can produce carbon dioxide and hydrogen gases with small portions of formic acid and alcohol;

and

2. The an aerobic granulobacter pectinovorum can yield butyric acid and alcohol with small amounts of carbon dioxide and hydrogen.

Figure 4. Decomposition of materials containing cellulose.

extensive bacterial action. Within days of applying cover, temperatures may be expected to rise. Peak temperatures of around 160°F can be achieved in anything from a few days to several weeks, after which there is a slow decline as the contents of the cell stabilize.

The bacteria which produce this initial burst of heat are aerobes, which liberate heat energy previously stored by the plant and animal material. As a result of this activity, the oxygen content of the landfill is drastically reduced and faculative and anaerobic bacteria begin to take over. This takeover is marked by a rapid decline in temperature and the beginning of methane production.

Water

Water is also a controlling factor in the bacteriological micro climate. It is essential to nearly all processes of decomposition. In the initial stages of landfill

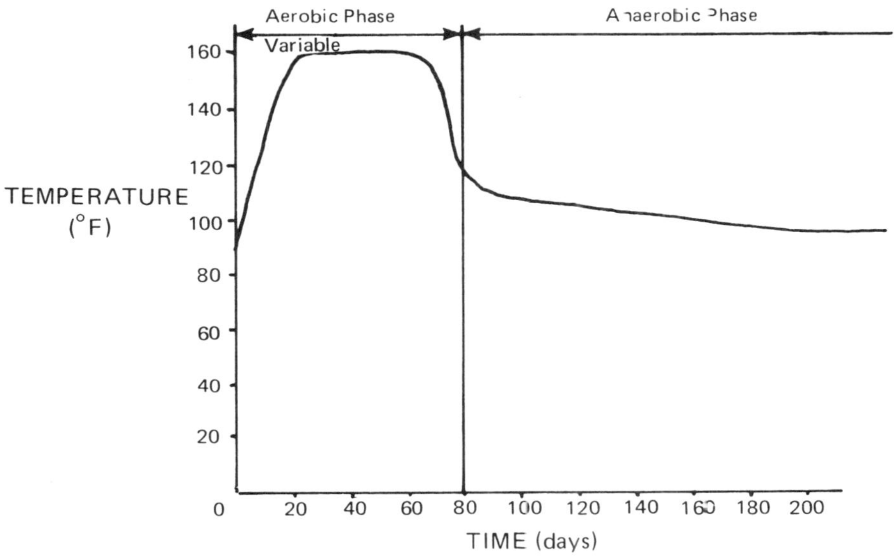

Figure 5. Average temperature range of landfill decomposition.

development, there is rarely adequate moisture available in the material. Moisture is added by rainfall and also by subsequent bacterial action. In general 25% moisture is the required minimum for decomposition to begin[4]. Residential refuse contains 15% to 20% moisture and thorough mixing on the site will reduce the overall percent moisture to less than 15%.

Moisture content also determines the rate of refuse decomposition. The maximum rate of microbial decomposition has been observed to occur at a moisture content of 60%.[4]

Of these three control factors under discussion (oxygen, temperature and water), water is the most difficult to maintain at an even rate. The problem is that unlike oxygen and temperature, failure to control water in the fill can result in environmental impact external to the landfill site.

A substantial amount of research has been undertaken in this field since water discharging from a landfill (leachate) has been characterized as highly contaminated.

In order to understand the mechanism of leachate generation and movement, we must start with a discussion of the natural processes involved in the whole cycle of water movement in the natural environment.

[4] "Notes on the Science and Practice of Controlled Tipping of Refuse", by R. E. Bevan.

The Hydrologic Cycle

It is important to distinguish between the various processes which occur as a result of rain falling on the earth's surface. *Infiltration, percolation, evaporation,* and *transpiration* are all part of the hydrologic cycle. They may be defined as shown in Table 3.

TABLE 3

Processes of the Hydrologic Cycle

Process	Definition
Infiltration	Passage of water into the soil surface.
Percolation	Movement of water through the soil profile.
Evaporation	Moisture returned to the air by the heat of the sun vaporizing the rainfall.
Transpiration	Moisture returned to the air as water vapor from the surface of plants.

Although it is not strictly part of the hydrologic cycle, the first thing to occur when rainfall hits the ground will be *runoff*. Simply stated, the rainfall will flow over the ground at a velocity determined by the *slope* of the land and limited by the *roughness* of the surface. In landfill design, the velocity of runoff is important because soil particles can be eroded from the surface by runoff if the slope is too steep or the soil easily eroded. The problems of erosion control will be discussed in more detail in the section dealing with landfill design.

The *quantity* of runoff is of major importance to landfill design. Careful design of the slopes of the finished site as well as the intermediate stages of site construction can significantly increase the amount of rainfall which will run off the soil surface. The proportion of a given quantity of rainfall which will run off is determined by the slope of the land and the characteristics of the soil. Design criteria suitable for estimating slope requirements for specific soil types are presented in the design section.

Of the water remaining, after runoff a certain amount will be lost due to *evaporation* and *transpiration*. Techniques to estimate these quantities for various geographic locations will also be discussed in the chapter dealing with detailed design.

The total hydrologic cycle may be represented by Figure 6.

The prime objective of good landfill practice is to minimize the amount of water infiltrating the soil and percolating through the refuse to produce *leachate.* However, some water inevitably soaks through and we must consider what happens to water which does infiltrate the cover material and percolate through the landfill.

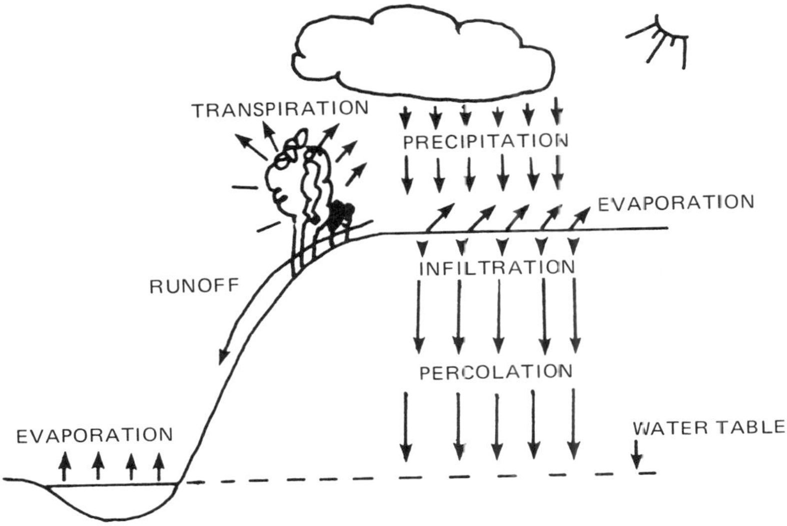

Figure 6. The hydrologic cycle.

During construction moisture is "trapped" in the landfill as a result of rainfall occurring during the operation. After the application of cover material, subsequent lifts are built up and final cover is applied. Where infiltration exceeds evapotranspiration and runoff, some additional water will penetrate the landfill. Initially the water is absorbed by the material in the landfill itself. While this absorption is taking place, there may be some additional compaction of the fill due to settlement. This compaction may slightly reduce the absorptive capacity of the contents.

After a period of 2 to 20 years (considerable differences are indicated in the literature between one site and another) the full absorptive capacity of the contents is reached. This is known as the field capacity. It is important to note that this is not the same as saturation. The following diagram of subsurface water zones will assist in defining terms.

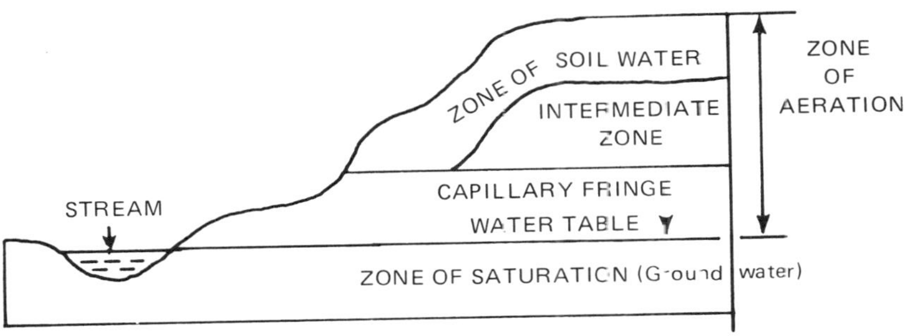

Figure 7. Subsurface water zones.

A sanitary landfill, by definition, should not be constructed below the water table. It should be located in the zone of aeration in the intermediate zone. A considerable volume of the landfill may remain rich in oxygen during the period of moisture absorption, since many of the spaces in the refuse will remain open. Aeration continues to take place during landfill operation, as a result of the draw down effect of water movement. When field capacity is reached additional percolation has the effect of displacing roughly the same amount of moisture from the landfill. The moisture displaced in this way is termed leachate. Leachate continues in its downward movement through intervening soil layers until it reaches the water table. At this point it passes into the ground water system.

We may construct a simple water budget for a sanitary landfill as shown in Figure 8.

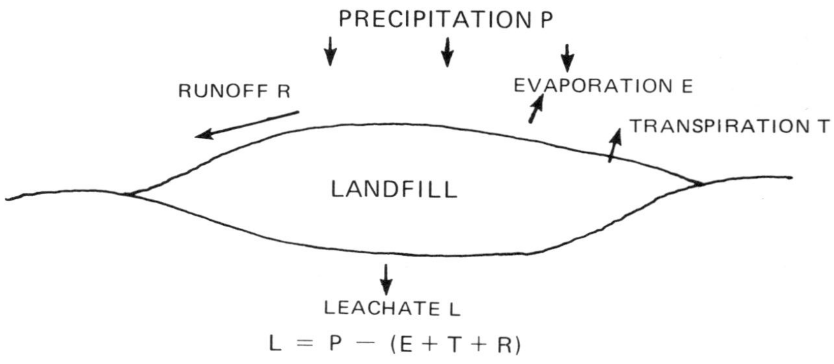

Figure 8. Water budget for a sanitary landfill.

Leachate

It should be clear from the foregoing that Leachate (L) is not always significant since many areas of the country lack sufficient rainfall to establish such a surplus. However, there are places in the nation and times of the year when liquid discharge will occur from some landfills. So we must turn our attention to what might be contained in this discharge (leachate) and what happens to it on its subterranean journey.

Many research attempts have been made to characterize leachate. In almost all cases a liquid leachate is collected directly from the base of a landfill or a simulated model. Subject to analysis the leachate characteristics are extremely varied. Commonly there is little control over the quantity and characteristics of input solid waste so it is perhaps not surprising that output characteristics vary widely. Also, disposal practice varies so widely from one test site to another that it is difficult to know whether the leachate characteristics are not more appropriately those of an open dump than a sanitary landfill.

All one can really say is that it seems clear from the literature that leachate taken directly from the base of a landfill may be abnormally high in the following constituents:

Iron (Ferric and Ferrous)
Chlorides

Leachate may also be high in:

Organic nitrogen
Phosphate
Sulphate
Hardness
Total Solids (Dissolved and Suspended Solids)
BOD (Biochemical Oxygen Demand)
COD (Chemical Oxygen Demand)
Trace Metals

It is an important principle of good landfill design that consideration must be given to the potential generation of a discharge of this type.

The main method of control is to limit infiltration by drainage design and suitable planting. The second line of defense depends upon natural renovation in the soil.

Natural Renovation

A number of researchers have shown that natural renovation can be an effective method of treating leachate, provided adequate care is taken in the siting and the subsequent operation of the fill. We can understand the concerns for careful siting and operation from the following discussion of the various mechanisms of leachate renovation.

Filtration – This describes the physical retention by the soil of suspended solids in the leachate. The soil separation layer between the landfill base and the water table simply acts as a large thick filter paper. The soil filters out solid particles above certain dimensions determined by the soil pore size. In addition, with the increase in solids retention by the soil, increasingly smaller solids in the leachate can be filtered out as the pore space is effectively reduced.

Solids retained in this way, which are largely organic in nature, are broken down by the action of soil bacteria. Thus the system is effectively self cleaning. However, the filtration system is only effective in "cleaning out" solid particles plus some colloids above a certain size. As a result of biological action in the landfill itself, the leachate contains organic and inorganic material in solution which quickly pass through the filtering system. Fortunately most soils have the capacity to soak up liquids and hold them for a period of time rather like a sponge.

Absorption – The ability of the soil to retain moisture in this way is not a treatment facility in itself, but the phenomenon does serve to retain moisture long enough for other treatment processes to occur. These treatment processes may depend upon the presence of suitable soil bacteria or they may depend upon the presence of chemicals in the soil particles which will react with the leachate to produce chemicals more amenable to breakdown. Another more important means by which dissolved pollutants may be retained by the soil structure until chemical or biological treatment can be effective is adsorption.

Adsorption – This phenomenon is often confused figuratively and literally with absorption. Adsorption describes the mechanism by which molecules may be "held" on the surface of a substance. This phenomenon occurs between liquids, gases and solids, but here we are only concerned with adsorption as it applies to liquid leachate passing over solid soil particles.

Adsorption occurs when a charged molecule (called an ion) in the leachate passes over a soil particle which contains a site possessing an opposite charge. Clays and organic materials possess a number of charged sites due to their molecular make-up. Positively charged molecules are referred to as cations and negatively charged molecules as anions. Strictly speaking, both positively and negatively charged sites can be found in the soil.

The dominant state of charge in clay minerals is a negative one. This means that any "free" negative sites in the clay will be filled by positively charged ions in the leachate. In practice, of course, these sites are often filled as a result of the normal percolation of moisture through the soil. However, the state of charge of these "natural pollutants" is usually very weak, and dissolved chemicals in leachate which have a greater ion strength will usually be preferentially adsorbed, replacing those of lesser strength.

The quality of positively charged (cation) particles which can be adsorbed by the negatively charged clay minerals and organic matter can be measured. The quantity of charge is called the "Cation Exchange Capacity". The usual measure is milli-equivalents per 100 grams. This is a number of grams (equivalent to the molecular weight) of singly charged negative sites in one hundred grams of the dried material.

The ability of clay particles, in particular, to hold charged particles in the soil, is very important, not only for leachate renovation, but also for soil fertility in general.

An example of an everyday use of adsorption, is the common water softener. Zeolite or permutit crystals are used as ion exchangers to take out calcium ions from the hard water, replacing them in the water with sodium ions. When most of the sodium ions have been replaced by calcium, the zeolite is regenerated by passing a concentrated salt solution over the zeolite. The ion exchange reaction is thus reversed because of the high concentration of sodium and the softener is ready to begin again.

In leachate renovation, adsorption is important because a soil with good cation exchange characteristics has the potential to take out and hold most of the pollutants common to a typical leachate stream.

It is important to note that many of the ions taken out of the leachate stream may only be "held" for a period before being replaced by another. "Other things being equal, the higher the valence of the ion, the greater is its replacing power and the more difficult it is to displace when already present on the clay."[5] However, during the retention period, chemical and biological renovation will have the opportunity to function. Also, some adsorption processes are almost irreversible. Potassium is the commonest ion that can be "fixed" in an almost non exchangeable state by the process of ion exchange.

A final word of caution should be added in this discussion of adsorption and its importance in leachate renovation. It does not automatically follow that the soil with the highest cation exchange capacity will necessarily provide the best landfill base. Strictly speaking, a high cation exchange capacity merely indicates a high clay content in the soil. If the clay content is too high, the soil will not allow adequate passage for moisture movement due to lack of sufficient permeability. Leachate movement will then occur laterally rather than vertically and any ability which the underlying soils may have for renovation will be lost.

Chemical Precipitation – As a result of chemical or bacteriological reactions which occur in the course of percolation through the soil, some constituents of leachate initially taken into solution may be precipitated out in an insoluble form. Insoluble salts are then readily filtered out by passage through the soil medium. The chief determinant of precipitation conditions is the pH. In general, the higher the pH of the soil/moisture complex the greater the tendency for precipitation to occur.

Bacteriological Action – Bacteriological activity begun in the landfill does not automatically cease if there is a net moisture movement out of the fill into the soil interface. Provided the environment is conducive to growth, bacteria will continue to develop in the soil interface and further the breakdown of complex substances in the manner described earlier in this chapter. The mechanisms of filtration, absorption and adsorption all assist in retaining nutrient material thus making it available for further breakdown by bacteriological action.

Dispersion and Dilution – Although not exactly mechanisms of natural renovation, dispersion and dilution also play an important role in determining the effects of leachate upon the environment.

Dispersion describes the combination of lateral and vertical movement by which net leachate flow may leave the landfill site. Ideally, the greater proportion of the movement will be vertically downwards through the underlying strata. If the sub-base has been carefully selected, natural renovation will reduce any impact upon the groundwater. Dispersion insures that the water flow will take place over a wide area like the release from a fine sieve as opposed to the release from a single source like a spigot.

---

[5] From "Clay Mineralogy", by Ralph E. Grimm, Copyright 1968 by McGraw-Hill Inc. Used with permission of McGraw-Hill Book Company.

It is important to note that for dispersion to occur, the ability of the sub-base to allow moisture movement equal to that in the fill is very important. In landfills where the underlying strata are not capable of taking the total flow, leachate may migrate laterally to areas where it is difficult to control. In situations where lateral movement is also restricted by low permeability, the water level may build up in the landfill. This is called a water "mound". The water level increases until sufficient hydrostatic pressure is built up to force artesian streams to occur at the base of the landfill. Little or no natural renovation can take place under such conditions.

Dilution occurs whenever the net leachate migration from the site meets the groundwater system. Frequently, the contribution of the leachate from the site is significant orders of magnitude less than the total groundwater quantity available in the area. Thus remaining concentrations of dissolved solids can be significantly diluted in the groundwater system. This dilution may not be achieved immediately since it is thought that leachate travels in laminar flow, there is therefore no turbulent mixing action to aid in the dilution process.

Chapter 3

SITE SELECTION

A great many factors influence the selection of a site for a sanitary landfill. Most site selection factors fall into two categories:

1. public attitudes
2. technical

The most perfectly suitable site technically, will never be established if the general public are strongly opposed to it. Neither will the best conceivable public information effort guarantee acceptance if the technical factors are strongly weighted against it. A balance of technical and public relations factors is necessary in order to ensure success — that is the art of landfill site selection.

The only means of ensuring that the job is done right is to conduct an organised and comprehensive site selection study.

Public Attitudes

At the outset, some indication of public attitudes may be gained from the regional studies mentioned previously. The results of the preliminary studies and associated public meetings will indicate general areas of need for disposal facilities.

In general, those areas which have the most clearly defined need for disposal volume, will be more favorably disposed towards the prospect of a new landfill. However, even in such cases of need, community leaders who may recognise the need may be among the last to favor a local site. Commonly they would prefer to talk about resource recovery or rail haul or any other fashionable trend in order to reserve commitment and presumably stay popular. Only rare individuals will side with a local site if they feel that their own political future is at stake.

Strictly speaking, this is not a *public* attitude and the covert opposition of local political figures need not be a problem. If there is any rule to follow in relation to such people, it is "beware October." Political attitudes are hardened in public on the month (October) or months before elections. This is a bad time to be looking for a site. At any other time, especially the day after the elections, you may find many friends ready to face the reality of the situation and further the case for a new landfill.

If possible, commitments are very good things to have in looking for a site. Where local communities need sites, obtain a commitment that your site will be used. The level of financial commitment on behalf of the site-owner and operator is becoming such that a use commitment of this sort from the local community is almost mandatory in order to obtain financing. If you are also in the business of collecting waste, you will need no introduction to the fact that this provides considerable leverage in favor of a new and local site. Non-players in this area take note.

To summarize the foregoing — it helps to have customers.

However, there is another commitment or partial commitment which is useful at this stage and that involves the State Agency responsible for landfill regulation. It helps to establish contact here as soon as possible. They can help in many different ways. To speak of a commitment here with respect to the State Agency is perhaps premature. Regardless of the needs for landfill, they will, or should, be careful not to favor one site over another until they consider applications. However, most states will, at some stage, judge the quality of a site by the quality of the written application and it helps to find out early what particular concerns they have about landfills.

Assuming the general area of search for a site has been determined at this stage based on knowledge of the regional plan, definition of major areas of need and potential customers, then a "windshield" search of the area will reveal likely looking sites. The criteria for this kind of search are generally as follows:

- Availability of land
- Good drainage characteristics
- Suitable soil characteristics
- Visually isolated location
- Above highest known or indicated flood levels
- Access to major transportation routes

Clearly some of these criteria can be conflicting. Others can be overcome by good design, but in the initial search it is best to aim high. There will no doubt be compromises to come.

Land Availability

A "windshield" search should not start blind. Even though the general area of search has been determined, there is little point in merely coasting around without defining some target areas.

The problem is that land availability is a changable phenomenon. The use of land is affected by zoning restrictions and a willingness to sell on the part of the owner. Zoning restrictions tend to be somewhat less final where a public agency is concerned as compared to a private individual. Zoning restrictions only pertain to incorporated areas though some non-incorporated areas may be residential and resistant to landfill development.

Zoning — Clearly, the first step is to establish the planning status of land in the target area. Zoning maps may be obtained from the local planner's office or a copy can be made covering the area of interest. Commonly, sanitary landfill is handled as a variance to a specific classification such as heavy industrial development. The terms heavy and light industry are generally used indiscriminantly and without qualification, but the general principle seems to be that "heavy" suggests "away" from residential areas. Areas in this category definitely suggest less "struggle" in order to establish a landfill, but areas outside this category should definitely not be ruled out. If the sanitary landfill planned contains a beneficial and impressive final use the prospects are at least reasonable that a variance may be granted.

Ownership — Armed with the zoning information, the next step is to establish ownership. Most counties have ownership plat books for sale which will provide reasonably up-to-date information on the extent of present patterns of ownership. Obviously, single ownership of a target area is preferable to multiple ownership, particularly where protracted negotiation is involved.

Ownership is only half the battle. The willingness to sell determines real availability.

This may be an appropriate assignment for a reliable realtor, preferably one experienced in the art of buying land for this purpose. (In fact, a realtor could be assigned to conduct a search as a preliminary to the windshield survey.)

In many areas, realtors have joint agreements which allow access to files on land available through a number of different realtors. This is not as all inclusive as it sounds since only "members" of the particular group of realtors are generally included. Those who charge a non-conforming commission or merely choose to work independently, are not necessarily included. However, the files of land available for purchase are still probably the most comprehensive available and there is commonly no charge for looking since the commission on the sale is the chief concern.

Other sources of land availability information are the files of major utilities, Federal and local government agencies.

Many utility companies (particularly electricity suppliers) have achieved the status of major land holding companies. Railroads, of course, fit the same category. A visit to the land or estates development offices of local utilities will often reveal a considerable number of land holdings of all sizes as well as a willingness (in most cases) to consider sale. This is only a first step, of course, and inquiries should be general rather than specific. However, within the definition of industrial development, a number of areas of interest can be defined.

A valuable source of information on land purchase price which may be referred to in order to obtain a good reading of what may be a fair price are the records of public agencies. This information is in the public domain and a matter of public record, though it may be necessary to push sometimes to see the data. Many local agencies actively pursue land acquisition policies as a land use control strategy and also for purposes of public development. The data on prices, areas and locations can provide a valuable guide to local costs. Among the various Federal Agencies, the Bureau of Land Management (BLM) may be the most likely to provide useful background information.

A visit to the BLM may reveal an unsuspected ownership of suitable areas by the Federal Government. Availability is another question, but it may be worthwhile to look for areas which are *not* Federally owned, and isolated by those which are. Look for isolated areas of Federal ownership. This information is all in the public domain and available in map form, except for some mid-western states which are only now going through the process of mapping areas of Federal ownership. The field office of the Soil Conservation Service may also help by adding local knowledge of land availability while the search is broadened to obtain soil information.

Drainage

The control of site drainage is an important element of site operation. The site must be carefully designed to be free draining both during the operation and upon completion. Areas "upstream" of the site which might naturally drain across the face of the landfill must be diverted around the new fill. Areas "downstream" of the site must be capable of carrying the additional flow to the nearest drainage system.

The prospective site must, therefore, be appraised for the ability to control "to" and "from" drainage.

The art of visual evaluation of a site for good drainage is relatively simple, but it is surprising how many people just do not bother. Stated very simply, imagine what happens to the site when it rains. In fact, if it is possible to visit immediately after a major down-pour, so much the better. Contour maps of the area are vital for this exercise and can be picked up at any supplier of U.S. Geological Survey (USGS) maps. Using the contour map, it is useful to outline the area of interest and trace the direction of surface flow which might be expected to flow under gravity. The area of interest extends beyond the area which might ultimately be used as a landfill. The adjacent area may well control water movement through the site. Major problem areas to look out for are:

- Areas of perched runoff collection
- Groundwater recharge areas
- Direct surface flow to major streams

Perched Runoff — As a result of existing natural drainage patterns, it is not unusual for substantial quantities of runoff to flow under gravity to a low point where it is prevented from further movement. After periods of heavy rainfall, water will remain perched in the area until it percolates through the subsoil or until it is dissipated by evaporation and plant usage.

This condition could prove troublesome to the siting of a sanitary landfill. The depressed area should be filled and raised to a free draining contour and the resultant drainage then channeled around the projected landfill. If the troublesome area is extensive, this could prove a costly beginning or maybe even an insurmountable problem depending upon the attitude of the neighboring owner.

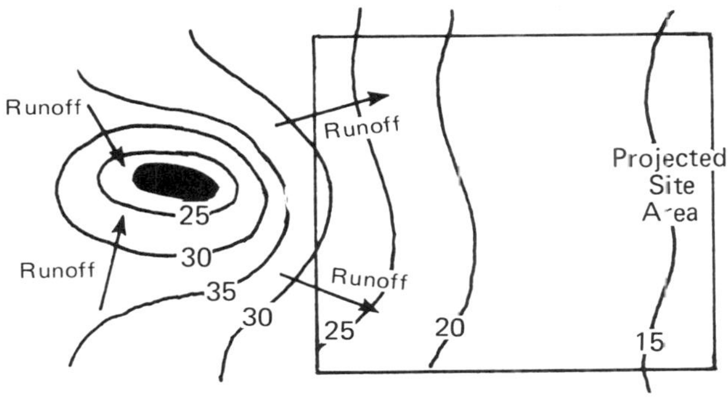

Figure 9. Perched runoff.

Whether this condition will give rise to groundwater contamination depends of course upon the general direction of ground water movement, but at this stage it would be prudent to assume the worst case. Also, it is a matter of geological record that groundwater movement will commonly occur in the same general direction as the natural surface water movement, so the chances are high that perched water is a "real" problem.

Strictly speaking, there are only two methods of handling the problem, direct or indirect. The direct method as described previously requires that the low spot up-gradient of the site should be filled and seeded to encourage runoff in the predominant direction. This will increase flow over the projected site, and that problem can be eliminated by the design of an appropriate diversion channel.

The newly filled area should be graded to encourage runoff at a minimum slope of 2%. Clearly this solution may be simply executed by including the adjacent offending area in the projected land purchase. Alternatively, direct negotiation with the owner of the adjacent land might include an agreement to fill this area prior to full landfill operation.

Indirect methods of control may be necessary if negotiations to gain some direct control over the adjacent property breakdown. Two indirect methods may be employed:

- Impermeable barrier
- Drainage way

In the example cited, the 35 ft contour is essentially the surface drainage divide between the two areas. Where possible, control at this divide presents the best option. An impermeable membrane of clay to a depth of 20 feet (down to the 15 ft contour which marks the end of the site area) would provide protection against groundwater flow across this divide. However, this may prove costly if clay is not available nearby. A less expensive alternative would be to construct a deep (20 ft) gravel drainage way to induce the groundwater movement to travel the path of least resistance. This latter solution will be particularly effective in soils which are slightly to moderately impermeable. In both methods it is important to realise that water movement is being controlled by changes in the direction of ground water flow. Thus, these expedients must be handled with care and an eye to where the new direction of flow will take the water. At this stage it is only necessary to realise that such expedients are possible and with skill and practice a rough visual assessment of the site area can yield valuable results.

Groundwater Recharge — Groundwater recharge areas are not always easy to spot, but there are some clues which may be useful. To a trained eye, the contour maps depict a great deal of history. In particular, the history of water movement. Clearly, natural drainage over time can etch significant channels into the topography and depending upon the general cross section of the channels, fairly accurate statements can be made about the types of flow which caused the channel. For our purposes, we are concerned here with natural drainage channels, which can usually be identified very clearly as a series of V-shaped contours. Depending upon the level of groundwater in the area, the channel may appear on the map with the familiar thin blue line in the center, depicting a natural stream. Whether a stream is shown or not, it remains true that as more runoff collects along the length of the stream or drainage-way, the flow in the channel increases and this will show up in the contours as a progressive rounding of the contour lines as you progress down stream. If anything should occur in the natural scheme of things to reduce the amount of flow in the channel, one would expect the channel to reduce in size. This would be shown as a series of more acute contours.

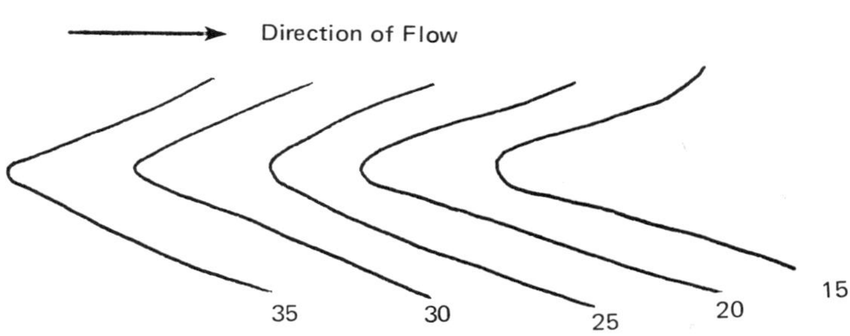

Figure 10. Natural drainage contours.

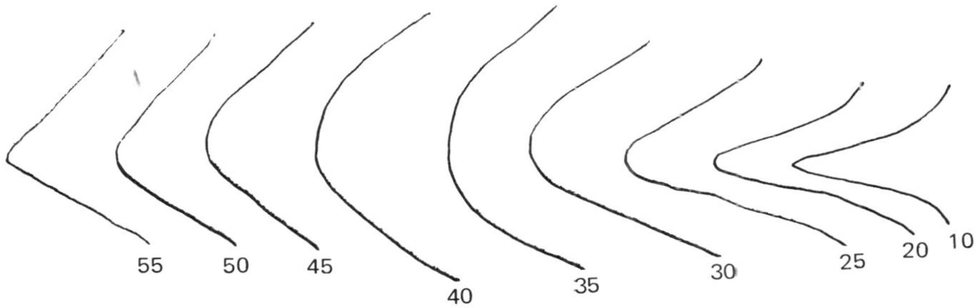

Figure 11. "Unnatural" drainage contours.

This succession of contours from acute to rounded and back to acute provides some important clues to the nature of the underlying geology. Clearly the volume of water which naturally flows in this channel during rainfall cannot sharply reduce in quantity, unless the underlying soil and geological structure is so permeable that the flow is diminished by loss to the ground. Be careful that the shape of the contours and the sudden acuteness in particular, is not due to increased velocity in the surface water stream. This would occur if the ground surface became suddenly quite steep downstream of the area where the contours are beginning to round off.

However, this situation would be very clear from the steepness indicated by the closeness of the downstream contour lines. Where this latter condition does not exist, it may be assumed that some loss to the groundwater system is indicated in the area of sudden reduction in channel size shown by the contours. Clearly this indicates permeable underlying structure and it should be a warning that the site is less than ideal, additional investment may be necessary to correct the condition. While you are simply looking for options, the area should go down as not a first choice.

Direct Surface Flow to Major Streams – The presence of streams in the area being surveyed will be clearly indicated by the characteristic blue lines on the USGS maps. If there is direct overland flow to the stream from the site, then there are two classes of problem:

- Surface flow
- Groundwater flow.

Surface flow will represent the biggest headache during the operation of the site. Clearly direct runoff during this period would need to be diverted since any liquid waste could easily be inadvertently carried to the stream before daily cover could be placed. The diversion ditches could be expected to be necessarily costly since the drainage required would be the reverse of the natural drainage pattern presently in existence.

Also, it will be remembered that in general, the ground water flow tends to follow the direction of surface water flow and thus the flow of ground water which will permeate through the fill may be expected to ultimately find its way to the stream.

It should be noted here that pollutant discharge to the stream is not of itself, necessarily untreatable. Under present laws wastewater treatment plants are in fact, granted licenses to discharge certain quantities of pollutant daily. The problem is that the present thrust of the law suggests that these permissable discharges may be reduced sharply in the next few years. Also, some treatment is usually required before discharge in the case of wastewater treatment facilities, and it seems very probable that similar requirements may be imposed upon landfills.

As will be discussed in detail under the section dealing with design, the ultimate purpose of the plan both during the operation and upon completion, must be to reduce the amount of rainfall which permeates the fill itself. This can be achieved by careful grading and selection of cover material. However, the surface water flow which results from this type of design can amount to a significant quantity of water during times of heavy rainfall, and the impact upon a nearby stream could therefore, be quite substantial. In other words, by trying to solve the leaching problem, heavy runoff conditions are created. During operations these heavy runoff flows could be large enough to carry debris from the fill area, not to mention the risks of heavy silt loads from overland erosion.

Clearly some intermediate treatment must be placed between the landfill and the stream and this could most effectively be a large siltation basin ideally situated on the far side of the landfill from the stream.

So far we have discussed drainage characteristics only in the context of map interpretation, but it will be obvious that most of the features discussed can also be seen on the ground. However, it is sometimes rather difficult to make these interpretations once you find yourself on the side of the road looking at a large chunk of real estate. The trick is to be quite simplistic about it. Literally, imagine a rainstorm and see what you would do if you were a light object lying on the surface at various locations. Think particularly of the area outside the selected location for the landfill. Remember that the drainage characteristics can be altered on the landfill, but you may have little control over what happens outside the area.

Suitable Soil Characteristics

Many experts judge soil to be the most important single criterion for judging landfill suitability but usually for the wrong reasons. Commonly soil is judged for its workability as a cover material though this is one feature of the soil which can be easily modified. The drainage characteristics of the soil, however, may be impossible to change.

Soil is important to landfill design for two reasons:

- **INTERFACE.** This is generally the material which lies between the completed landfill and the groundwater system.
- **COVER.** This is the material used to cover the landfill (daily and finally).

Good landfill practice requires underlying soils which will a low moisture from the landfill to percolate *slowly* through the soil structure before it reaches the ground water system.

Soils required for cover material should ideally be less permeable than the soil under the landfill. The function of cover soils is to prevent or limit the passage of moisture to the fill material. The cover soil should be capable of developing a good stand of vegetation since ultimately the finished landfill will be contoured to encourage runoff and vegetation will reduce the risk of erosion.

These are *ideal* requirements. Some compromise may be necessary in the design of each site.

A considerable amount of information is available on soils in most areas of the country. The United States Department of Agriculture, Soil Conservation Service (SCS) in cooperation with a number of agricultural experimental stations at state universities provides many detailed publications dealing with soil characteristics and properties. These include individual county soil surveys, irrigation and drainage guides and soil interpretation sheets.

However, before we can talk about specific properties, to look for, we should begin at the beginning and consider how soils are actually formed. Many factors influence the formation of soil. In order of importance, they can be listed as shown in Table 4.

Table 4

Factors Which Influence Soil Formation

Nature of Parent Materials
Climate
Native Vegetation
Drainage
Time
Man

Parent material is the original geologic formation from which the soil is produced by various natural processes of disintegration and decomposition. These parent materials or the soils which result from their decomposition need not necessarily be representative of the immediately underlying geology, since they may have been considerably moved around by glacial, wind or water action.

Climate largely determines the type of weathering which the parent material undergoes in the process of conversion to soil. Climate also determines the type of vegetation which contributes to soil development through root action and also plant decay. Temperature and rainfall are the chief components of climate to influence soil production. In some areas, wind action may also contribute significantly to soil development.

Native vegetation contributes to the soil development in many different ways. Root action may physically aid in breaking down large particles. Plant decay contributes to the organic content and structure of the soil. Plant growth and moisture usage directly influence the quantity of water which percolates through the soil, leaching out important constituents of the soil in its travels.

Drainage is generally controlled by topography. The quantity of moisture in the soil at any one time directly influences the amount of drainage, the rate of weathering and also the development of color in the soil.

All of these variables are of course, a function of time. Soil formation is generally a function of geologic time, though the final factor, man, can bring about changes in soil structure very rapidly.

In general, the process of soil formation creates a characteristic series of layers which are often considered conceptually the same for most soils. This conceptual layering of the soil in its natural process of formation is referred to as the soil profile. Each layer is referred to as an horizon and a vertical section through the soil showing the various horizons is referred to as the soil profile. An idealized soil profile is shown in Figure 12.

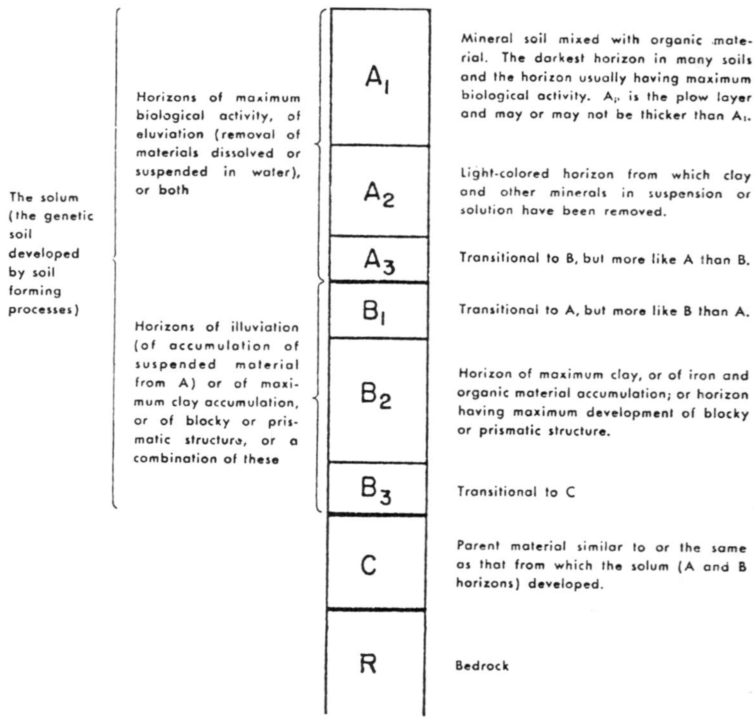

Figure 12. Principle horizons of upland soils.

Source: "Wastewater Management for Lake Michigan – A Study of C-SELM" by U.S. Army Corps of Engineers, July 1971.

It is important to realize that many areas of the country have soils with very similar profiles. When this similarity exists, the soils which have similar characteristic profiles are said to be in the same soil series. Often each of these soil series is named after a nearby geographic feature or town. When similarities in soil series are found to derive from similarities of parent material and color, several soil series may be grouped together as soil associations.

Soil association maps are developed by the Soil Conservation Service (SCS) and these will be found the most valuable tool available for preliminary survey. However, it should be noted that soil association maps are produced by the SCS largely on the evidence of 6 feet test pits which are opened up for examination by SCS field personnal on approximately 500 ft centers. Clearly this grid basis of investigation is very necessary in terms of mapping large areas, but it should be kept in mind that significant variations can occur at lesser intervals than 500 feet. Also a depth of investigation of 6 feet is a good average indicator and ideal to obtain agricultural data, but not nearly deep enough to provide the best information about landfill suitability.

This general information has enabled the SCS in some areas of the country to produce maps showing limitations to sanitary landfill siting. These maps and limitation criteria should be used with discretion. Not all of the limitations may be viewed as significant constraints. Nevertheless, this kind of information, if used selectively, can provide an excellent beginning for landfill investigation. In particular, indicators of areas with limited depth to seasonal high water table, are absolutely invaluable. Depth to bedrock is also well worth looking at in addition to flood hazard indicators and texture.

Having set the scene on soils, their formation and general characteristics, let us return to the two chief functions of soils which may be briefly summarised as follows:

- interface
- cover

Soil As Interface

The ideal requirements of soil as interface are poorly understood though the later design chapter in this book attempts to establish some general guidelines.

However, we are here concerned with what characteristics can be identified in the course of a search for a site. Since the nature of the soil interface is such that it is chiefly in the region of the C-horizon, there can be little data obtained from a visual inspection of the site.

Soils data available from the Soil Conversation Survey will be the chief source though soil information at depth is only available by inference. The chief concerns at this stage will be to limit the risk that the underlying strata may prove highly fissured or excessively permeable.

Borehole examination of a tentatively selected site is imperative in order to determine interface conditions. The general guidelines outlined in the chapter on detailed design will provide adequate information to assess results of borehole investigations.

Soil As Cover

As stated previously, the literature is replete with guidelines on this subject. The Soil Conservation Service have developed a general listing of soil limitations which is presented here for general guidance:

TABLE 5

USDA Soil Conservation Service Classification of Soil Limitations for Sanitary Landfill (Typical of Pennsylvania Area)

Limiting Factor	Degree of Limitation: Slight	Moderate	Severe
Depth to seasonal high water table[1]	Deeper than 4′ below surface	1½′ to 4′ below surface	Less than 1½′ below surface
Slope	0–8 percent	8–15 percent	15+ percent
Depth to Bedrock[2]	More than 5′	3′ to 5′	Less than 3′
Stoniness	Nonstony to stony	Very stony	Extremely stony to rubble land
Surface soil texture	Sand, loamy sands, sandy loams, loams, silt loams	Silt, clay loam	Silty clay loam, clay, much, peat
Flood hazard[3]	Seldom	Occasional	Frequent
Permeability[4]	Acceptable if less than 2″/hour		
Limestone Bedrock[5]			

[1] Seasonal high water table will prevent proper landfill operations during certain seasons and seepage can cause contaminated liquids to flow out on the lower banks.

[2] This is depth to hard unrippable bedrock, and on-site investigation should be made to determine actual depth to bedrock.

[3] This is an estimate of dominant condition, and on-site investigation to determine actual overflow frequency should be made.

[4] Soils with rapid permeability (greater than 2.0 inches per hour) have slight limitations, but are hazardous to ground-water quality.

[5] Soils underlain by cavernous limestone are hazardous to ground-water quality.

In my own opinion, the strictly soil related items in this list are the most important. (Depth to high water table will be discussed later in this section. Slope has already been discussed in the section on drainage.)

Depth to bedrock is important to the requirement for soil as interface, but the criteria contained in Table 5 offer little assistance in planning that particular element of landfill design. The data presented according to these criteria can only indicate whether or not bedrock occurs above or below 6 feet. A site with less than 6 feet depth to bedrock should be eliminated unless a very limited refuse volume is required or cover material may be obtained elsewhere. Data regarding depth greater than 6 feet can only be obtained from borehole investigation.

Stoniness is a useful parameter since it provides direct information regarding the workability of the soil as cover material. We will see later that there is some direct correlation of permeability and stoniness and ease of percolation is not strictly a sort after quality of cover material.

Texture provides by far the best direct clue to the value of a site since a field investigation can quickly ascertain where the soil fits in a general classification.

Soil is primarily made up of sand, silt and clay. The percentage distribution of each of these constituents in addition to the particle sizes, determines the texture. Variations in the distribution of the three primary constituents and the general terminology for soil type can be shown very clearly in a classification chart.

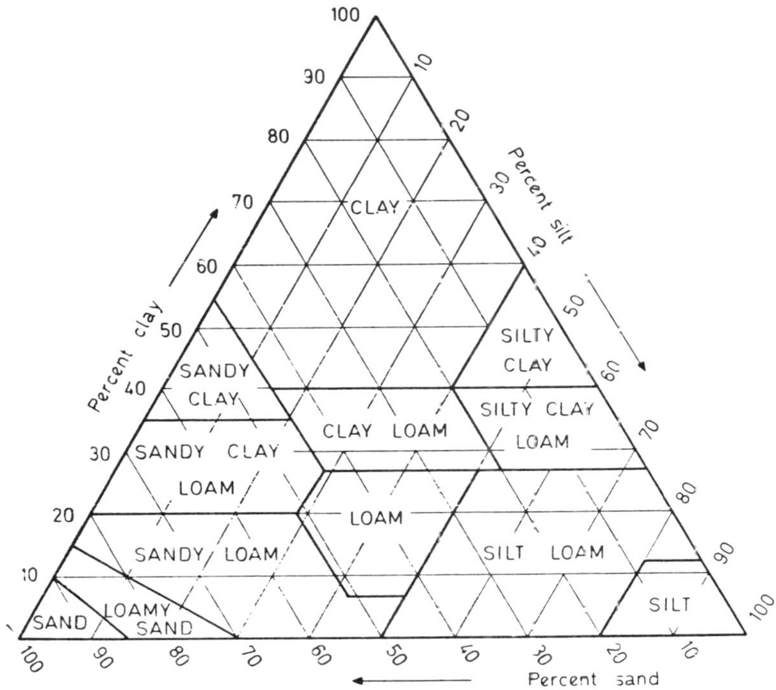

Figure 13. Soil classification chart.

This is called a ternary chart. It can be used very easily to locate the general classification of a particular distribution of clay, silt and sand. Conversely, it shows very clearly the composition of each common designation of soil type.

On-Site Tests for Soil Texture – The texture of a soil can be easily determined on site by first moistening and then estimating the proportions of clay, silt and sand as it is worked between finger and thumb.

TABLE 6

Soil Texture Class Description[1]

Sand – Soil consisting mostly of coarse and fine sand and containing so little clay that it is loose when dry and not sticky at all when wet. When rubbed, it leaves no film on the fingers.

Loamy sand – Consisting mostly of sand, but with sufficient clay to give slight plasticity and cohesion when very moist. Leaves a slight film of fine materials on the fingers when rubbed.

Sandy loam – Soil in which the sand fraction is still quite obvious, which moulds readily when sufficiently moist but in most cases does not stick appreciably to the fingers. Threads do not form easily.

Loam – Soil in which the fractions are so blended that it moulds readily when sufficiently moist, and sticks to the fingers to some extent. It can with difficulty be moulded into threads but will not bend into a small ring.

Silt loam – Soil that is moderately plastic without being very sticky and in which the smooth, soapy feel of the silt is the main feature.

Sandy clay loam – Soils containing sufficient clay to be distinctly sticky when moist, but in which the sand fraction is still an obvious feature.

Clay loam – The soil is distinctly sticky when sufficiently moist, and the presence of sand fractions can only be detected with care.

Silty clay loam – This contains quite subordinate amounts of sand, but sufficient silt to confer something of a smooth, soapy feel. It is less sticky than silty clay or clay loam.

Silt – Soil in which the smooth, soapy feel of silt is dominant.

Sandy clay – The soil is plastic and sticky when moistened sufficiently, but the sand fraction is still an obvious feature. Clay and sand are dominant and the intermediate grades of silt and very fine sand are less apparent.

Clay – The soil is plastic and sticky when moistened sufficiently and gives a polished surface on rubbing. When moist, the soil can be rolled into threads, and it is capable of being moulded into any shape and takes clear fingerprints.

Silty clay – Soil which is composed almost entirely of very fine material, but in which the smooth, soapy feel of the silt fraction modifies to some extent the stickiness of the clay.

[1] From the SCS Field Handbook, 1969.

In general, soil with good workability and compaction characteristics is the most desirable cover material. A soil containing large quantities of clay may present an operational problem during wet weather and, under certain conditions, may tend to shrink and crack, thereby making it difficult to maintain a tight cover. Clayey soils may also present a difficult excavation problem during cold weather.

Sands and loamy sands are too permeable and non-compactable, and, to a lesser degree, similar problems occur with pure silts. Therefore, the mixed loams with low clay contents have been determined to be the best cover material.

Selection of a loam as the most desirable cover material has been recommended in a number of standards for sanitary landfill, ASCE Manual No. 39 recommends a soil containing 50 to 60 percent sand, with the balance being approximately equal amounts of clay and silt. A U.S. Public Health Service publication on landfills states that sandy loam is considered to be an excellent cover material as is a sandy loam or sandy silt (up to 50 percent silt). Pennsylvania Department of Environmental Resources generally considers the mixed loams with low clay contents as the most suitable cover material, in general agreement with the ASCE (1959), U.S. Public Health Service (1968) and the New York State Department of Health (1969).

Some controversy surrounds the boundary lines between acceptable and unacceptable cover materials and there is considerable divergence of view when it comes to including the clay loams and the silty and sandy clays. Where there are strict regulations, these should be followed. However, in my own opinion, the argument which excludes these soil types primarily on the grounds that they will give trouble in working is not strictly within the perview of government responsibility. Where clay loam, for example, is readily available, a good case can be made for using it if a good friable soil material is scarified into the loam. The argument that it cracks upon drying when used as final cover can also be blunted by the addition of 2 to 3 inches of top soil which will keep the surface moist and prevent drying.

On balance a wide range of soils may be used as cover materials provided the appropriate precautions are taken. I would redraw the ternary diagram to show the soils suitable for cover material as shown in Table 7.

Visual Isolation

The matter of choosing a site which is not overlooked by someone from some vantage point is more difficult than it seems.

That the matter is important should not be in question. During the application of daily cover a working face is always open. Garbage trucks are continually unloading their contents onto the working face and one or more dozers are continually toing and frowing across the compacted waste. None of this is a particularly attractive intrusion on the local landscape. It is a short step from visual intrusion to local dissent and hostility.

Clearly, a site should be as visually isolated as possible without the place becoming ridiculously distant from the source of waste. Natural vegetation is, of course, ideal as a barrier and where significant wooded areas are indicated on the 7½ minute quad sheets these areas should receive special attention during the windshield survey. In the last resort, a planting program may be necessary to establish some visual cover, but there is clearly no substitute for existing vegetation. Even hardy quick-growing hybrids such as maples or poplars take 7 to 9 years before they present decent visual cover.

TABLE 7

Ternary Diagram Showing Soils Suitable for Cover Material

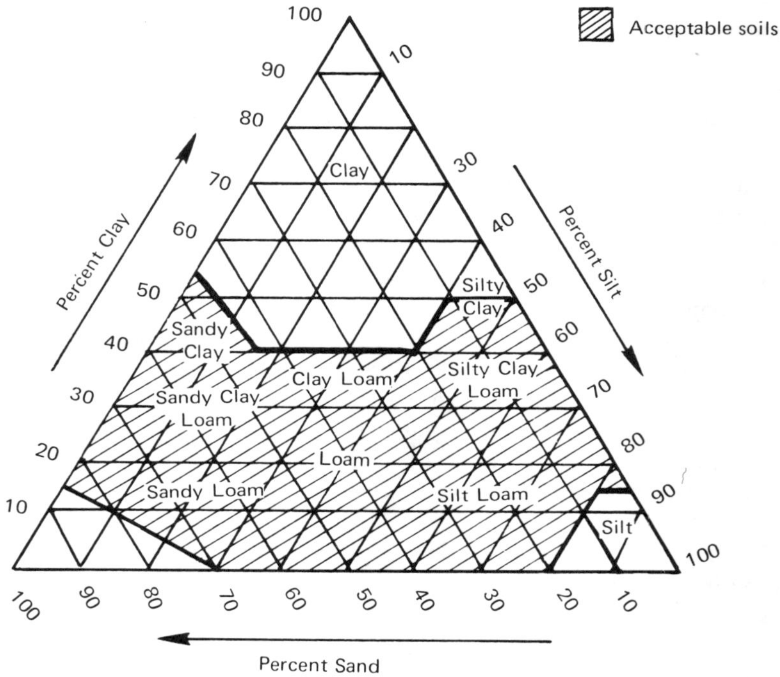

Trees do, of course, provide an additional advantage when considering litter control, but to make best use of them, it is necessary to have a good idea of the prevailing wind patterns in the area. It is worth taking the trouble to obtain a detailed chart of annual wind direction from the local weather station. An example is shown in Figure 14. A chart of this type is called a wind rose.

It takes little or no time to obtain one and it gives a clear indication of the prime location for tree cover. The greatest concern for visual cover should be to screen vision from the highways. The map will show clearly where all roads (major and local) might overlook the site and when the search has narrowed down to a few options all roads should be traversed to check overview.

The search for visual isolation will need to be tempered by the fact that easy access is also necessary. The economics of adding a long all-weather access road will, therefore, govern the decision at some stage. Typically, the long access road will pay off as a design strategy if the search for a site is taking place near a large town or city. Under these circumstances, urban development may be very widespread and

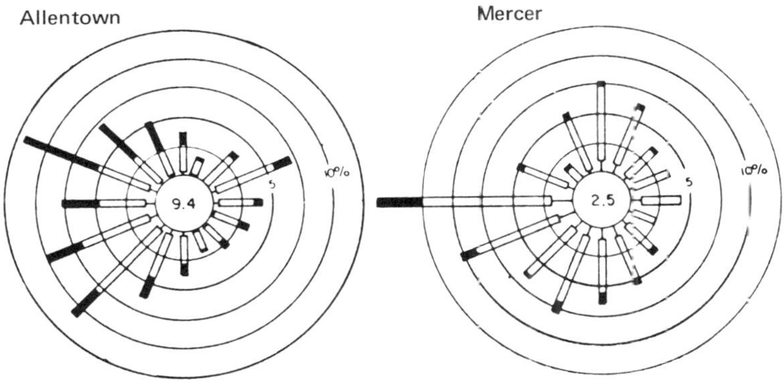

Figure 14. Typical wind roses. Showing wind direction and frequency.

finding a site not overlooked by anyone may be comparatively difficult. A long access road may achieve such isolation, though such sites will be particularly hard to find, near areas of high residential density.

Paradoxically, in the rural areas where concern for visual isolation may be much less, there are many areas where long access roads may be established though in these areas they are less necessary.

So far, we have been concerned with road level or lateral isolation because this is the most easily attained. By locating the site 200 feet or more from the major highway, substantial visual isolation may be achieved by careful design of the access road. See the later chapter on design for further details. It can be argued that this isolation may be achieved by careful design of the landfill topography later, but it should be clear at this stage what the options are. Often by careful selection of the site, the prospects for such design can be greatly improved.

An additional feature of visual isolation which can improve the design potential of the site is topography.

The most important topographic considerations are those discussed earlier as functions of site drainage. However, within the limits of good drainage requirements, there are many features of natural topography which may improve the visual aspects of the site. Areas not easily seen from the highway are ideal.

This may occur where highways are low lying or where the site itself is low lying compared to highways.

Visual isolation of the working area may even occur where the site itself is higher than the surrounding area.

High lying areas are sometimes neglected on the assumption that they are easily seen from the surrounding area. By careful siting of the working area, this may be avoided and the site "problem" turned to advantage.

All the foregoing rather obvious precautions may count for nothing, of course, if the local planning office is not consulted at some stage. In particular, the Depart-

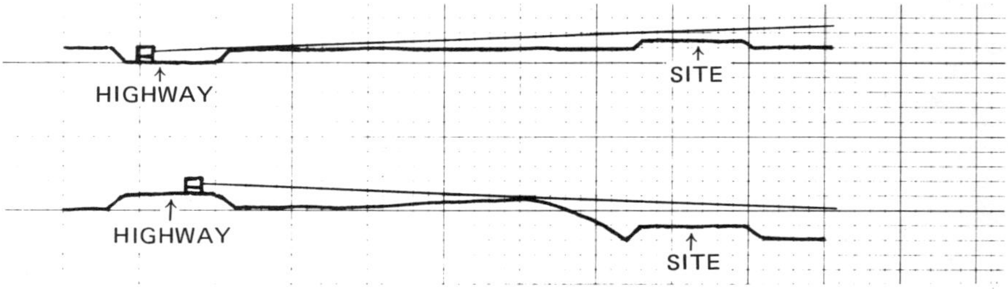

Figure 15. Highway overview.

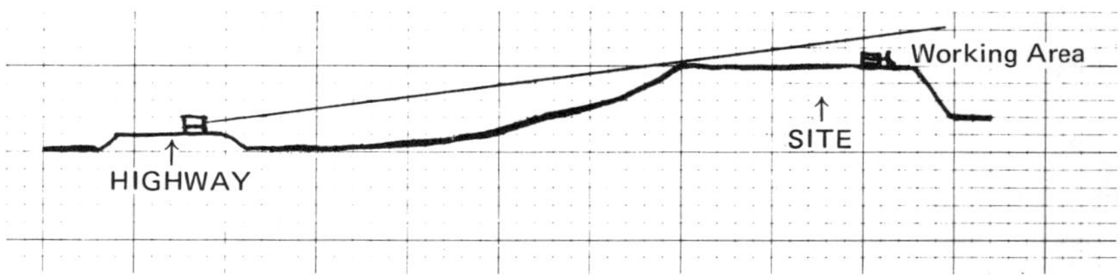

Figure 16. Strategic location of working face.

ment of Transportation of a Regional Planning Commission or the State DOT should be consulted regarding future transportation plans. This is a very necessary precaution to reduce the risk of a potential site being clearly seen from a highway not presently in the picture.

Depth to Water Table

The water table is the fluctuating boundary between the zone of aeration and the underlying zone of saturation. The seasonal high water table (which occurs at the end of the ground water recharge season in the Spring) should be as deep as possible at a potential site. A deep water table allows maximum advantage to be taken of the renovative properties of the unsaturated soils.

The soil limitations data from the SCS is of very little assistance on this subject, but the field officers may be able to fill the gap. Commonly, field officers on the SCS have a wealth of experience in the local area and may be in a position to give general assistance regarding areas of high water table.

Water well users in the area can also provide valuable assistance by permitting simple line measurements to be taken on local wells to measure depth. In fortunate circumstances the original well drilling logs may be available. However, data from such sources often requires interpretation by an expert.

Ultimately, the depth to water table (with adjustment if necessary for seasonal fluctuation) will determine the life of the fill. The following simple cross section may help to explain this seemingly sweeping statement:

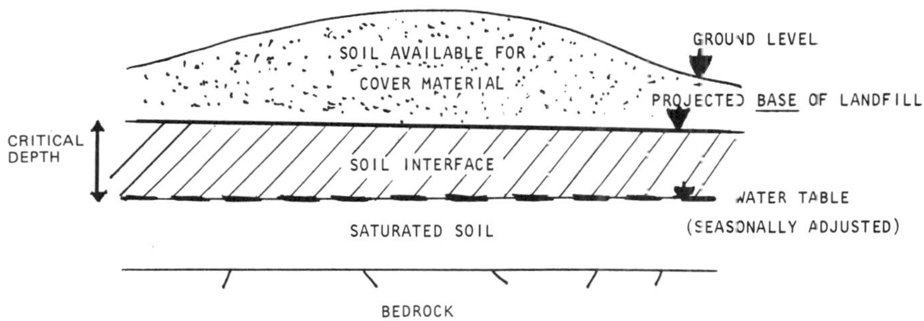

Figure 17. Soil interface diagram.

As we have discussed earlier, the ability of the soil to provide renovative capacity is an important support feature to good landfill design. One measure of this capacity in quantitative terms is the depth of soil between the landfill base and the high water table. This margin of soil between landfill base and water table is termed the soil interface. The actual depth of this strip of soil is the subject of much debate. Some guidance to the factors involved in making decisions about depth of interface will be offered in the design chapter. It will suffice here to observe that this thickness need only be subtracted from the total depth to high water table to reveal the soil depth available for cover. I believe there is a more scientific basis for this judgment than State edict, but as a practical matter that may not count for very much if the local regulations are rigidly enforced. All that aside, it should be clear from the foregoing the scale of importance of the depth to water table.

If the SCS or local water wells do not reveal depth to water table data, it is imperative that additional deep drilling should be conducted in order to check this depth. This is obviously particularly necessary where there may be some question whether or not the depth will be adequate. There is only one way to be sure and that is to drill and find out.

When the depth is known, the tables contained in the detailed design section can be used to determine the potential volume of refuse the site will hold.

Access to Transportation

It is a matter of some importance that the selected site should be relatively near reliable transportation routes. Obviously, this is particularly necessary in those areas of the country which have the doubtful pleasure of snow in winter. In these conditions, access to considerable stretches of minor road may prove a liability in winter even though they may afford superb visual isolation. It is necessary to strike a balance.

This particular element of finding a site usually takes care of itself in as much as good road maps are used in planning a windshield survey and access by road tends to limit examination of remote areas.

However, once again a visit to the local planning office may prove helpful if a map of proposed new transportation routes can be examined during the visit. It should be understood, of course, that probably about 70% of all conceptual transportation plans never result in any useful activity in the way of actual construction. Detailed plans and specific schedules are a reasonably reliable prelude to meaningful plans. Projected new routes should be carefully reviewed. In some cases, they may effect substantial changes in projected access routes for alternative sites. For the landfill operator who is in the picture early, knowledge of projected new routes may even make a difference in land costs.

Summary

In summary, the chief criteria for finding a site can be presented in the form of a table to assist in making the final selection.

TABLE 8

Decision Table for Landfill Site Selection

SITE NAME	COMMENTS	SCORE
Land Availability		
Drainage		
Soils		
Isolation		
Depth to Water		
Transportation		

Any comparison methodology may be used. Factors may be scored as a fraction of 10 (say) and then rated according to importance. So long as this method is applied by the same evaluator that is perfectly acceptable. Personally, I would rate them in order of importance in the order in which they appear. (Land availability is clearly the most important, followed by drainage, followed by soils, etc.)

Alternatively, some people prefer mentally juggling the various factors and intuitively coming up with a judgment. People with a good record at that sort of thing should probably stick to it.

Chapter 4

SITE INVESTIGATION

Assuming the results of the site selection process yield a good potential site, detailed site investigation is necessary in order to verify previous data and to provide detailed information for the design process.

The first basic requirement is a more detailed knowledge of the soils and geology than previously obtained in the course of site selection. A good geologist (preferably with some knowledge of groundwater hydrology) should be engaged to examine the site, plan a borehole examination program and obtain sufficient detailed information to be able to draw several critical cross-sections of the site.

Very often, the need for economy begins to impose a constraint at this point since substantial expenditure has been incurred to date and not one pound of waste has been landfilled. However, economies at this point are totally illusory. Lack of detailed knowledge of the underground structure of the site could be very costly in terms of lost potential to exploit all the resources available on the site. In addition, it is also true that mistakes made at this point (based on inadequate information) are certainly very costly.

Whether or not more information is, in fact, required is strictly a matter of judgment, though there are few sites which can be adequately planned without detailed borehole investigation. If the data search during the site selection process was sufficiently successful that early well drilling data was obtained from a number of sources, there may be a possibility that no more borings will be required. In most cases, however, this kind of information does not provide enough evidence to reliably reconstruct site geology though it is invaluable in planning the boring program.

A good test of the adequacy of existing information is to draw critical cross sections through the site and identify the underlying structure and groundwater location. Where there are clearly voids in the data and knowledge of local geology suggests greater variation in subsurface structure than limited data can describe, then more data is needed.

The borehole information should provide enough basic information about the site that subsurface cross sections can be drawn with reasonable assurance of accuracy. Figures 18 and 19 provide an excellent example of a well-researched landfill site with excellent detail in the cross sectional diagrams

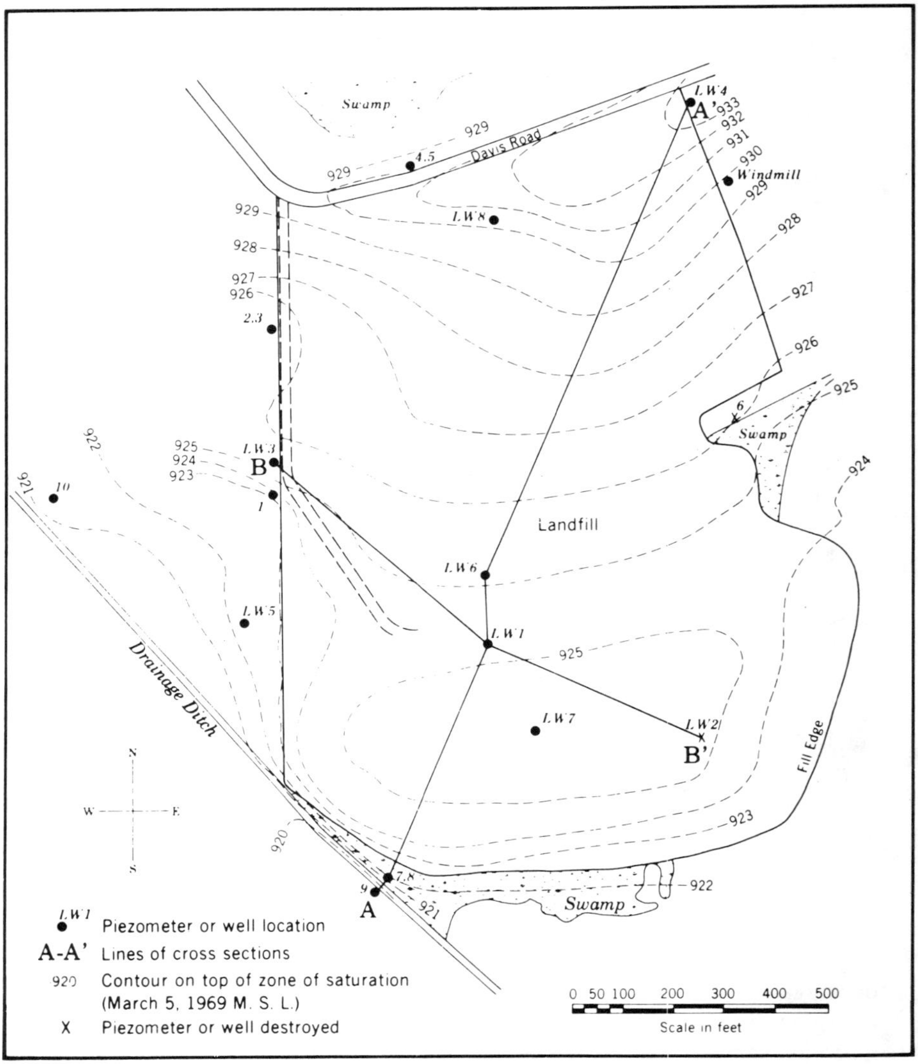

Figure 18. A good example of a well cross-sectioned landfill. Plan view of the Woodstock landfill and surrounding area, showing locations of borings and the contours of the top of the zone of saturation. Gradients are away from the upland of the northern part of the landfill in all directions. In the older part of the filled area, the gradient is southward to swampy areas bordering the landfill or to the drainage ditch west and southwest of the landfill. Some influence of the landfill is shown by a steepening of gradients on the southern edge. This steepening suggests that a small ground water mound lies beneath the landfill.

Source: "Hydrogeology of Solid Waste Disposal Sites in Northeastern Illinois", by G. M. Hughes, R. A. Landon and R. N. Farvolden of the Illinois State Geological Survey. Published by the U.S. EPA, 1971.

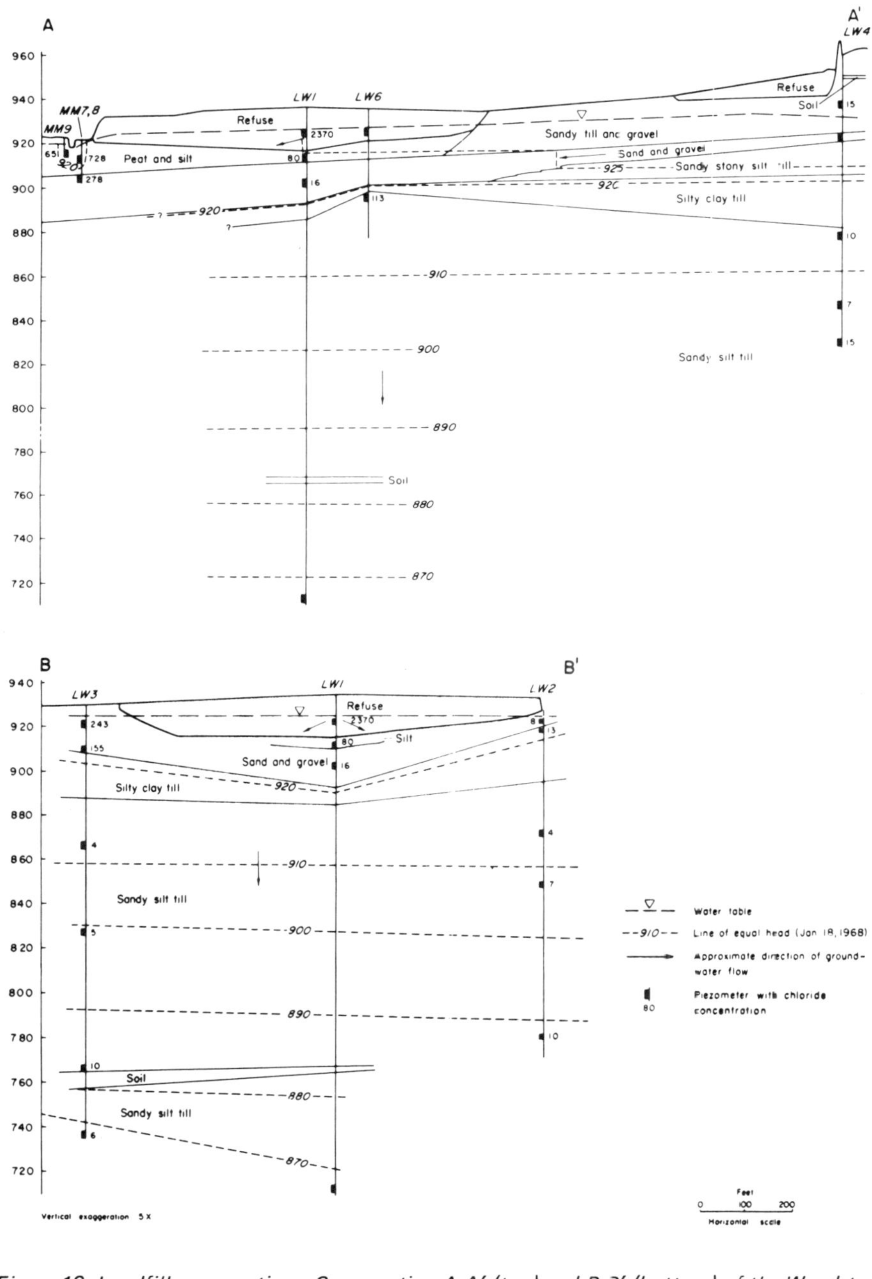

Figure 19. Landfill cross-sections. Cross section A-A′ (top) and B-B′ (bottom) of the Woodstock landfill with selected chloride concentrations. A vertical gradient in the silty clay till and a strong lateral flow of ground water in the shallow materials above the till are evident. Ground water discharges into the drainage ditch near MM 7 and MM 8, cross section A-A′.

Source: "Hydrogeology of Solid Waste Disposal Sites in Northeastern Illinois", by G. M. Hughes, R. A. Landon and R. N. Farvolden of the Illinois State Geological Survey. Published by the U.S. EPA, 1971.

Number of Boreholes

As a general rule, state regulatory agencies do not delineate the borehole numbers required or the sequence in which the investigation should proceed. In general, the number required and their location should simply be whatever is adequate to support the plan (enough to draw reliable cross sections).

In recent years, there has been a trend in some states towards specifying the number of boreholes according to the size of site. This seems to me to be unfortunate since the requirement will vary considerably dependent more upon geological variations than acreage.

As a general rule, even the smallest site with a suspected straightforward geological structure would probably need at least three boreholes to draw a suitable cross section. This might be appropriate for 10 acres or less. For a larger site up to say 50 acres, a more appropriate number might be 5 or 6 to describe two cross-sections over the site. Larger sites may require 8 or more.

These general guidelines may be summarized as follows:

TABLE 9

Borehole Requirements – General Guidelines

Site Size	Approximate Number	General Configuration
Up to 10 acres	3	
10 to 50 acres	5 - 6	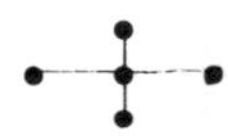
50 to 100 acres	8 - 9	
100 acres or more	11-15	

These numbers are only intended as general indicators. Where the underlying geology is very complex, it is possible that a substantially greater number of borings may be required.

In the final analysis, the distribution of boreholes over the site area can only be determined in relation to the actual site and the accumulation of information about the underlying structure. In practice even when a decision is made about numbers of borings, the field geologist and the design engineer working with the drill operator can adjust the working plan as information is received from each borehole. The quality of interpretation depends to some extent upon the capability of the equipment utilized.

Choices of Drilling Technique

In some areas of the country there may be no choice at all since the local contractor may only have one piece of equipment. In areas where geological conditions are not well documented, it is often appropriate to hire the local contractor. This strategy makes full use of his local knowledge and experience in the area and also reduces mobilization charges to and from the site.

In areas where some choice of drilling rig may be exercised, it is worth having the geologist who will conduct the investigation discuss options with the contractor.

In general, the choice of method of drilling is governed by the expected nature of the site geology, equipment availability and economics.

Figure 20. A typical lightweight drilling rig.

Courtesy: Soiltest, Evanston, Illinois

Coreboring — As the name implies this method has the ability to take out a solid continuous core which may then be taken away for detailed examination. The drilling bit itself is hollow with the cutting edge around the periphery of the tube. Core barrels are mounted in sections (usually 2 feet sections) above the cutting bit and as the boring progresses additional sections may be added. In landfill evaluation work, it is usually unnecessary to retain a long continuous core of the entire geological structure. Commonly, cores may be required at 3 to 5 foot intervals or wherever a significant change in geology occurs. The core of material is retained inside the barrel by a flap or core retainer in the end near the cutting edge. As the drill is retracted, the weight of the core closes the flap and prevents the loss of the core. However, the extent to which the sample remains intact is a function of the geological material. In clay, the percentage core recovery may be as much as 90 percent, but in dry sand or weathered rock recovery may fall to as little as 30 percent.

Figure 21. Casing shoe bits for core drilling.

Courtesy: Soiltest, Evanston, Illinois

Of all the boring methods available, core boring tends to be the most costly though in most instances, it does have the capability to provide the most complete information.

However, on sites where the geology is thought to be fairly uniform, it may not be necessary to go to the expense of coring and the auger drill may be adequate.

Auger – As the name implies, the auger drill is merely a long continuous helix or corkscrew. It is usually mounted on the same kind of rotary drilling rig used for core boring and as the drilling goes deeper, additional sections (usually 5 foot in length) are added to the drill head.

The system has the advantage of relative simplicity and ease of operation though it can only be used in unconsolidated material. It has the disadvantage that the sampled material is not withdrawn intact, but in a continuous stream of cuttings from the drill head.

However, an experienced geologist can often provide adequate field interpretation and preserve representative samples for further investigation.

Hollow Stem Auger – This type of drill combines the speed and simplicity of the auger with the advantages of core boring. A continuous core of unconsolidated material can be retained in the hollow stem of the auger wherever core samples may be thought appropriate. Five feet sections are added exactly as described previously for the auger except that these sections are hollow.

Piezometric sampling tubes may even be inserted down the hollow stem, but the limited core diameter reduces the value of this expedient. In general, only a one inch casing will pass down the hollow stem and this is unduly narrow for groundwater sampling. Also, due to the limited space remaining in the hollow stem, it is virtually impossible to pass sand or gravel down with the piezometer and groundwater flow into the instrument is, therefore, often restricted. To reduce this problem somewhat, a number of slots are often cut into the piezometer tube for a length equivalent to the depth of the aquifer.

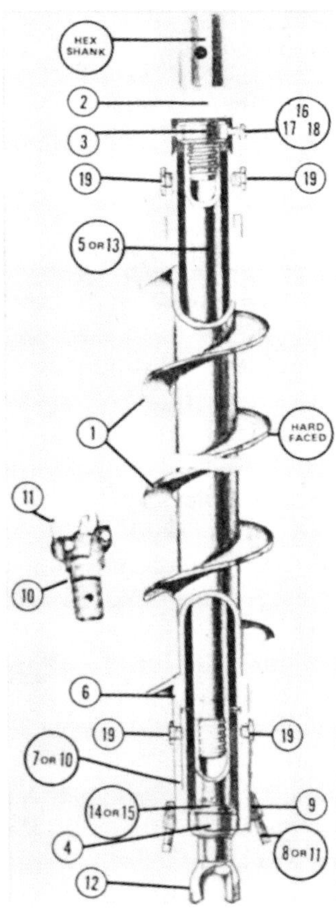

1. Auger
2. Adapter Cap
3. Adapter Center Stem
4. Center Plug
5. Center Rod
6. Complete Lead Assembly
7. Carbide Insert Cutter Head
8. Carbide Insert Fingers for Cutter Head
9. Pins for Fingers
10. Conical Finger Cutter Head
11. Finger Bits for above Cutter Head
12. Drag Bit
13. Center Stem
14. Drive Clip Type Pin
15. Socket Set Screw
16. Steel Bolt
17. Lock Nut
18. Lock Washer
19. 7/8" Lock Screws

Figure 22. Hollow earth auger.

Courtesy: Soiltest, Evanston, Illinois

In spite of the disadvantages in placing sampling tubes hollow stem augers are very economical for the continuous sampling of unconsolidated materials. However, neither of the two augers discussed here is appropriate for rock boring.

Mud Rotary – The use of the mud rotary is often suggested by a suspected shallow depth to rock. A large number of contractors use either the mud or the air rotary type and it is, therefore, very competitive.

The rotary bit itself is usually of the roller or tri-cone type. Rotation of the bit in the bottom of the borehole sets up a chipping and crushing action as the cones rotate. Teeth are usually steel, often surfaced with tungsten carbide or other hard facing. Full carbide inserts are used for the hardest and most abrasive formations.

Figure 23. Rotary tri-cone bits. These bits are used in hard (Type H) and medium (Type MS) formations with drilling fluid or compressed air. Used in exploration and geophysical work. Advantages of this type bit includes: avoiding gumming in shale; removes large cuttings and longer life in difficult formations.

Courtesy: Soiltest, Evanston, Illinois.

The mud rotary derives its name from the use of a liquid slurry as a cutting agent. The cutting fluid is fed to the rotary bit by hollow tube down the center of the rig. The fluid (commonly a bentonite or kaolinite slurry) keeps the drill head cool and floats cuttings up to the surface. Here lies the greatest disadvantage of the system since there is very poor definition of geological sequence. An experienced driller can approximate the location of major changes in structure by the subtle changes in drilling action, but sampling is complicated by the fact that different materials exhibit different rates of flotation in the cutting liquid.

Nonetheless, with care and experience samples can be retained by this method and taken back to a laboratory for analysis. A bio-degradable lubricant is often used when drilling wells by this method as the regular slurry has a tendency to cake the hole and prevent water movement into the well-point.

Air Rotary – This method of drilling is similar to the mud rotary except that compressed air is used as the lubricant instead of mud. The compressed air tends to eject cuttings from the borehole more efficiently and for this reason sample definition is somewhat less fuzzy than the mud rotary method.

In both methods, the efficiency of boring (which is a direct measure of drilling cost) tends to be very high. It is not unusual to achieve as much as 100 feet of drilling per day where the number of hole changes and mobilization is kept to a minimum.

A summary of the foregoing discussion of alternative drilling techniques is presented in Table 10.

TABLE 10

Summary of Landfill Site Alternative Drilling Techniques

Material	Technique				
	Coring Drill	Auger	Hollow Stem Auger	Mud Rotary	Air Rotary
Sand — Dry	X		X	X	
Wet	X		X	X	
Silt	X	X	X	X	X
Clay	X	X	X	X	X
Rock Weathered	X			X	X
In situ	X			X	X

Boring Plan

Assuming the choice of boring system is made, a boring plan should be drawn up to guide the drill operator. The field geologist will stand by and make adjustments in his plans as necessary based on his interpretation of the results at the drill head. The importance of a good and experienced field geologist cannot be overstressed. Field investigation by drilling should be a one shot effort wherever possible. To return later with the rig for more data because the wrong decisions were made in the field can be very costly and time consuming.

The importance of previously collected boring information may become apparent at this point since the choice of new borings should be made in a manner which makes maximum use of previously collected data (where it is felt to be reliable).

Previously obtained information regarding groundwater levels and potential hydraulic gradients is also valuable here since careful setting of the borehole positions at this stage may permit some of them to double later as monitoring wells. The surrounding groundwater quality data is also important to the monitoring process, and it should be born in mind while establishing the boring plan that up-gradient monitors (not necessarily on the property) are also important.

Sampling and Analysis

Soils — During the course of the boring operation, soil samples should be taken at 5 foot intervals or at each major change in soil type whichever occurs first. When a transition zone is reached, the borehole is usually pumped clear of cuttings (and lubricant if the mud rotary is used) and a soil sampler is used to obtain an undisturbed sample.

The split tube or split spoon sampler is one of the most popular soil samplers. It is merely a long tube with a cutting edge at one end split in two down its length and

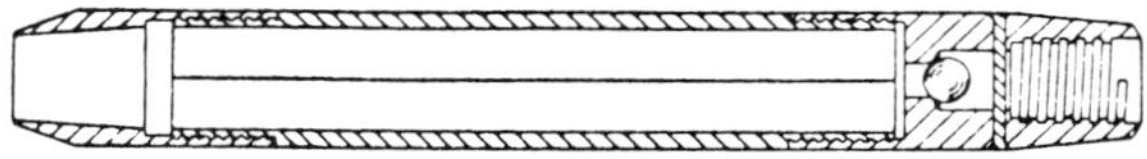

Figure 24. Split tube sampler. One of the most popular types of soil samplers is the split tube or split spoon sampler. This sampler is a tube split lengthwise and held together by a head and a hardened shoe. The sampler is forced into the soil by jacking, hydraulic pressure or driving. The sampler is equipped with a ball check valve in the head which prevents washing out of the sample while being hoisted from the hole. The ball check can be removed if not required. When withdrawn from the hole the sampler shoe and head are unscrewed and the sampler opens like a book.

Courtesy: Soiltest, Evanston, Illinois

held together with a head and a hardened shoe. The sampler is driven into the soil and the sample itself drives air out through the ball check valve. When the check valve is closed, it prevents the sample being washed out while the sampler is pulled from the borehole. The sample may then be removed by unscrewing the head and the shoe and the split tube opens like a book.

The samples obtained in this way should be carefully preserved for laboratory analysis. Each sample should be labeled to indicate the location on the site where the sample was obtained and also the depth at which each sample was taken. Dependent upon the depth of sampling, the soil will ultimately provide cover material for the compacted refuse or it will form the crucial interface between the refuse and the groundwater table. In either event a sieve analysis is important in order to determine the general soil classification on the ternary diagram. For soils which may ultimately form the interface, a reliable laboratory should be charged with the responsibility of determining:

- soil permeability
- pH
- cationic exchange capacity

As we will see later, these three factors are critical to the design of the soil interface.

For soil as cover material, the sieve analysis is very important. In addition, a determination of the specific soil type will prove useful. Soil types in categories A and B of Appendix C will generally prove to have better characteristics for cover material since they resist infiltration and encourage runoff.

Chapter 5 contains detailed information on a method by which soil pH and cationic exchange capacity may be used to determine the necessary interface thickness between high water table and landfill base.

Thus far, we have discussed drilling only in relation to soil investigations. There are two additional aspects of landfill design which on site drilling can assist in evaluating and they are depth to rock and groundwater location.

Rock — Auger drilling will not normally penetrate rock unless the rock structure itself is quite soft and friable. We are concerned to know at what depth rock is reached. On many sites, this is a comparatively simple matter since rock is generally synonymous with drill resistance or refusal.

However, in some cases, such as sites in the northern states, where the site may contain materials transported by glacial action (morraine) many large boulders may give false first refusal readings of depth to rock. Obviously, site interpretation based upon this kind of false reading of depth to rock can result in a significant waste of landfill volume since the soils underlying the rock fragments would be left intact and never exploited as cover material.

Depth to rock as measured by drill refusal is not necessarily the limit of moisture movement. In igneous and metamorphic rock particularly the weathered zone may be quite resistant to drilling and yet it may represent several feet of permeable material before an impermeable rock surface is reached. Some additional rock drilling may be necessary on some sites to determine the extent to which any rock surface is rendered permeable by extensive fracture cracks. If the rock is highly fractured, there is a risk that leachate would move through the fissures in the rock and establish a path to the groundwater system. In this situation, the landfill designer is concerned to establish an adequate soil separation between the rock and the refuse in order to ensure adequate renovation. In cases where the rock may prove permeable to water movement, the rock structure is logically an extension of the soil separation zone. Commonly, rock has reduced capacity for leachate renovation except for filtration capacity and some limited characteristics which may extend the process of chemical treatment.

Groundwater — If the site itself has been carefully chosen, the depth of groundwater will be quite adequate to allow an appropriate depth of soil interface and still leave room for an economic life. All that is required at this stage is to confirm the results of the original investigation.

However, it is here that, with forethought, the requirements for leachate monitoring can be anticipated and the drilling necessary to check depth to groundwater can be carefully selected to double at a later stage as a sampling point.

The direction of groundwater movement should be established wherever possible though on some sites this can be complicated by complex geological formations. Movement will occur from points of high groundwater level in the direction of points at lower groundwater level. In other words, a groundwater gradient line may be drawn to identify the direction of flow.

The presence of the drilling rig on site should also prompt questions about water supply. If a new source of water may be required for maintenance purposes, it is wise to establish a water well on the site and mobilization charges for the rig will only have to be paid once.

Contour Map — Once the soils and geological examinations are complete, the

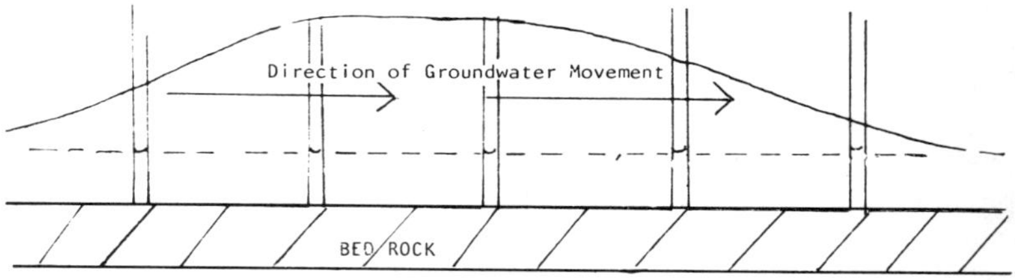

Figure 25. Groundwater gradient.

next basic requirement is a detailed contour map. The USGS sheets used earlier for site investigation are fine for general purposes, but a more detailed knowledge of the site topography is now required.

Many states specify the contour interval required (this may be anything from 1 foot intervals to 5 foot intervals). Where no specific requirements are made 2 foot contour intervals are usually adequate.

Commonly, contour maps to this level of detail are not available and they must be specially prepared for the site. Before this relatively expensive task is begun, it is not a bad idea to consider the past history and general location of the area as contour mapping may have been completed for some other purpose. The following table provides some possible reasons why the area may have been mapped in detail and gives an indication of the source of the detailed information.

TABLE 11

Potential Sources of Detailed Contour Maps

Source	Original Reason for Mapping
Regional Planning Commission	Comprehensive regional mapping
City Planning Office	General planning
Department of Transportation	Alternative highway mapping
State Department of Natural Resources	Statewide natural resources inventory
Local Department of Economic Development	Once considered for industrial development

It should be considered a stroke of good luck if detailed contour maps are found for the area selected as a landfill site, but the possibility should not be discounted altogether and a preliminary search should be made.

If none of these sources yields contour information, a detailed contour map will have to be produced. There are two options available:

- Surveying
- Aerial survey

Surveying – The survey work will require the services of an experienced surveyor and a crew of at least one or two assistants. Also a good surveying level, staff, transit chain and pins and a knowledge of the whereabouts of the nearest bench marks.

It may be that a legal survey performed at the same time will reduce the total surveying costs.

Compared to aerial survey contouring, survey work can be quite costly and time consuming. However, in the winter months during periods of extended snow cover, there may be no choice if the contouring needs cannot wait.

Aerial Survey – Aerial contouring can be an extremely fast means of obtaining accurate contouring information. In particular, the method is unsurpassed where there are areas of difficult terrain. Some ground surveying is required, of course, in order to establish ground control. The chief disadvantage is that weather conditions do not always permit flying. Snow cover and dense vegetation prevents accurate interpretation of the air photographs from which the contours are drawn.

Whichever method is chosen, the final result should be a detailed (reproducable) contour map which can be used with confidence through the many phases of the design process.

Chapter 5

DESIGN

Interface Design

The first and most important question in the design process is to determine what will be the average elevation of the base of the landfill. To establish this elevation it is necessary to know what the soil separation will be between landfill base and water table.

There is no simple answer to this question. In fact, there is considerable disagreement. In many instances, the characteristics of the soil separation are ignored and statements are made in some texts which suggest that depth of separation is the only important characteristic. Many states rather arbitrarily specify a certain depth of separation. Without analysis of the subsoil, there is no scientific basis for such requirements. Many texts on the subject of sanitary landfill ignore the question completely. Indeed, soil suitability is often discussed as though the only critical factors were those related to the use of soil as cover material.

This book attempts to establish criteria for both aspects of soil use:

- as landfill/water table separation
- as cover material

However, it should be plainly stated and clearly understood that the criteria suggested can only be used as a guide. The subject is not yet sufficiently well understood that figures can be given without the need for qualifiers. Also, in the final analysis all sites are different and there is no substitute for the judgement of an experienced professional with an intimate knowledge of a particular site.

The concept presented is perhaps more important than the suggested range of figures. I believe that only by analyzing specific characteristics of the soil interface can intelligent statements be made about how thick the interface should be.

What are the Governing Criteria? – In the chapter on Fundamentals, we discussed the process of leachate renovation. If we can identify the soil characteristics most active in the process of renovation, I believe these are the governing criteria for the design of the soil interface.

We may summarize the previously discussed mechanisms of leachate renovation as follows:

- Filtration
- Absorption
- Adsorption
- Chemical Precipitation
- Bacteriological Action

If we are to design the soil interface, we need to identify the soil characteristics most conducive to the development of each of these mechanisms.

Understanding the complex and varied nature of each of the mechanisms of renovation makes it extremely difficult to establish a single parameter which can be used as a key design factor. However, we may identify the chief mechanism of renovation, and design to maximize its contribution to the overall characteristics of a soil complex.

In truth, the established science of sanitary landfill ends here since there is no fixed body of agreement about the chief mechanism of leachate renovation. As in all sciences there is a substantial body of research but it is somewhat early to make a definite judgement on the subject.

Recent work on the use of soils to renovate mine drainage seems to support a growing body of opinion that the most important renovation mechanism is probably adsorption. The real answer probably depends upon the constituents of the landfill itself. However, if we are dealing with a typical sanitary landfill containing predominantly household refuse then we would be chiefly concerned that iron and other metal salts should be taken out of solution before any net moisture movement to ground water takes place. This might tend to support the general thesis that adsorption may be the most important mechanism.

However, having labored that thesis to death, it may be said that in some respects it does not matter. That statement requires extensive explanation but it may be initially summarized in the following brief paragraph.

The chief enhancer of adsorption capacity appears to be a high pH. This factor also aids in chemical precipitation and also provides a conducive environment for bacteriological growth. Thus, if we proceed on the assumption that adsorption is the chief mechanism, we will aim for high pH soil interface and in so doing improve some of the other main mechanisms of renovation.

The Importance of pH – Strictly speaking, pH is simply a shorthand method of referring to the quantity of free hydrogen ions (H+) in a solution. For convenience, since the numbers in a liter of solution are very very small, pH is the logarithm of the reciprocal of the hydrogen ion concentration.

$$pH = \log \frac{1}{H+}$$

pH is measured on a scale of 0 to 14. When the pH is less than 7, the system is acid. When the pH is above 7, the system is alkaline.

Due to the logarithmic relationship, a small increase in quantity of hydrogen ions is deceptive. A solution of pH equal to 4, for example, has 10 times more hydrogen ions than one of pH equal to 5, or 100 times more than one of pH equal to 6.

In solutions, pH is a direct measure of hydrogen ion concentration. In soils, the determination is complicated by the fact that the soil consists of two parts:

- soil water
- soil fabric

The pH of a soil/water system only approximates the hydrogen ion concentration since hydrogen ions are held on some of the cation exchange sites of the soil fabric.

We have said earlier that adsorption appears to be the most prominent mechanism of leachate renovation. The measure of adsorption is cation exchange capacity. It is defined as the quantity of positive ions necessary to neutralize the negative charge of a unit quantity of soil under a given set of conditions. Cation exchange capacity (CEC) is usually expressed as milli-equivalents of cations required to neutralize the negative charge of 100 grams of soil at pH equal to 7. It has been noted that in the B and C horizons, the dominant source of CEC is due to the presence of clay minerals. It is important to realize that clay minerals also possess pH dependant charges. In fact, the CEC for some clay minerals increases dramatically as the pH increases.

In general, as the pH increases above 6, the pH dependent charge increases progressively, reaching a maximum pH of around 10.

The relationship between CEC and pH for clay minerals and organic matter can be illustrated schematically as in Figure 26.

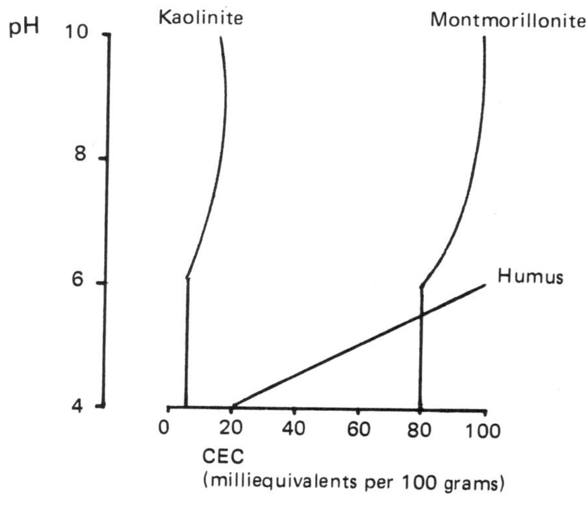

Figure 26. Relationship between pH and CEC for clay minerals and humus.

Source: Yearbook of Agriculture, 1957, U.S.D.A., Washington DC.

This brings us to two important design requirements:

- aim for highest attainable CEC
- enhance pH as much as possible

It will be clear from the foregoing that the second of these requirements also has the effect of increasing the CEC of the soil interface.

Table 12 gives a general indication of CEC values of various clay minerals.

TABLE 12

Cation Exchange Capacity of Clay Minerals[1]

Clay Minerals	**CEC (milliequivalence per 100 grams)**
Kaolinite	3–15
Halloysite $2H_2O$	5–10
Halloysite $4H_2O$	40–50
Smectite	80–150
Illite	10–40
Vermiculite	100–150
Chlorite	10–40
Sepiolite – Attapulgite – Palygorskite	3–15
Montmorillonite[2]	70–100
Zeolite (Chabazite)[2]	400

[1] Main Source: From "Clay Mineralogy", by Ralph E. Grim Copyright 1968 by McGraw-Hill, Inc., Used with permission of McGraw-Hill Book Company.

[2] Supplemental information from "The Encyclopedia of Geochemistry and Environmental Sciences". Edited by Fairbridge. Published by Van Nostrand Reinhold, (distributed by Halsted Press, New York City).

Infiltration, Percolation and Permeability

So far, we have been concerned with defining the characteristics of the soil interface which will most effectively renovate leachate. We have also identified (rather loosely) the fact that adsorption appears to be the chief mechanism by which dissolved solids may be taken out of solution as the leachate moves through the soil.

However, the actual movement of water into and through the soil also plays a most important role in the renovative process.

Earlier, we defined infiltration as the passage of water into the soil surface and percolation as the downward movement through the soil. Percolation is strictly a phenomenon which occurs in *unsaturated* soil. The downward flow is carried by the numerous noncapillary channels in the soil structure. Permeability defines the rate of movement of water through a *saturated* soil medium. Permeability can be measured in either a horizontal or a vertical plane.

The rate at which water moves through soil can be expressed as a percolation rate or permeability according to whether the soil is saturated or unsaturated. However, the limiting factor to infiltration and percolation is ultimately permeability. Unless the permeability of the subsoil is adequate to carry away water passed down the soil horizons first by infiltration then by percolation, the system will become water bound.

Permeability is influenced by particle size, void space and soil structure. Table 13 provides illustration of the influence of particle size on the coefficient of permeability (k).

The coefficient of permeability (K) is defined by Darcy's Law:

$$Q = K\,I\,A$$

Q = Quantity of water discharged per unit time
K = Coefficient of permeability
I = Hydraulic gradient
A = Cross sectional area through which water percolates

The void space in the soil is proportional to the particle size. The smaller the particle size, the smaller the void space.

Soil structure also influences permeability particularly in fine grained soils. When the soil structure is aggregated, the large channels created increase the permeability of the soil. Fine grained soils which are not aggregated generally have a much lower permeability.

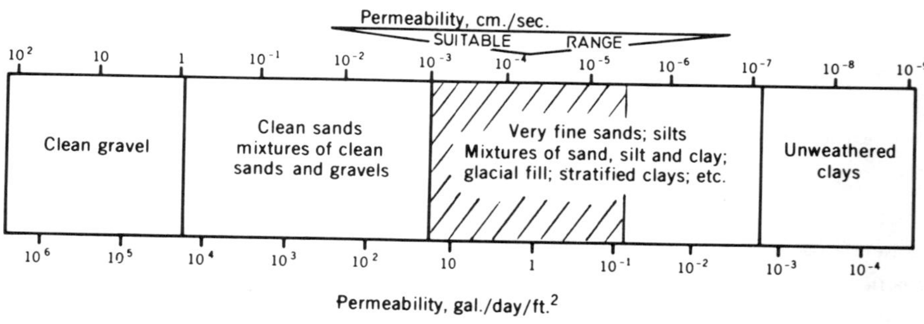

Figure 27. Range in permeability of different soil classes (modified from Todd, 1959, p. 53).

Permeability is presented in most soil science publications in descriptive terms. The terminology used in the Illinois Drainage Guide is fairly standard. The definitions are as follows:

TABLE 13

Influence of Particle Size on Permeability[1]

	Particle Size Range				"Effective" Size		Permeability Coefficient – *k*		
	Inches		Millimeters						
	D_{max}	D_{min}	D_{max}	D_{min}	D_{20} in	D_{10} mm	Ft/yr	Ft/mo	Cm/sec
Derrick STONE	120	36	–	–	48	–	100×10^6	100×10^5	100
One-man STONE	12	4	–	–	6	–	30×10^6	30×10^5	30
Clean, fine to coarse GRAVEL	3	1/4	80	10	1/2	–	10×10^6	10×10^5	10
Fine, uniform GRAVEL	3/8	1/16	8	1.5	1/8	–	5×10^6	5×10^5	5
Very coarse, clean, uniform SAND	1/8	1/32	3	0.8	1/16	–	3×10^6	3×10^5	3
Uniform, coarse SAND	1/8	1/64	2	0.5	–	0.6	0.4×10^6	0.4×10^5	0.4
Uniform, medium SAND	–	–	0.5	0.25	–	0.3	0.1×10^6	0.1×10^5	0.1
Clean, well-graded SAND & GRAVEL	–	–	10	0.05	–	0.1	0.01×10^6	0.01×10^5	0.01
Uniform, fine SAND	–	–	0.25	0.05	–	0.06	4000	400	40×10^{-4}
Well-graded, silty SAND & GRAVEL	–	–	5	0.01	–	0.02	400	40	4×10^{-4}
Silty SAND	–	–	2	0.005	–	0.01	100	10	10^{-4}
Uniform SILT	–	–	0.05	0.005	–	0.006	50	5	0.5×10^{-4}
Sandy CLAY	–	–	1.0	0.001	–	0.002	5	0.5	0.05×10^{-4}
Silty CLAY	–	–	0.05	0.001	–	0.0015	1	0.1	0.01×10^{-4}
CLAY (30 to 50% clay sizes)	–	–	0.05	0.0005	–	0.0008	0.1	0.01	0.001×10^{-4}
Colloidal CLAY ($-2\,\mu > 50\%$)	–	–	0.01	10Å	–	40Å	0.001	10^{-4}	10^{-8}

Typical values of permeability coefficients.

[1] From "Wastewater Management For Lake Michigan – A Study of C-SELM", by U.S. Army Corps of Engineers, July, 1971.

1. Rapidly permeable (more than 6 inches per hour) and moderately rapidly permeable (2 to 6 inches per hour).
2. Moderately permeable (.6 to 2 inches per hour).
3. Moderately slowly permeable (.2 to .6 inches per hour).
4. Slowly permeable (.06 to .2 inches per hour) and very slowly permeable (less than .06 inches per hour).

For landfill design moderately slow permeability is desirable so that movement of leachate will be retarded, and renovative processes will have maximum time to operate. On the other hand, soils that are almost impermeable, such as clays, will not permit downward percolation of leachate and may cause perched water conditions and concomitant surface seepage problems. Soils that have moderately rapid permeability (greater than 2.0 inches per hour) are generally classified as hazardous to ground-water quality since leachate could move rapidly to the water table.

In general, the soils in the moderately permeable or moderately slowly permeable class would be preferable for the soil interface between the landfill base and the water table. These soils allow sufficient contact time for renovation action while preventing the build up of water head to a point where horizontal movement of the leachate may take place.

Interface Design Criteria – At this point, it may be useful to summarize the features of the soil interface which appear to be most important in the renovation process:

- High cation exchange capacity
- Moderate permeability

In addition, we have learned that a high pH either in the leachate or in the soil interface will enhance cation exchange capacity whilst also increasing chemical precipitation.

It should also be clear at this point that specific design criteria for a soil interface can only be established in relation to drainage design and final use. The amount of infiltration and percolation which the surficial design will allow to enter the compacted refuse is essentially the amount which the soil interface must be capable of renovating, aside from evapo-transportation.

From a hydrogeological point of view, the development of specific design criteria to achieve this end amounts to a very tall order. We have talked thus far without much acknowledgement that the earth upon which we are obliged to build landfills rarely exists in uniform sandwiches of geological material. The qualities of the varied sub-base we have to work with can, of course, be estimated by taking samples in a few locations. These samples can be analyzed for a few specific criteria and those criteria applied to a design process.

The State of the Art makes this possible, but it must be said that in the final analysis each site must be judged on its merits. The information presented here may be used as a general guideline to establish a reasonable design strategy.

To establish a methodology to design a soil interface for maximum renovation, we initially require a set of leachate characteristics. As we have observed earlier these tend to vary considerably.

However, for the purposes of illustration, I propose to select from the leachate characteristics which the EPA makes most use of in their own design literature[1]. If we put together a list of leachate characteristics from the high side of the quoted analysis, we should have a fairly representative indication of characteristics typical of an initial leach. (It is noted later that after the initial period, there is a decline in leachate strength.)

The hypothetical leachate then has the following characteristics.

TABLE 14

Hypothetical Leachate Characteristics

		mg/litre
Hardness	$CaCO_3$	8,000
Alkalinity	$CaCO_3$	9,500
Calcium	Ca	2,500
Magnesium	Mg	500
Sodium	Na	2,000
Potassium	K	2,000
Iron (Ferric)	Fe	1,500
Iron (Ferrous)	Fe	500
Chloride	Cl	2,500
Sulfate	S	750
Phosphate	P	100
Ammonia Nitrogen	NH_4–N	500
BOD		30,000
Zinc	Zn	150
Nickel	Ni	100

What is known about these particular parameters which will assist us in determining their respective amenability to renovation?

We know whether cations or anions are formed in solution – this can be readily added to the table. We also know the valence or the number of ions which will be formed per atom and hence the respective ability to be adsorbed by soil particles – this can be shown by the number of + or – signs next to the chemical symbol.

We also know the equivalent weight of each parameter (i.e. atomic weight divided by valence) – this will enable us to determine the milli-equivalent weight. (Remember that cation exchange capacity is measured in milli-equivalents per 100 grams.)

---

[1] "Sanitary Landfill Design and Operation", D. R. Brunner & D. J. Keller, U.S. EPA 1971.

If we know the milli-equivalent weight of each parameter in addition to the concentration of that parameter in the leachate, we can work out the quantity of milli-equivalents per litre of leachate.

For example: Calcium has an equivalent weight $= 20$

or a milli-equivalent weight $= \dfrac{20}{1{,}000}$

In a liter of water this much calcium would produce a concentration of $\dfrac{20}{1{,}000}$ grams/litre or 20 milligrams/litre

Now, the leachate contains 2,500 milligrams/litre of calcium.

Which is equal to $\dfrac{2{,}500}{20}$ milli-equivalents/litre

or 125 milli-equivalents/litre

Each parameter can be analyzed in this way to produce the following table:

TABLE 15

Leachate Characteristics and Cationic Exchange Capacities

Leachate Constituent	Ion Charge	Equivalent Weight	Cationic Milli-equivalents per litre
Hardness	$CaCO_3$	*	—
Alkalinity	$CaCO_3$	*	—
Calcium	Ca++	20	125
Magnesium	Mg++	12.15	41
Sodium	Na+	23	87
Potassium	K+	39.09	51
Iron (Ferric)	Fe+++	18.62	81
Iron (Ferrous)	Fe++	27.92	18
Chloride	Cl-	**	—
Sulfate	S--	**	—
Phosphate	P---	**	—
Ammonia-Nitrogen	NH_4 *N+	17	29
BOD		***	—
Zinc	Zn+	32.7	5
Nickel	NI++	29.4	3
		TOTAL	440 Milli-equivalents

* These are non-specific indicators of hardness. The causes of hardness and alkalinity are, of course, the dissolved solids measured below.

** These negatively charged ions (cations) can only be removed by positively charged soil particles (anions). Most soils have relatively low anionic exchange capacity. However, there are characteristically few cations in leachate and the soils necessary to achieve high cation exchange rates will generally contain sufficient anionic exchange capacity to remove cations from the leachate.

*** Not applicable. Organic fractions are renovated by chemical and bacteriological action.

Total cations present per litre of leachate expressed as mill -equivalent = 440. Now we may assume soil CEC = 15 milli-equivalents per 100 grams of soil. This is a good average for the correct soil type though it is impo tant that the f gure should be derived from test results in all cases. The cation exchange capacity at the proposed site should be measured at the projected depth of the landfill base. An average of the measured CEC would then be used for this evaluation.

At the C horizon where the soil interface will usually begin, some of the exchange sites may already be "occupied" by ions leached downward in the normal course of soil development. However, at this depth, the percentage of occupied sites is commonly very low. Also, weakly held cations will be displaced by cations possessing a stronger charge and a higher valence. Later in the process of decomposition newly charged sites will actually be created as a result of the natural breakdown of the soil structure.

Now we have 15 milli-equivalents of cation exchange capacity for every 100 grams of soil, and we require enough soil to renovate leachate containing 440 milli-equivalents. Therefore, we need 440/15 × 100 = 2933 grams of soil/litre of leachate. In other words 2933 grams of soil are required to renovate 1 litre of leachate.

Changing units to those in most common use we have:

2933 grams to renovate 1 litre of leachate
Equivalent to 0.0624 × 2933 pounds to renovate
1 cubic foot of leachate

or 183 pounds to renovate 1 cub c foot of leachate.

Now, the density of a good clay loam (oven dried for 24 hours) ranges from 70 to 100 pounds per cubic foot. Taking 85 pounds per cubic foot as average, the *volume* of soil required for renovation may be written as follows

$\frac{183}{85}$ cubic feet of soil to renovate 1 cubic foot of leachate

or 2.15 cubic feet of soil to renovate 1 cubic foot of leachate

or 2.15 *feet* of soil to renovate 1 *foot* of leachate

More generally for any location

$$R = \frac{6.25\ L_{cec}}{S_{cec} \cdot S_d}$$

Where R = Renovative thickness of soil in feet of soil per foot of leachate.

L_{cec} = Cationic Exchange Capacity of Leachate in Milli-equivalents per litre.

S_{cec} = Cationic Exchange Capacity of Soil in Milli-equivalents per 100 grams of soil.

S_d = Density of soil (oven dried for 24 hours) in pounds per cubic foot.

Or where my suggested leachate characterization is acceptable:

$$R = \frac{2{,}812.5}{S_{cec} \cdot S_d}$$

If we know the total amount of leachate generated per year, we can estimate the ideal depth of soil interface between landfill base and water table.

From the earlier discussion of the fundamentals of sanitary landfill operation, we established the following general equation:

$$L = P - (E + T + R)$$

Where L = Leachate
P = Precipitation
E = Evaporation
T = Transpiration
R = Runoff

Later in this section on design, simple methods will be presented to establish evapotranspiration and runoff. Thus, we can obtain all the necessary unknowns in the above equation. Take New York City rainfall as an example:[1]

Average Annual Precipitation (P) = 49 inches
Average Annual Evapotranspiration (PE) = 26 inches[2]
Runoff (R) = 17 inches[3]

$$L = 49 - (26 + 17)$$
$$L = 6 \text{ inches per year}$$

Thus, on the basis of 2.15 feet of soil for 1 foot of leachate, about 1 foot of soil interface would be adequate to renovate the total leachate quantity estimated for the first year.

For each subsequent year, the leachate will be quantitatively, but not qualitatively the same. Assuming continual breakdown of the landfill contents the leachate will undergo an improvement in overall quality.

The results of experimental work performed by R. C. Merz in California as reported in an EPA publication[4] on landfill design provide some clues to leachate improvement with time.

[2] Derived from the section dealing with evapotranspiration.

[3] Derived from a later calculation in the section dealing with runoff determination. Total Annual Runoff = 16.8 inches.

[4] "Sanitary Landfill Design and Operation", U.S. EPA – 1971.

The onset of leachate occurred after about 1 year and the cation concentration declined rapidly to about half the concentration each subsequent year until the end of the third year. However, reports of other experiments vary considerably. At this point, the conscientous designer can only make what appears to be a reasonable approximation of a worst case and use that as a basis for design until more information comes to light on the rate of decline of leachate concentration.

In any event, the importance of this section rests chiefly on the development of a methodology. Where known local data are more reliable than the examples presented here that data should be used.

Where no good data is available, the assumptions presented here are offered as a good approximation considering the state of the art at the present time.

It is proposed to consider leachate concentration at a maximum immediately the landfill reaches field capacity. (In fact, some minor leaching may occur earlier due to channeling or as a result of a *wetting front* moving through the fill.) The initial concentration usually declines rapidly during the first year. However, we may simplify later computation by assuming constant cation concentration during the first (and each subsequent) year of leaching. We will assume 50 percent concentration in the second year of leaching and 25 percent in the third. During the fourth year and each subsequent year, we will assume the leachate to be at 12½ percent of the initial concentration. Students of leachate studies will recognize that this is a fairly generous approximation of *leachate* concentration since the decline in concentration is usually considerably greater than indicated. This idealized leachate concentration is illustrated in Figure 28.

In the New York City example already quoted we found that one foot of soil (assuming average CEC = 15) would be adequate to renovate first year leaching. Thus, the second year would require 0.5 feet, the third year 0.25 feet and each subsequent year 0.125 feet.

I will assume we may safely take 20 years as the period of interest as far as leaching is concerned. A well designed site will achieve a substantial age before field capacity is reached, sometimes as much as 20 years. With the addition of the 20 years of leaching, that represents a total period of 40 years. After some years of leaching the decline in leachate concentration is substantial. After 40+ years leachate concentration from a well designed site should be negligible.

To provide *renovation capacity* over a twenty-year period of leaching, the soil depth required in our example would be 3.875 feet.

It should be noted that the cell depth in the California study was a little over 19 feet. Therefore the soil interface thickness of 3.875 feet is appropriate only in the case of a 19 feet cell. (Or 0.2 feet of soil is required per foot of cell depth.) For the more common example of eight foot lifts, one lift would therefore require 1.6 feet, two lifts 3.2 feet and three lifts 4.8 feet. Table 16 summarizes this.

I think it may be necessary at this point to back up a little and make some general observations about these figures. I think it is important to emphasize how

*The concept of renovation *capacity* is critical to this statement. A thorough understanding of this capacity will make it possible to establish a design life for landfills according to their capacity to renovate (and no more).

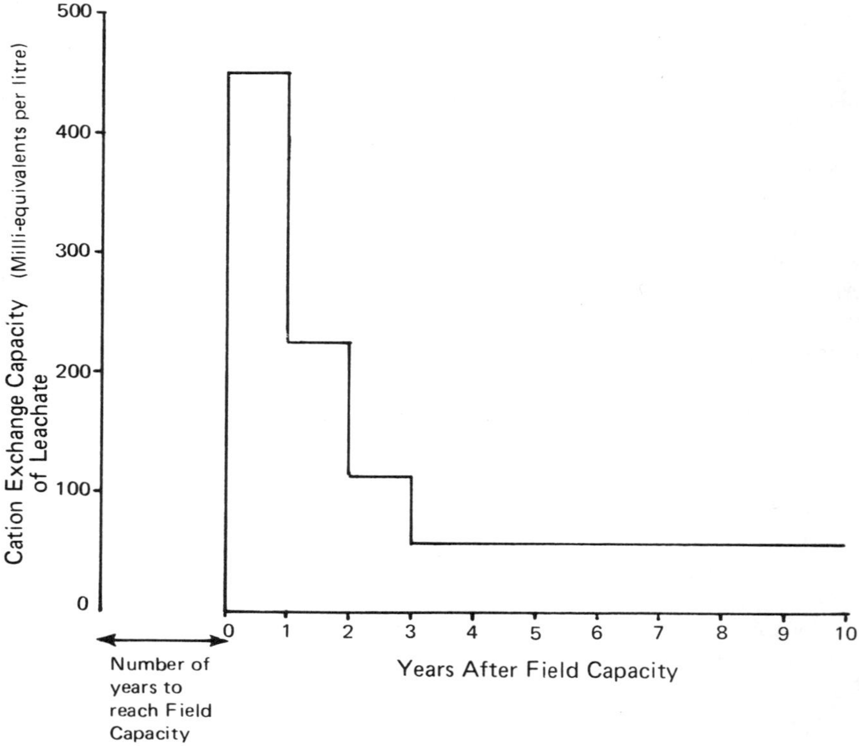

Figure 28. Idealized leachate concentration curve.

TABLE 16

Depth of Soil Interface for Various Depths of Fill

Number of Cells (Average Depth 8′/Cell)	Depth of Soil Interface (feet)
1	1.6
2	3.2
3	4.8
4	6.4

these figures were arrived at rather than these specific results. It is important to re-emphasize that these are only rough estimates based on available information. If the methodology presented here can be taken and refined in light of new data, I will be more than satisfied.

For the present, the state of the art is such that we need to consider including a safety factor in the above analysis.

We have observed the importance of cation exchange capacity (CEC) in the preceding analysis such that this point need not be re-emphasized. We also noted two independent elements of cation exchange capacity in the soil:

- constant CEC
- pH dependant CEC

I suggest the latter may be very easily exploited in landfill design by simply laying down a 2 feet to 3 feet bed of demolition debris at the base of the landfill. The percolation of leachate through this layer of calcium-carbonate will raise the pH substantially, thus potentially increasing the CEC of the soil interface by as much as 20 percent.

I do not suggest that this should be included in earlier computations because of the uncertain and varied nature of the action, but I do suggest that this could be an important safety factor in the landfill operation.

Landfill Method

Once the determination of soil interface thickness has been made, the specific method for landfilling may be selected. In theory, if the soil interface depth is added to the high water table elevation, the new elevation represents the base of the landfill. The volume of suitable cover material which remains above the base elevation determines the life of the site unless cover material is imported from elsewhere.

Most texts refer to two different methods of landfilling, namely:

- trench method
- area method

Some texts throw in an extra method for good measure (the ramp method — a transition between the two).

In truth, it is often hard to determine where one definition starts and the other finishes.

Trench Method — Basically, the trench method involves digging a deep trench usually with a drag line and systematically filling the trench by the accepted layering method. The method has a number of limitations, not least the fact that a substantial volume of cover material remains unused between the trenches. Also only one lift is possible. When the trench is filled, there is no potential for further lifts unless artificial trench walls are built from "spare" cover material (not a recommended procedure). The remaining finished surface is often subject to differential settlement with little or no settlement between the trenches.

The method may have uses in some rare circumstances, (in particular where containment may be practiced for hazardous material) but it does not use any site to its fullest potential. I will, therefore, concentrate on the area method.

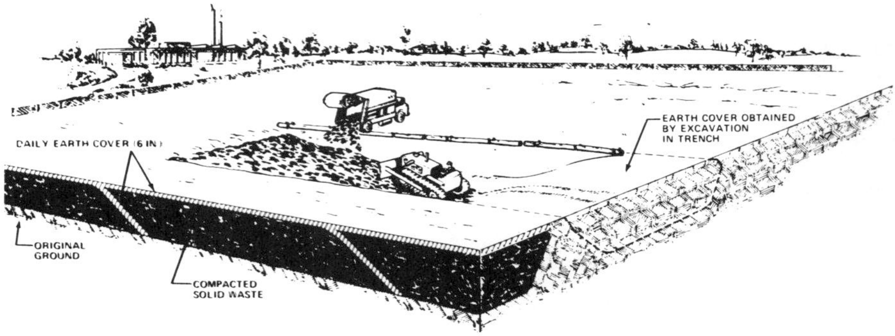

Figure 29. Trench Method. The waste collection truck deposits its load into the trench where the bulldozer spreads and compacts it. At the end of the day the dragline excavates soil from the future trench; this soil is used as the daily cover material. Trenches can also be excavated with a front-end loader, bulldozer, or scraper.

Source: "Sanitary Landfill Facts", by Sorg & Hickman, U.S. Department of Health, Education and Welfare, Bureau of Solid Waste Management, 1970.

Area Method – The area method may be very simply visualized as the complete excavation of a site down to the base elevation followed by the systematic building up of fill areas according to a predetermined plan. The complex pattern of cutting cover from one area and filling refuse in another which is typical of the area method may perhaps be one reason for interest in the simpler techniques of the trench method. However, with care and a good plan, there is no reason to loose track of the area method.

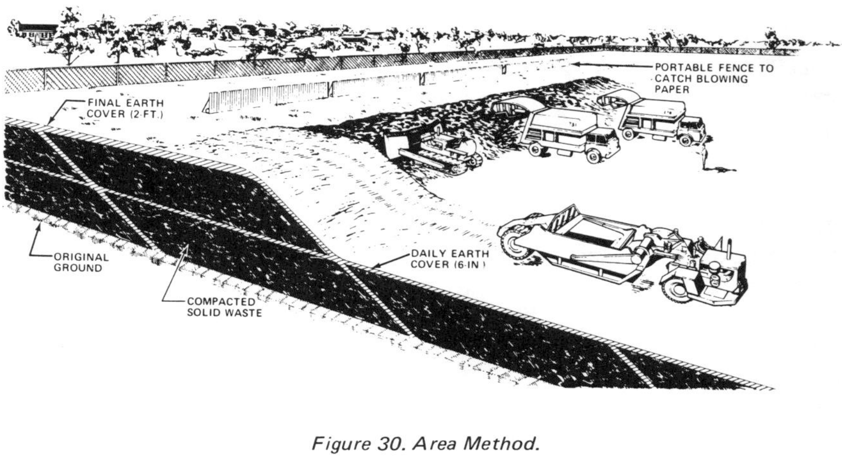

Figure 30. Area Method.

Source: "Sanitary Landfill Facts", by Sorg & Hickman, U.S. Department of Health, Education and Welfare, Bureau of Solid Waste Management, 1970.

Volume Calculations — At this point we may reassess the projected refuse volume which the site will take. Having established the refuse/water table separation it is a simple matter to calculate the volume of soil remaining. This may be achieved by the application of the Prismoidal Formula. In my own view, any more complex technique is inappropriate considering the many changes which the site will go through before final cover is laid. An estimate within 10–15% is all that is required at this stage.

Prismoidal Formula — In order to apply this method of volume calculation it may be necessary to break up a complex site into a series of approximately regular prismoidal shapes. Figure 31 illustrates the way in which a site may be reduced to fairly simple prismoidal volumes.

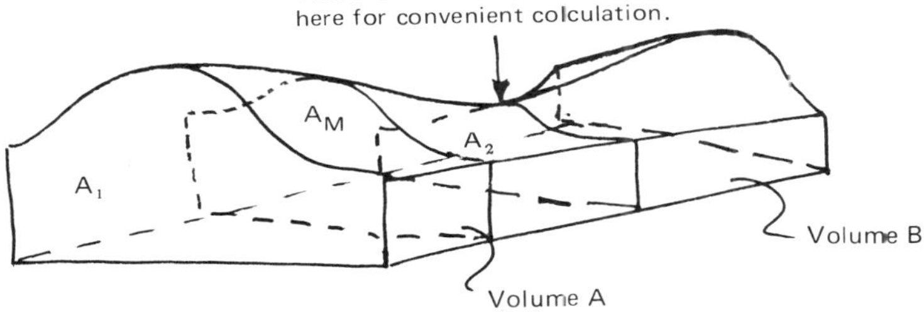

Figure 31. Sectioning site for volume calculations.

Now each section can be handled using the formula:

$$V_A = \frac{L}{6}(A_1 + 4A_M + A_2)$$

Where V_A = Volume of prismoid A
and A_M = Area of the midsection between A_1 and A_2

Clearly the application of this formula depends upon the availability of a simple method of calculating the areas at each cross section.

There are two simple methods available:

- planimeter
- give and take

If a planimeter is available the area may be obtained accurately provided the instrument is properly calibrated and handled with care.

Alternatively, the areas may be approximated by using the give and take method. As shown in Figure 32, the area is first drawn on squared paper. The more complex the shape the smaller the graph scale.

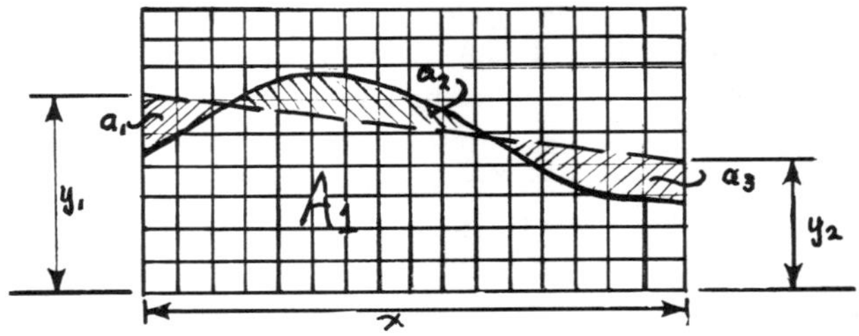

$$a_2 = a_1 + a_3$$

Figure 32. Give and Take Method.

Place a transparent straight edge across the area in such a way that the area above the line is approximately equal to the two areas below the line. With a little trial and error the straight edge can be moved until the areas are equal. Use the squares for reference to check the areas on either side of the line. Now the area of A_1 can be simply computed as:

$$A_1 = x\left(\frac{y_1 + y_2}{2}\right)$$

The midpoint area A_M and the end area A_2 can each be calculated in this way and the prismoidal formula applied very simply.

The result of these computations will yield a prismoidal volume approximately equal to that of the section of the site being considered. Obviously the same technique is used for all the volume sections and the aggregate yields the total volume of cover material available.

At this point many authorities suggest multiplying the cover volume by some simple factor to obtain the projected volume of the refuse which can be taken at the site. This is a good idea so long as it is appreciated that the figure is intended as a rough check only. The idea that cover material is 20 to 25% of the total volume of the site as stated in some texts is based upon an estimate of six inches of daily cover plus two feet of final cover. It is perfectly correct, except that the calculations provide the designer with no clear idea of *when* the bulk of the cover material is required.

The truth of the matter is that the bulk of the cover material is not required until the site is filled. In planning the fill operation from the beginning it is more useful to know the sequence of cover requirements in order to be able to plan an orderly stripping operation.

The cover requirements are a function of daily refuse intake since the compacted refuse is required to be covered only once a day. Thus it is a simple matter to convert the projected daily refuse intake in tons to a projected cover requirement in cubic yards. For convenience Table 17 may be used. Note that an in-place density of 1000 lbs. per cubic yard is assumed, this should be a good average. If the site yields much less in practice some reevaluation of the operation and/or the equipment may be necessary.

TABLE 17

Daily Cover Material Requirements (Based on 6 inch daily cover)

Refuse Intake (Long Tons per day)	In Place Volume @ 1000 lbs per yds^3 (yards^3)	Total Area of Daily Cell (yards^2) Cell Depth in Feet			Total Daily Volume of Cover Material (yards^3) Cell Depth in Feet		
		6 ft	8 ft	10 ft	6 ft	8 ft	10 ft
25	56	28	21	17	5	4	3
50	112	56	42	34	9	7	6
100	224	112	84	67	19	14	11
200	448	224	168	134	37	28	22
500	1120	560	420	335	94	70	56
750	1680	840	629	502	140	105	84
1000	2240	1120	839	671	187	140	112

In order to set aside sufficient cover material for the final 2 feet which is applied upon completion it is now necessary to consult Table 18.

TABLE 18

Final Cover Material Requirements

Acreage	Final Cover Material Required (yards^3)
25	80,700
50	161,400
100	322,800
150	484,200
200	645,600
250	807,000
300	968,500

For intermediate acreages the final cover requirement is equal to 3228 yd^3 per acre.

Now we have sufficient information to plan the total life of the site. Assuming the total volume of cover material is V and the life of the site is L in years then the following equation holds true:

$$V = \text{Final Cover per acre} \times \text{Acreage} + \text{Number of operating days per year} \times \text{Daily cover} \times L \text{ yrs.}$$

$$\text{or } V = 3228A + 300^{*}\, DL \quad \text{where V is in cubic yards.}$$

To take a specific example let us assume we have a 25 acre site available for landfill and plan to take in about 500 tons per day laid down in eight foot cells

$$V = 3228 \times 25 + 300 \times 70^{**} \times L$$

Assume we have previously calculated that the total cover material available on the site amounts to 500,000 cubic yards. Then we have

$$500{,}000 = 80{,}700 + 21{,}000\ L$$

$$L = \frac{500{,}000 - 80{,}700}{21{,}000}$$

$$L = 20 \text{ years}$$

This will serve as a general estimate for planning purposes. There are, in fact, other constraints to landfill volume, though the cover availability is commonly the major constraint.

(It should be noted, of course, that if a nearby source of good cover material is available then this additional volume should be added to the estimate of cover volume available from the site itself).

Final Site Use

Once the total life of the site is determined, the detailed site planning can begin. The approximate volume of refuse used to estimate the life of the site will provide a good basis for designing the topographic features of the site. The first consideration at this point is the final site use. Commonly this is considered last. Sometimes it is not even considered at all.

In my own opinion it is the key to acceptance in many areas. This is particularly true where there is major public concern. If a local landfill can be shown to represent a significant future benefit the public may be ready to put up with the

---

*Assuming 6 day week less holidays

**From table – 500 TPD requires 70 yds^3 daily cover

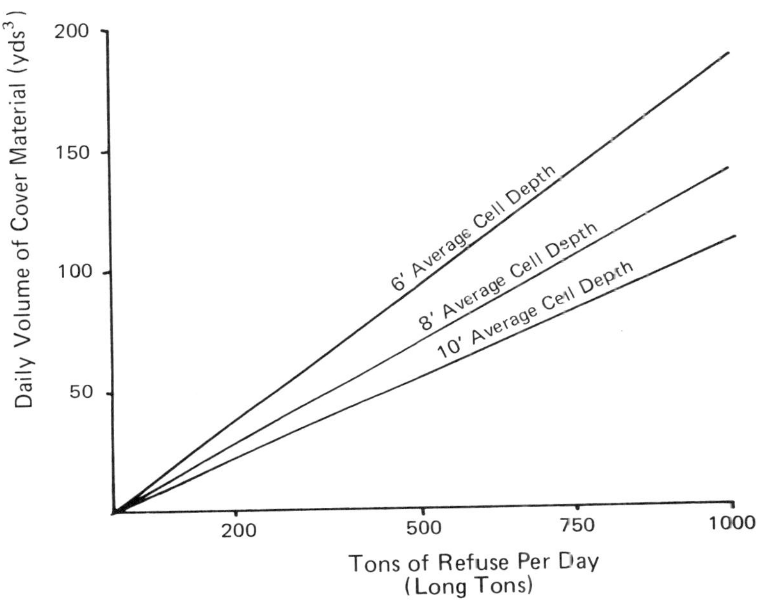

Figure 33. Daily cover material requirements.

inconvenience. This is particularly true if the planned use can be phased-in after only two or three years of operation. What future use would be acceptable?

Consult local plans. In many areas local or regional planning offices have recreational plans which generally delineate the minimum recreational requirements for the region. Often these plans indicate what types of outdoor recreational activities are lacking and the general location of deficient areas.

Recreational needs which may be appropriately met from landfill design fall into the following general categories:

- Play areas
- Golf
- Winter activities

Play Areas – General criteria for some popular outdoor play activities may be found in Appendix A. In general, activities with limited flat area and foundation requirements should be selected since runoff from extensive flat areas is hard to control. In addition, even on a well run landfill a certain amount of differential settlement can be expected and a hard surface flat area such as a tennis court could well be subject to a certain amount of heave and differential settlement. However, if an area on the working landfill is set aside for materials such as fly ash or building rubble even these sensitive uses may be considered.

Fit the activity to the area and any local plans which might be available. If there are open play areas nearby either at school grounds or local parks there will be little use for more, and your plan will consequently have less appeal.

Golf – The general standard for golf courses is about 8 acres per hole. Thus an eithteen hole course would require 150 acres. Another useful yardstick is that the population of about 3000 persons per hole is often used. For a large landfill in an area deficient in this type of facility the golf course is an ideal final use for a landfill.

A number of examples of conversion to golf courses may be identified. The rolling nature of the required terrain allows for good drainage control, while the careful development of vegetative cover is ideal in controlling erosion and maximizing evapo-transpiration.

If the site lends itself to this kind of development but lacks the acreage for a full scale course – consider a miniature course complete with hills, valleys and greens, but scaled down for family use. Remember that a golf course is an expensive investment and should be properly planned by a specialist if it is to be popular and well used. See Appendix B for design criteria.

Winter Activities – Particularly in the flat mid-western states winter activity oriented development of a landfill can be very appropriate. There are so few suitable sledding slopes available that an area incorporating sledding tracks would undoubtedly be very popular in most parts of the mid-west. In 1972 the Chicago Metropolitan Sanitary District conducted a feasibility study to determine if a ski mountain development would be feasible for solid waste from Chicago. In 1975 a revised development along these lines began in the western suburbs of Chicago.

There is a general paucity of data regarding design criteria for sledding slopes, but the following design notes may be helpful. The development of a good native grass cover on the steep side slopes is an absolute necessity to reduce erosion and also increase evapotranspiration. However, the heavy snow compaction resulting from sled traffic can be very damaging to underlying vegetation and some rotation of slope use would seem to be a necessary precaution. Alternatively the actual sledding surface may be compacted gravel with native grasses developed on the less frequently used areas.

The increasing popularity of the snowmobile suggests that a suitable track for these vehicles might also be an appropriate development, particularly in the northern states. This activity would need to be very popular in the local community before such a plan is proposed or a snowmobile plan may turn out to be as unpopular as a landfill plan. Also the ground compaction which results from snowmobile activity is notoriously destructive of natural vegetation, so some care must be exercised in preparing the track area.

A summary of the foregoing with additional notes is shown in Table 19.

Road Layout – Planning the access road and internal road layout is the next step. When this is done, minor corrections can be made to the volume calculations if any special features of the site reduce the availability of cover material. However,

TABLE 19

General Summary of Recreational Features Suitable for Landfill Areas

Activity	Advantages	Disadvantages
Play areas	Popular. Small area requirements	Flat areas. Somewhat difficult to drain
Court games	Popular	Large flat areas required – differential settlement may be a problem
Putting green	Popular	See above
Golf course (18/9 or miniature)	Popular. Good natural drainage characteristics	Settlement creates maintenance problems
Nature/hiking trails	Popular	Some tree cover required – care required in tree selection
Golf driving range	Very popular. With careful contouring could be ideal for drainage control. No heavy traffic to impair grass growth	Expensive. A great deal of pre-planning may be necessary
Archery/Rifle range	Fairly popular	Flat area requirements pose drainage problems
Model Airplane flying	Shortage of areas set aside for this use	Inclined to be noisy – some flat areas required
Garden plots	Growing in popularity	Slope requirement presents a problem. Landfill gas and settlement could be problems
GoKart track	Popular	Flat – Noisy – Not conducive to maintaining grass cover
Motorcycle scramble	Very popular. Good natural drainage	Noisy – Not conducive to maintaining natural cover
Open air zoo	Very popular. With careful design good natural drainage could be achieved	Extensive paved areas could be a problem due to differential settlement. Maintaining natural vegetation might be difficult in some penned areas

it should be clear at this stage that cover material represents the major natural resource of the chosen site. It should not be wasted.

Key decisions made earlier about final site use will begin to impact here but they should not be allowed to interfere with the efficient use of cover material.

Access – We need to concern ourselves first with the problems of access since this may impact upon road layout. The chief goals in designing the point of entry to the site should be:

- minimize traffic interference on the main road
- minimize turn around time
- assist in screening the site from the public
- present attractive point of entry

To minimize turn around time the access road must be clean and free of traffic. Ideally separate entry and exit roads should be used. Some further notes on this problem will be found in the Chapter dealing with special precautions. If the access easement is wide enough it is often useful to bend the road at an angle of 20° or more at a point about 50 ft. from the main road. This has the effect of eliminating a direct view of the site from the roadway.

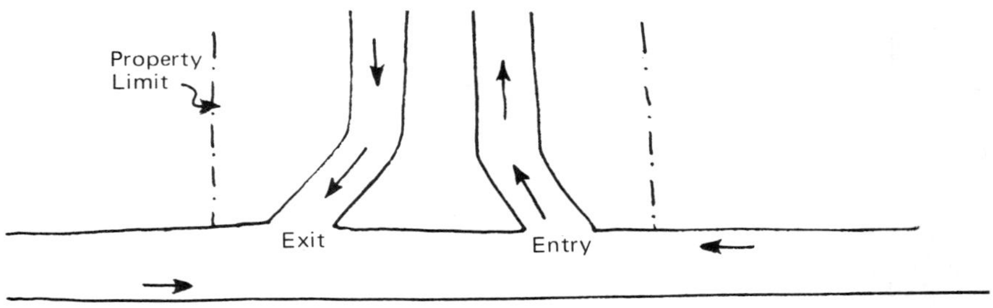

Figure 34. Separate entry and exit roads.

In many states a minimum buffer zone is specified in the regulations for landfill design. Where there is none a separation of at least 100 ft. from a busy highway is a good guide. If the buffer zone is already heavily wooded and there is good visual cover (at all times of the year) this may be reduced to 50 ft. Screens are particularly important for the access road since drivers following packer trucks on the main highway will tend to visually follow turning traffic.

Sites with narrow access easements present quite different problems. In a situation like this it is often a good idea to forgo the advantages of a wide separation between entry and exit road and simply plan a wide road access with a central island. Suitable planting on the island can be used to limit the view into the landfilling area.

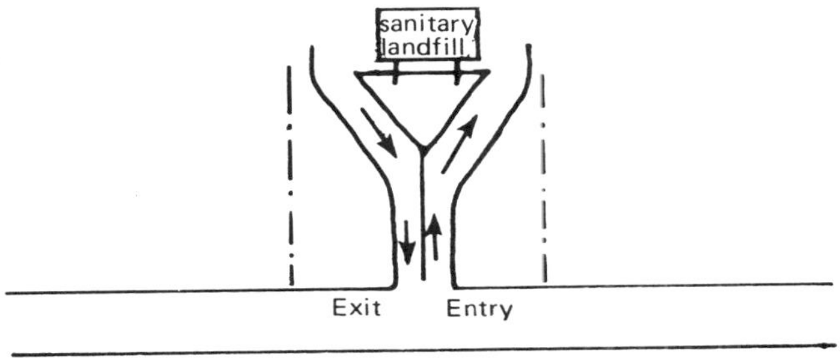

Figure 35. Single entry/exit road.

The island scheme can also be a good way of displaying signs indicating ownership, permit status, charges, liquid and hazardous waste limitations, etc. The advantage of a large sign in the center can be that in the winter when foliage offers little visual screening the sign can do just as well.

The requirement of an attractive and clean entrance and exit sound like rather fringe concerns in a chapter dealing with basic design. However, rightly or wrongly, the public image of a landfill site is raised enormously in stature by a clean and attractive exterior. For most members of the public the entrance is all they will ever see of a landfill. Frankly, if the entrance looks bad the assumption will follow that a bad landfill lurks behind it.

Trees and shrubs will provide the most attractive frontage at least cost and provided they are carefully selected they will last longer than the landfill. Choose hybrid varieties since they are generally more hardy and fast growing. Avoid maples since they are extremely sensitive to saline groundwater conditions and in the winter some road salting may be necessary to keep the access road safe.

Site Roads – The roads inside the site represent the framework or skeleton of the entire plan. If the roads are poorly designed the value of the whole site could be drastically reduced.

The goals in planning internal site roads should be:

- to provide the most permanent, trouble-free road surface possible
- to minimize turnaround time
- to lay down a road which may be useful in the final plan

Obviously, above every other consideration, a site road should get the truck to the working face and out as quickly as possible. Yet the whole question of planning a site road is often left to chance. On many sites packers make their way across unconsolidated fill in all weathers and struggle back to firm ground like drunken dinosaurs where they shed mud and refuse at such a rate that the entire township must be alienated by the mess. A site like this saves hundreds of dollars by not planning a proper road and loses thousands of dollars in lost business and poor public relations.

It is difficult to generalize about site road requirements since sites can vary so much in topography and general layout. But, in general, on most sites, I feel consideration should be given to having two separate operating areas.

- An area near the point of entry should be designated the wet weather area
- The remaining area should be designated the dry weather area

During bad weather when site roads may become quite treacherous this arrangement can help minimize turnaround time and keep mud and refuse off the roads.

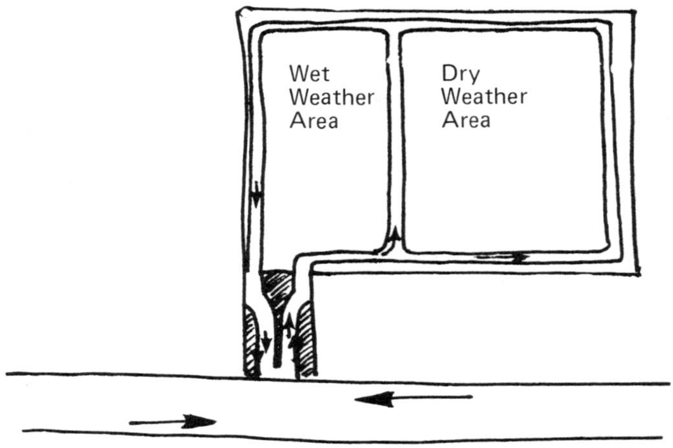

Figure 36. Simplified site layout showing wet and dry areas.

This road configuration also provides some useful flexibility in that most areas of the site may be approached by two alternate routes. Thus in case of an accident continuous operation can be maintained.

Roads in Conformance with the Final Plan – Municipal operations may be more concerned with this problem than private operators. However, the latter would do well to consider the prospects of sale upon completion. The availability of a good access road base would be a useful asset.

It will be clear from the earlier discussion on site use that the most popular end product for finished landfills is as recreational areas. The road network requirements for recreational areas are not complex or extensive and with a little forethought and imagination they can often be built into the landfill design.

At a minimum, a recreational area will require access routes to within 30–100 yards of the recreational facility plus parking space. A municipal landfill operation might therefore be wise to construct an oversized personnel parking area close to the main landfill area. A large maintenance area could also be part of the complex.

For the private operator the problems are a little different. Without an assured buyer for the site and an assured final plan (unless he goes into the recreation business) probably the minimum should suffice (i.e. a good peripheral road).

Drainage Plans

Thus far the following constraints upon the topographical design of the landfill have been discussed:

- base landfill elevation
- cover volume
- site planning

A general structure plan for the site should have begun to emerge by this stage. The over-riding criteria will commonly be the site planning requirements particularly if extensive final use is planned.

If no such final use is contemplated, the surrounding land forms should be taken as a clue to an acceptable topographical design for the site. Reflection of the same hill formation in the site design is the major consideration if there are no other criteria. In a flat area, the hill slopes should be shallow and unassuming. The final land forms should not impose harshly upon the landscape.

Now an approximate landform plan showing approximate slopes should be drawn rather like a many facetted series of inclined planes over the face of the site.

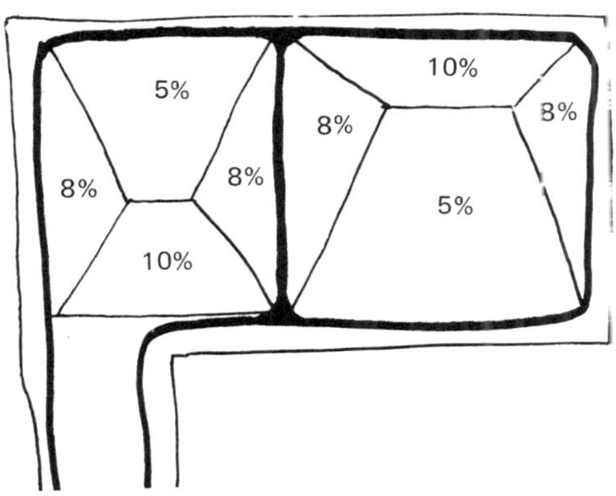

Figure 37. Drainage landform plan.

This represents the first approximation of drainage surfaces. Now the problem is to establish their efficiency under local rainfall conditions.

Drainage Design – In order to adequately design drainage structures for the landfill area, it is necessary to establish what the peak rate of runoff or discharge will be. The peak rate of flow is a function of:

- rainfall
- antecedent condition
- land treatment
- soil type
- slope

Rainfall – Annual data on rainfall at the site is rarely directly available unless the selected site is close to the home of a meteorological enthusiast. However, the U.S. Weather Bureau's Rainfall Frequency Atlases cover the entire United States. They provide rainfall frequency data for areas less than 400 square miles for durations of 24 hours and for frequencies from 1 to 100 years.

When dealing with landfill sites, the data from the Atlases need not be adjusted with respect to specific geographical area since the drainage area is so small. However, quite significant differences occur across the country with respect to the distribution of rainfall with time. For the purposes of simplified analysis, the SCS has identified[5] two major regions across the United States.

- Type 1 – Maritime Climate
- Type 2 – Summer Thunderstorm Region

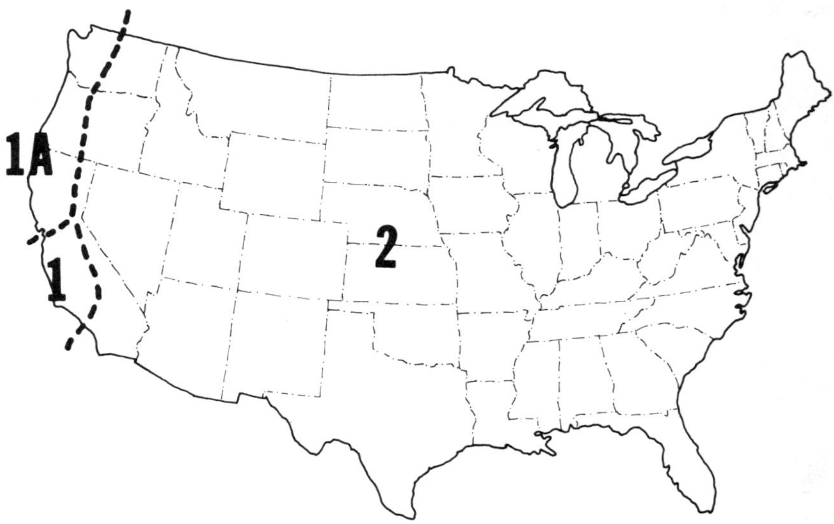

Figure 38. 24 hour rainfall distribution types.

[5] "A Method for Estimating Volume and Rate of Runoff in Small Watersheds", K. M. Kent (retired), Chief, Hydrology Branch, Soil Conservation Service.

The type 1 and Type 2 regions are based upon generalized rainfall data from Weather Bureau records. The variations which occur in using these generalized regions are within the accuracy of rainfall records themselves. The 25-year 24 hour storm record is generally chosen as the basis for runoff calculations. The use of generalized Type 1 or Type 2 regions is adequate correction for the variation in storm intensity during the 24 hour period. Thus, the initial rainfall data for the general vicinity of the site can be obtained from the U.S. Weather Bureau map in Appendix C. (For the unique storm distribution (Type 1A) representative of the coastal side of the Sierra Nevada and the Cascade Mountains in Oregon, Washington, northern California and the coastal regions of Alaska graphs may be obtained from the West Regional Technical Service Center, SCS. Portland, Oregon.)

Antecedent Conditions – The Soil Conservation Service has conducted research on a number of experimental plots of land to establish imperical formula for use in determining the effect of these factors upon runoff quantities. The experimental values for different soils and types of land treatment are made available as Curve Numbers (CN).

Obviously the quantity of runoff will vary according to whether or not the land has been previously wetted by rainfall. The wetted state of the land is known as the antecedent condition. Extreme conditions exist when the antecedent condition is either very wet or very dry. In the nomenclature used by the SCS these conditions are respectively AMC III and AMC I.

An average antecedent moisture condition is represented by AMC II and this is generally used as the basis for runoff calculations. However, where weather conditions suggest the incidence of the extreme antecedent conditions the given CN's can be prorated from Table 20.

Curve Numbers – A table of suitable CN values for landfill design is shown in Table 21 and also repeated in the Appendix.

Soil Type – Appendix C lists more than 8,000 soils and their hydrologic group. The results of the soil survey will establish soil types which can then be identified in the table. In general, the four hydrologic groups are described in Table 22.

Slope – Slope is the final determinant of runoff quantity. In the curves produced by the SCS reproduced in Appendix D, three general slope factors are included in Table 23.

In general the direct use of the curves will give good approximations of peak runoff. For slopes beyond these ranges a method of interpolation is suggested by The Source Document, but it is unlikely that computations will be required for sanitary landfill beyond this range. We now have all the ingredients to establish an estimate of runoff quantity from any given area.

- Soils data can be used to identify soil group and CN value
- % slope can be estimated based on ultimate site use In any event finished slope should be in the moderate or steep range to maximize runoff quantity.
- The drainage area can be obtained from the site drawings

TABLE 20

Curve Numbers (CN) for Wet (AMC III) and Dry (AMC I) Antecedent Moisture Conditions Corresponding to an Average Antecedent Moisture Condition (AMC II)[1]

CN for AMC II	Corresponding CN's	
	AMC I	AMC III
100	100	100
95	87	98
90	78	96
85	70	94
80	63	91
75	57	88
70	51	85
65	45	82
60	40	78
55	35	74
50	31	70
45	26	65
40	22	60
35	18	55
30	15	50
25	12	43
20	9	37
15	6	30
10	4	22
5	2	13

[1] AMC I. Lowest runoff potential. Soils in the watershed are dry enough for satisfactory plowing or cultivation.
AMC II. The average condition.
AMC III. Highest runoff potential. Soils in the watershed are practically saturated from antecedent rains.

Source: "A Method for Estimating Rate of Runoff in Small Watersheds", U.S.D.A. SCS, 1973.

TABLE 21

Runoff Curve Numbers for Hydrologic Soil-Cover Complexes

Land Use and Treatment	Hydrologic Condition[1]	Hydrologic Soil Group			
		A	B	C	D
Contoured, grassed area	Poor	47	67	81	88
	Fair	25	59	75	83
	Good	6	35	70	79
Wooded area	Poor	45	66	77	83
	Fair	36	60	73	79
	Good	25	55	70	77
Roads					
Dirt		72	82	87	89
Hard surface		74	84	90	92

[1] Condition with respect to runoff quantity i.e. Poor = Maximum Runoff, Good = Minimum Runoff.

Source: "A Method for Estimating Volume and Rate of Runoff in Small Watersheds", U.S.D.A. SCS, 1973.

TABLE 22

General Soil Hydrologic Groups

A. (Low runoff potential.) Soils having high infiltration rates even when thoroughly wetted. These consist chiefly of deep, well to excessively drained sands or gravels. These soils have a high rate of water transmission in that water readily passes through them.

B. Soils having moderate infiltration rates when thoroughly wetted. These consist chiefly of moderately deep to deep, moderately well to well drained soils with moderately fine to moderately coarse textures. These soils have a moderate rate of water transmission.

C. Soils having *slow infiltration rates* when thoroughly wetted. These consist chiefly of soils with a layer that impedes downward movement of water or soils with moderately fine to fine texture. These soils have a slow rate of water transmission.

D. (High runoff potential.) Soils having *very slow infiltration rates* when thoroughly wetted. These consist chiefly of clay soils with a high swelling potential, soils with a permanent high water table, soils with a claypan or clay layer at or near the surface, and shallow soils over nearly impervious material. These soils have a very slow rate of water transmission.

Source: SCS Engineering Field Manual for Conservation, Chapter 2, 1970.

TABLE 23

Slope Factors for Peak Discharge Computations

Slope Factor	Slopes for Which Computations were Made (Percent/Slope)	Average Slope Range (Percent Slope)
Flat	1	0 to 3
Moderate	4	3 to 8
Steep	16	8 to 30

Example

Drainage area = 10 acres
Soil group = C
Land use treatment = Contoured grassland
Hydrologic condition = Good

From Appendix C the CN = 70

Assume that from the U.S. Weather Burea Rainfall-Frequency Atlas the 24 hour rainfall = 12 inches
Assume Type 2 Storm Distribution
Assume STEEP Slope
Use Sheet 17

Enter graph with 10 acres
Sight along vertical to meet 12″ rainfall line

Peak discharge = 105 cubic feet per second for
a slope range from 8–30%

For greater accuracy than this general range the reader is referred to the original source material where an extrapolation method is described. However, in most cases this level of accuracy should be adequate.

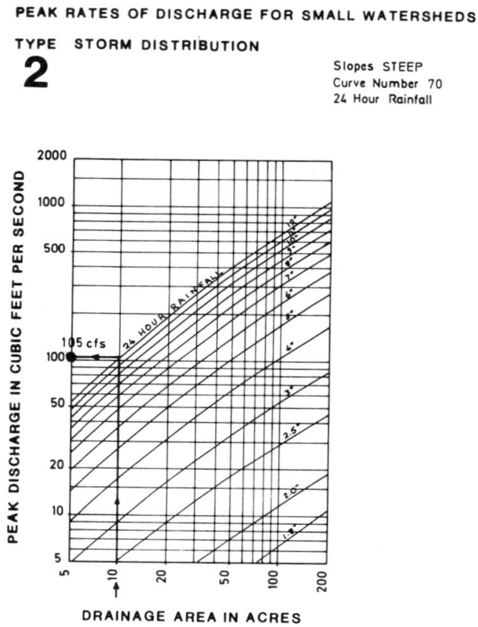

Figure 39. Example showing use of runoff curves.

Source: "A Method for Estimating Volume and Rate of Runoff in Small Watersheds" U.S.D.A. SCS, April 1973.

Cumulative Runoff

It should be stressed that thus far, we have been concerned only with establishing the maximum runoff. For this purpose, we have used the maximum 24 hour rainfall figures. The runoff thus generated must be capable of being conveyed off the landfill as efficiently as possible by a series of open channels or pipes. The supposition is that if we design these open channels for the maximum flow, they will be adequate for lesser flows. Later, an appropriate method to design the drainage channels will be presented.

However, at this point, we have some additional concerns that develop from the limited nature of the runoff calculations presented so far. There are two important reasons why we need to have some knowledge of the total annual runoff quantity:

- We need to estimate the total annual leachate generated.
- We may need to make provision for a settling pond in order to prevent eroded sediment flowing to an adjacent water course.

Unfortunately, most hydrological concerns center around the issue of maximum flow. Thus, there is no well established methodology for establishing annual runoff. We know that if we have peak 24 hour rainfall data, we can fairly accurately estimate runoff, but what do we do for total annual runoff when the rainfall data presents a multiplicity of rainfall figures?

Fortunately, there is a fairly simple methodology available which is accurate enough for our purposes. The rainfall equation which serves as a basis for the SCS curves presented in the appendices was developed by Victor Mockus and others around 1947.

Basically, the equation is:

$$Q = \frac{(P - Ia)^2}{(P - Ia)\ S + S}$$

Where: Q = accumulated direct runoff
P = accumulated rainfall
Ia = initial abstraction including surface storage, interception, and infiltration prior to runoff
S = potential maximum retention

Experimentally, it can be shown that the initial abstraction (Ia) is a fairly constant proportion of the maximum soil retention. In fact:

$$Ia = 0.2S$$

Substituting 0.2S for Ia in the Mockus equation we have:

$$Q = \frac{(P - 0.2S)^2}{P + 0.8S}$$

In order to show the rainfall/runoff relationship graphically, S values can be expressed as curve numbers by using the following equation:

$$CN = \frac{1000}{10 + S}$$

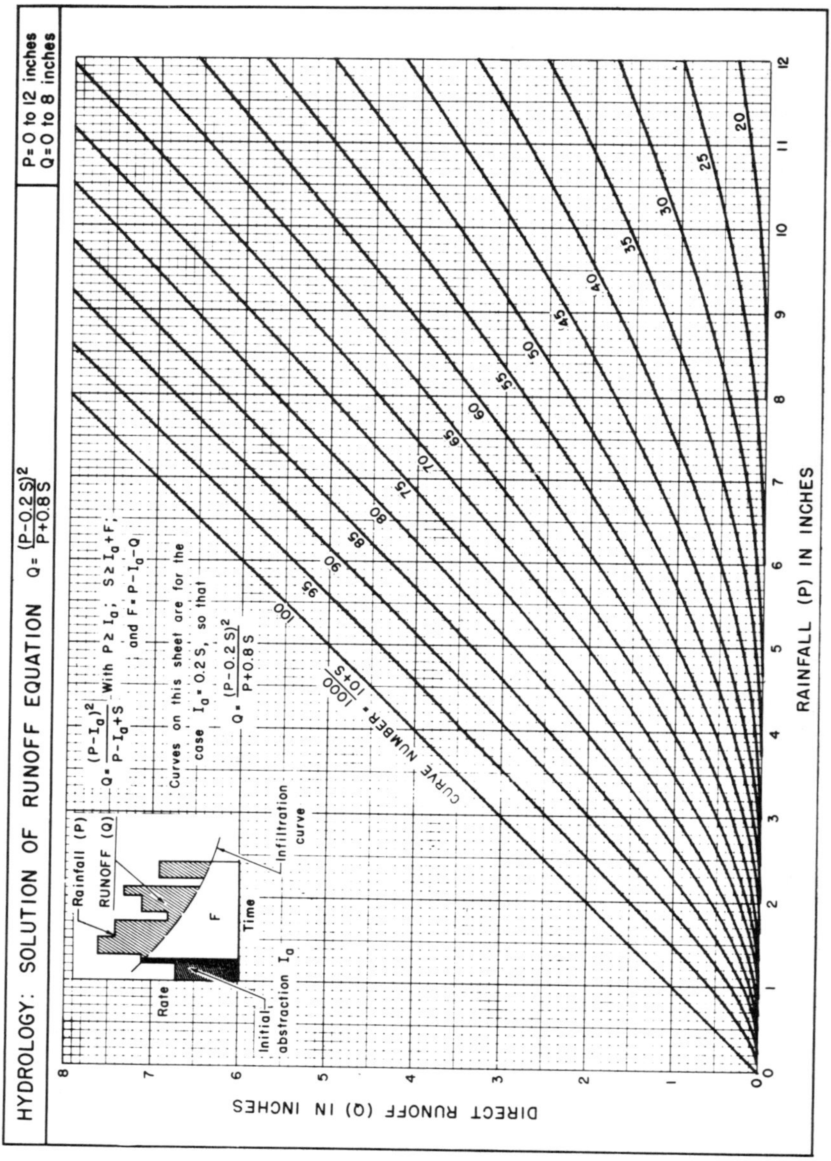

Figure 40. Solution of the runoff equation, $Q = \dfrac{(P - 0.2S)^2}{P + 0.8S}$

These are the curve numbers which we used earlier in the maximum runoff determinations for particular slopes and specific soil types.

If we dispense with the acreage evaluation and establish some average curve number to represent the whole site, we can establish a fairly simple methodology to arrive at an annual average runoff figure. We will use Figure 40 which gives a reasonable approximation of the solution of the Mockus equation.

We need daily rainfall figures for the area closest to the landfill. These figures may be readily obtained from the local Weather Bureau or SCS Field Office.

The method requires that we simply "eyeball" the last year of record (or an average year if such data is available) and underline all rainfall above 1 inch. We are making the rough assumption that rainfalls under 1 inch will give rise to a negligible amount of runoff. When this job is finished, we may have a chart rather similar to Table 24, the column for Q will be explained later.

TABLE 24

Example Chart to Determine Annual Runoff

Month	Frequency	Rainfall (P)	Runoff (Q)
January	2	2 inches	2.2
February	1	1 inch	.3
March	3	1 inch	.9
April	2	1 inch	.6
	2	2 inches	2.2
	1	3 inches	2.0
May	1	4 inches	2.4
September	3	1 inch	.9
	1	2 inches	1.1
October	1	1 inch	.3
	3	2 inches	3.3
November	2	1 inch	.6
		TOTAL	16.8 Inches

Now we need only enter the graph in Figure 40 with the appropriate curve number and fill in the column left blank for direct runoff (Q). Suppose we use a curve number 90 (soils in this range have the requisite low infiltration rates) then the Q column can be filled in as shown in Table 24. Note that the number in the Q column represents Q X Rainfall Frequency so we may add the Q column to obtain total annual runoff.

Total Annual Runoff = 16.8 inches
The Chart is typical of an area with
an annual rainfall = 49 inches
Therefore, runoff (Q) as percentage
of total annual rainfall = 34 percent

With the added sophistication of slope information and piecewise evaluation area by area, this percent runoff could be substantially increased, but for our purposes this estimate is adequate.

The reader will recognize that the estimated runoff figure of 16.8 inches was used earlier in this chapter to estimate total annual leachate quantity.

Drainage Channel Design – Careful design of the surface slopes to limit erosion and minimize rainfall infiltration is a crucial phase in the landfill design process. However, all this counts for nothing if the runoff from the slopes is not collected in drainage channels and diverted away from the landfill.

This process of collecting surface runoff and diverting it from the landfill is important both during the operational phase and in the final design. In large sites, this invariably means that temporary drainage channels must be installed, used for a year or so and then moved on or covered over as the landfill operation progresses. On smaller sites of 100 acres or less, it may be possible to design the site in such a manner that the drainage channels built early in the operation continue to be functional right through to the final design. Obviously, this latter arrangement is most economical. However, the basic requirement is that surface runoff should be quickly and efficiently transported from the slopes away from the site and even on some smaller sites temporary drainage channels may be necessary.

Open Channel Design – The most widely used formula for determining velocity of flow in an open channel is Manning's Equation:

$$V = \frac{1.486}{n} r^{2/3} s^{1/2}$$

Where V = mean velocity of flow in feet per second
r = hydraulic radius in feet
s = slope
n = coefficient of roughness

Manning's equation has the advantage of simplicity and also reasonable accuracy compared to experimental results.

For our purposes provided appropriate estimates of "n" can be made this equation is all we need to determine the design characteristics of any one of the many alternative types of drainage channels used on a landfill site. Manning also has the advantage that countless nomographs are available and these can be used with relative ease instead of battling through the formula. Appendix E shows a typical nomograph.

Some preliminary computation is necessary however.

The elements of the cross section of an open channel are as follows:

a = cross sectional area of flow
p = wetted perimeter
r = hydraulic radius = $\frac{a}{p}$

TABLE 25

Elements of Channel Sections

Section	Area a	Wetted Perimeter p	Hydraulic Radius r	Top Width T
 Trapezoid	$bd + zd^2$	$b + 2d\sqrt{z^2+1}$	$\frac{bd + zd^2}{b + 2d\sqrt{z^2+1}}$	$b + 2zd$
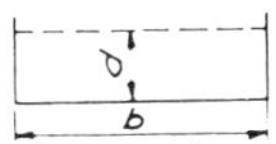 Rectangle	bd	$b + 2d$	$\frac{bd}{b+2d}$	b
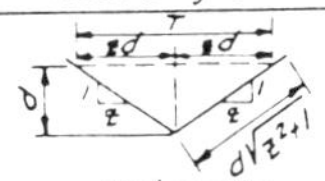 Triangle	zd^2	$2d\sqrt{z^2+1}$	$\frac{zd}{2\sqrt{z^2+1}}$	$2zd$
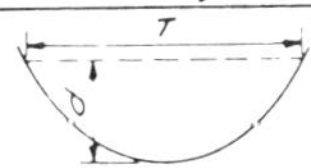 Parabola	$\frac{2}{3}dT$	$T + \frac{8d^2}{3T}$ [1]	$\frac{2dT^2}{3T^2+8d^2}$ [1]	$\frac{3a}{2d}$
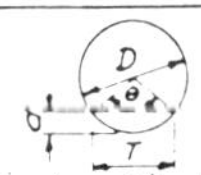 Circle - < ½ full [2]	$\frac{D^2}{8}\left(\frac{\pi\theta}{180} - \sin\theta\right)$	$\frac{\pi D\theta}{360}$	$\frac{45D}{\pi\theta}\left(\frac{\pi\theta}{180} - \sin\theta\right)$	$D \sin\frac{\theta}{2}$ or $2\sqrt{d(D-d)}$
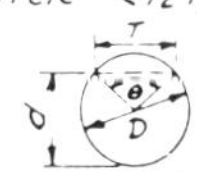 Circle - > ½ full [3]	$\frac{D^2}{8}\left(2\pi - \frac{\pi\theta}{180} + \sin\theta\right)$	$\frac{\pi D(360-\theta)}{360}$	$\frac{45D}{\pi(360-\theta)}\left(2\pi - \frac{\pi\theta}{180} + \sin\theta\right)$	$D \sin\frac{\theta}{2}$ or $2\sqrt{d(D-d)}$

[1] Satisfactory approximation for the interval $0 < \frac{d}{T} \leq 0.25$
When $d/T > 0.25$, use $p = \frac{1}{2}\sqrt{16d^2 + T^2} + \frac{T^2}{8d}\sinh^{-1}\frac{4d}{T}$

[2] $\theta = 4\sin^{-1}\sqrt{d/D}$
[3] $\theta = 4\cos^{-1}\sqrt{d/D}$
Insert θ in degrees in above equations

Source: "SCS Engineering Handbook Hydrology, 1971.

General formulas for obtaining area, welted perimeter, hydraulic radius and top widths of trapezoidal, rectangular, circular and parabolic sections are shown in Table 25.

Suitable values of Manning's "n" are shown in Table 26.

TABLE 26

Roughness Coefficients (Manning's *n*) for Open Channels

Channel description	Manning's *n* Good construction[a]	Manning's *n* Fair construction[a]
Open Channels, Nonvegetated Lining, Straight Alignment[b]		
Concrete, with all surfaces:		
Formed, no finish	0.013	0.017
Trowel finish	.012	.014
Float finish	.013	.015
Some gravel on bottom	.015	.017
Gunite, good section	.016	.019
Gunite, wavy section	.018	.022
Concrete bottom, float-finished, sides of:		
Dressed stone in mortar	.015	.017
Random stone in mortar	.017	.020
Cement rubble masonry	.020	.025
Plastered	.016	.020
Dry rubble (riprap)	.020	.030
Gravel bottom, sides of:		
Formed concrete	.017	.020
Random stone in mortar	.020	.023
Dry rubble (riprap)	.023	.033
Brick	.014	.017
Asphalt:		
Smooth	.013	
Rough	.016	
Planed wood, clean	.011	
Concrete-lined excavated rock:		
Good section	.017	.020
Irregular section	.022	.027
Flumes (steep slope)[b]		
Open Channels, Excavated, Straight Alignment[c], Natural Lining		
Earth, uniform section (best):		
Clean, recently completed	0.016	0.018
Clean, after weathering	.018	.020
With short grass, few weeds	.022	.027
Gravel, uniform section, clean	.022	.025

(continued)

Table 26 *(continued)*

Channel description	Manning's *n*: Good construction[a]	Manning's *n*: Fair construction[a]
Earth, fairly uniform section:		
No vegetation	.022	.025
Grass, some weeds	.025	.030
Dense weeds or aquatic plants in deep channels	.030	.035
Sides clean, gravel bottom	.025	.030
Sides clean, cobble bottom	.030	.040
Dragline-excavated or dredged:		
No vegetation	.028	.033
Light brush on banks	.035	.050
Rock:		
Based on design section	.035	
Based on actual mean section		
a. Smooth and uniform	.035	.040
b. Jagged and irregular	.040	.045
Channels not maintained, weeds and brush uncut:	Fair condition	Poor condition
Dense weeds, high as flow depth	0.08	0.12
Clean bottom, brush on sides	.05	.08
highest of flow	.07	.11
Dense brush, high stage	.10	.14

Source: "Engineering Manual", Editor in Chief, Robert H. Perry. Copyright 1959, 1967, by McGraw-Hill Inc. Used with the permission of the McGraw-Hill Book Company.

The Manning Nomograph in Appendix E is all that is necessary to complete the array of design tools required to handle any open channel design problem which may be encountered on a landfill.

It is, therefore, appropriate to consider at this stage just exactly what each type of channel is intended to accomplish on a working site.

Drainage Channels

Temporary Drainage Channels — Almost anything which will temporarily convey surface water runoff could fall under this category. However, the two types in most common use are:

- earthen swales
- split corrugated metal pipe

The former are probably most common where the drainage requirement is short lived and the growth of the landfill site is very rapid. The latter are more appropriate where the temporary drainage requirement will probably last for a year or more or where erosion problems could be potentially more severe. Also, split corrugated metal pipe may be used on larger sites where the runoff volume can be considerable and earthen channels would be subject to erosion.

Earthen Channels – Temporary earthen channels should be designed in accordance with the soil limitations. In general, the following permissible velocities apply:

TABLE 27

Permissible Velocities for Bare Earthen Channels

Soil Texture	Maximum Velocity ft/sec
Sand and sandy loam (non-colloidal)	2.5
Silt loam (also high lime clay)	3.0
Sandy clay loam	3.5
Clay loam	4.0
Stiff clay, fine gravel, graded laom to gravel	5.0
Graded silt to cobbles (colloidal)	5.5
Shale, hardpan and coarse gravel	6.0

Source: "SCS Engineering Field Manual for Conservation", 1970.

In any given problem involving the construction of an earthen channel, it would be appropriate to start with an evaluation of the capability of site equipment to construct a channel. Commonly, if a dozer is used, it would obviously be a simple matter to construct a triangular cross section. This would be the first assumption.

If we assume a modest slope of 1:4 then z = 4 in the formula in Table 25. Assuming depth = 2 feet.

Example: Hydraulic Radius

$$r = \frac{zd}{2\sqrt{2^2 + 1}}$$

$$r = \frac{4 \times 2}{2\sqrt{4^2 + 1}}$$

$$r = \frac{8}{2\sqrt{17}} = \frac{8}{2 \times 4 \cdot 12}$$

Hydraulic Radius $r = 0.97\ ft.$

Let us assume the soil is a sandy clay loam.

Therefore, the maximum velocity = 3.5 ft/second and Manning n = 0.022.

On the nomograph in Appendix E a line through 3.5 ft/second on the Velocity scale and 0.022 on the Manning Scale yields a point on the turning line. A line through the point on the turning line and 0.97 ft. on the Hydraulic Radius scale yields a slope requirement of S = 0.1 ft/ft and a discharge capacity of 12 cfs. Obviously, at this point, if neither S = 0.1 ft/ft or Q = 12 cfs is reasonable given the slope of the ground and the peak flow previously computed the cross section of the channel can be changed in order to arrive at a reasonable compromise.

Split Corrugated Metal Pipe — Obviously, in design terms, the above methodology is the same regardless of how the channel is formed. In practical terms, however, the split channel method requires some further explanation.

Clearly, the method involves cutting standard corrugated drainage pipe into two parts. The quickest method is probably to use a blow torch. This provides two open channel sections which are then "bedded down" in a roughly formed channel wherever the design calls for a temporary channel. Sections should be installed with the top section overlapping the lower section by at least 2 feet.

Weak mix concrete or roofing cement can be used to join each section since junction points will otherwise become major points of erosion. A very durable temporary drainage channel can be formed by this method with the additional advantage that it will withstand accidental impact from a passing dozer without sustaining too much permanent damage.

When the original location is no longer required, the split sections are easily raised and moved into a new location to be used again. Finally, when a permanent channel is required, they may be used as the basis for the final drainage swales.

Final Drainage Swales — In most respects final drainage is no different from temporary drainage design. The methodology is identical though the practical details may be different.

The major distinction is that the final swales should be designed to last. Commonly, they will be grassed over in the final plan. It may also be necessary for maintenance equipment to gain easy access and travel across the swales. The cross section most favored is the shallow parabola.

The final site plan will have a major impact upon the specific characteristics of the final drainage channels. There are many different types available of which the following is a selection of the most frequently used:

- grassed swale
- stone center ditches
- concrete ditches

Grassed Swale — The grassed swale is by far the most popular in spite of the initial preparation and planting problems. A protective binding mulch is sometimes used to protect the seeds until they establish sufficient root structure to withstand drainage flow velocities. Several propietary mesh products are available to protect the initial growth and the net slowly decomposes as the seeds become less vulnerable to erosion.

In extreme cases where the drainage swale is in fairly constant use, it may be necessary to lay down a complete sod cover in order to ensure good protection. Where this is contemplated, the stone center options should be considered.

Stone Center Ditches – In sections of the country where field stones or other sources of rock may be plentiful a stone center ditch may provide the best protection against erosion resulting from prolonged and frequent flow.

Stone center ditches may be installed as shown in Figure 41 either V-shaped or rounded. The required stone size in relation to slope of channel and depth of flow can be determined from the nomograph in Figure 42.

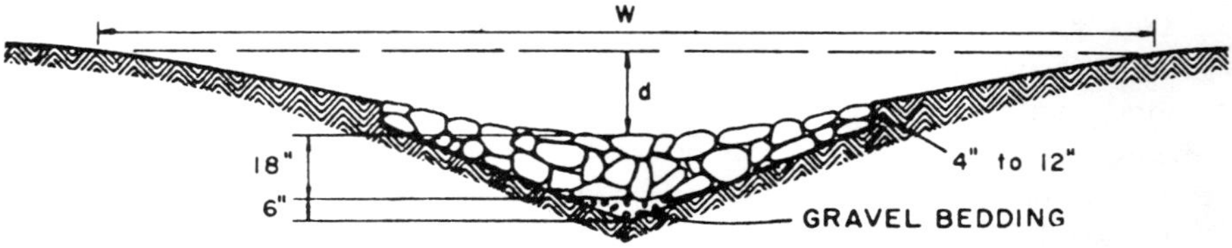

Waterway with stone center drain
V section shaped by bulldozer

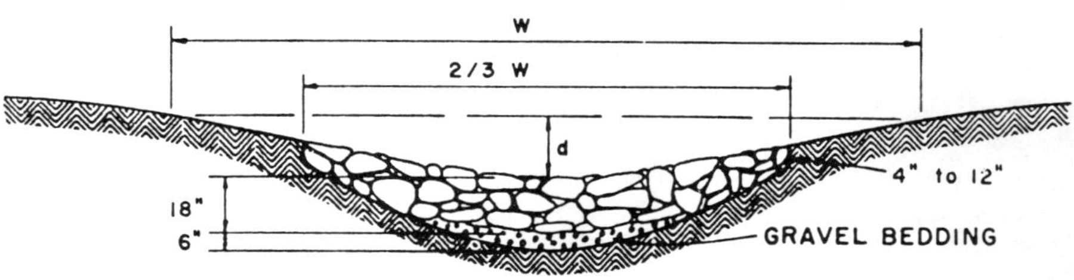

Waterway with stone center drain
Rounded section shaped by bulldozer

Figure 41. Waterway with stone center.

Source: "SCS Engineering Field Manual for Conservation", 1970.

Concrete Ditches – In situ concrete ditches may be the only feasible option for flows much above 6 ft. per second. They are not commonly used on sanitary landfills since the method is more costly than the previously mentioned alternatives. However, where the drainage design is extensive, the accumulated flow at the downstream end of the drainage system can be very considerable and a concrete final section may be the least cost option from a long term maintenance point of view.

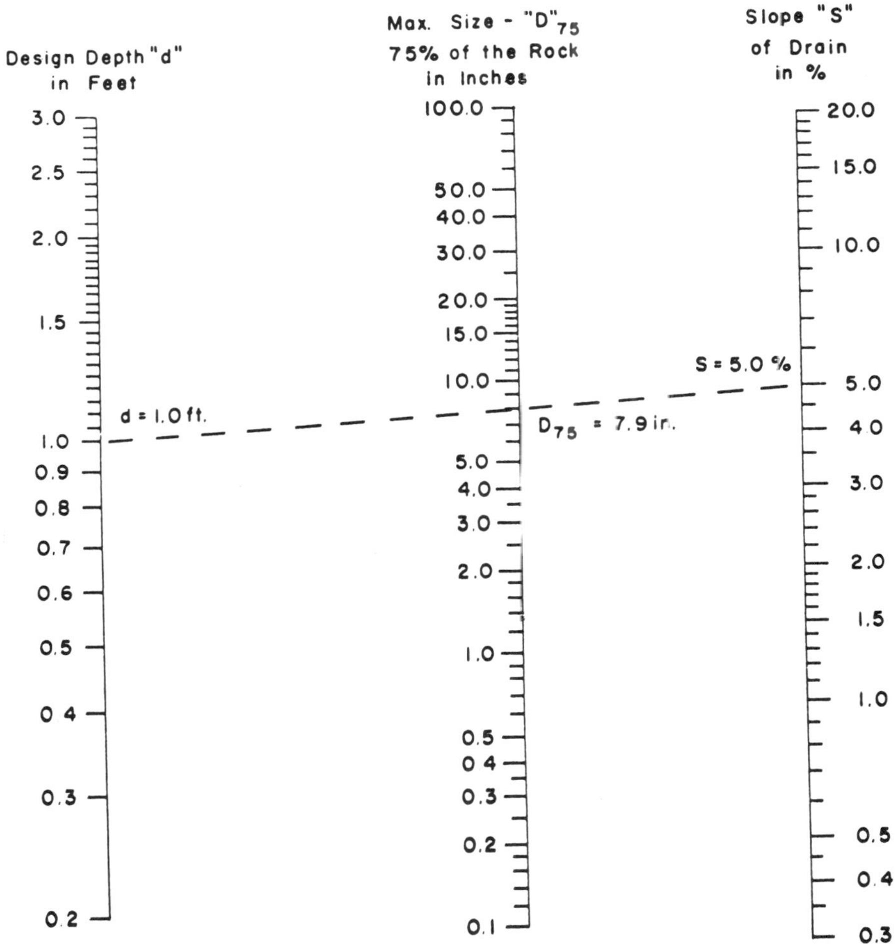

EXAMPLE: "d" = 1.0 Feet "S" = 5 %

Place straight edge at "d" value in Design Depth column and at "S" value in Slope column. Read rock size in middle column 7.9 inches. Say 8 inches.

FOR DESIGN:

25 % of the rock by volume should be in sizes of 8 inches or slightly larger. The remaining 75 % or less should be of well graded material, smaller than 8 inches, including sufficient sands and gravels to fill the voids between the larger rock.

Figure 42. Determination of rock size for stone center waterway.

Evapotranspiration

An important source of moisture loss from the surface of a landfill is the loss due to evaporation. An equally important loss associated with evaporation is that due to natural transpiration of grasses and other vegetation planted in the final cover.

However, the determination of the quantity of moisture loss in this manner is extremely complex. In itself, this may be one reason for the common neglect of this element of landfill design. Like many natural phenomena, the problem of quantification is actually more difficult than it seems at first glance.

The problem is that the combined phenomenon of evaporation and transpiration (commonly referred to as evapotranspiration) varies considerably from year to year. Consider for a moment the chief parameters of evapotranspiration:

- available energy (chiefly solar radiation)
- capacity of air to remove evaporated water (wind speed, turbulence, vapor concentration)
- vegetation type (root system, transpiration efficiency, moisture requirement)
- soil type (especially moisture retaining efficiency in the area of the root system)

Clearly, each parameter is literally as variable as the weather. As any farmer knows there will be times when there is too much rainfall to fulfill all the ideal requirements and times when there is too little.

However, sufficient information is available about the ideal conditions that relatively good estimates of potential evapotranspiration can be made. Considerable work in this field has been done by C. W. Thornthwaite and J. R. Mather and it is to Thornthwaite that the concept of Potential Evapotranspiration (PE) is credited. Stated very simple Potential Evapotranspiration is defined as the loss of moisture which occurs when all the needs of vegetational cover are met. Given this assumption an extremely detailed month by month analysis of climatological data can be conducted in order to obtain an estimate of PE. Unfortunately, few, if any, landfills are blessed with extensive climatological data and even where approximate data could be made available it is difficult to justify the extensive balance sheet procedure merely to determine potential losses due to evapotranspiration.

Fortunately, Thornthwaite evaluated PE for about 3500 weather stations in the United States and published a map of average annual PE in the Geographical Review of 1948. The map is reproduced in Figure 43 with the permission of the American Geographical Society of New York.

It is proposed that the approximate location of the site should be found on this map to obtain the estimate of PE required in later evaluation of total water budget for the landfill.

It will be clear from the way in which the water budget impacts the design considerations that the requirements for accuracy are not stringent. This fact, alone, justifies the use of average annual figures from Figure 43.

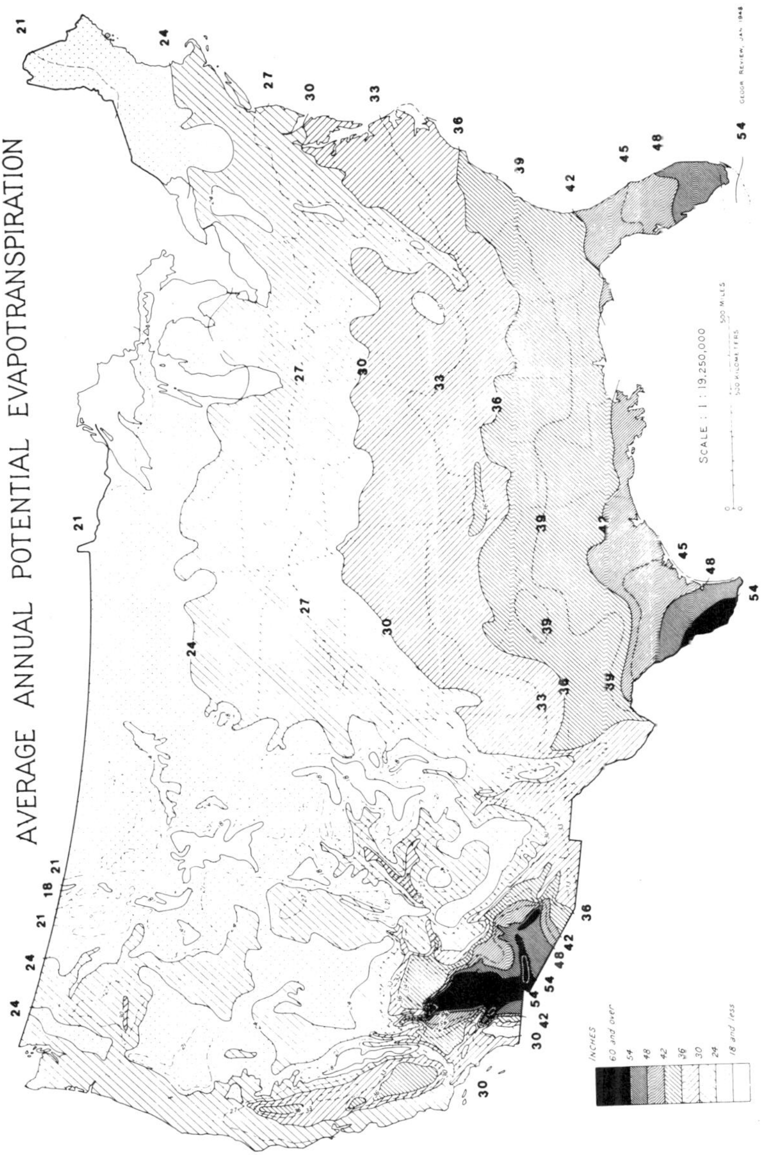

Figure 43. Average annual potential evapotranspiration over United States.

Source: Reprinted from the "Geographical Review", Vol. 38, 1948. Copyrighted by the American Geographical Society of New York.

In addition, it should be self-evident that the specific microclimatic conditions on most landfill sites are not those of fully vegetated areas (particularly in the early years). Vegetation tends to have a moderating influence on temperature and moisture loss due to evaporation. Heat retained by the poorly vegetated soil cover will have a tendency to increase evaporative loss after summer rains. Thus, in the summer months when real evaporation is commonly found to be less than PE, the nature of the landfill site itself may tend to reduce that discrepancy.

Erosion

In establishing the final design for the site, we also need to take into consideration the potential for soil loss due to erosion. Up to this point, we have considered increased site slope as a substantial virtue since in most respects it reduces the quantity of rainfall which will permeate through the soil and cause problems later in the life of the fill.

However, nothing is for nothing. As slope is increased, the velocity of runoff is increased along with the quantity of runoff. As the overland velocity increases, so does the tendency for fine particles in the soil to be washed away. Unchecked, erosion can cause serious problems on a finished landfill particularly if gullies are formed. In the V-shaped bottom of erosion gullies, the runoff velocity increases substantially over "sheet" flow and the resulting rapid erosion can produce gullies which go right through the final cover. Thus, the value of final cover protection can easily be completely eliminated by the work of natural forces.

To understand what might be done to check this problem, we need to understand the causes of erosion. It is generally agreed that the most important cause of erosion is the kinetic energy of the rainfall itself. Strictly speaking, rainfall provides the energy which detaches soil particles from the surrounding material while runoff merely provides the carrier necessary to transport the soil particles away.

Obviously, some soil types are less susceptible to the impact of rainfall. Equally obviously, careful management can make up for defficiencies in the soil. The amount of erosion or total soil loss which occurs as a result of a given rainfall can be represented by the following diagram:

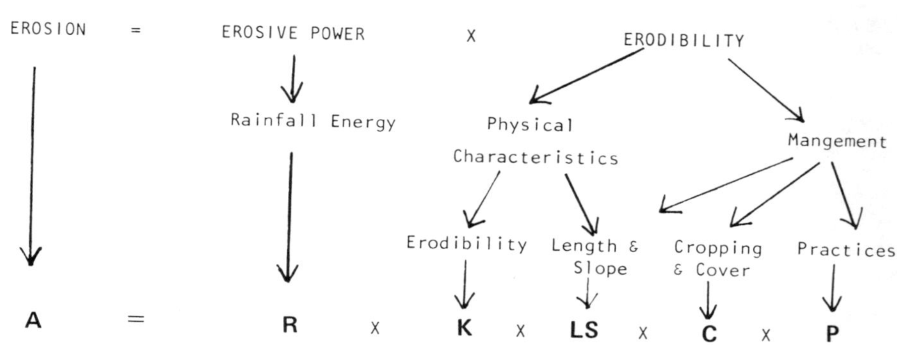

This produces the so called Universal Soil Loss Equation which is usually credited to Wischmeir:

$$A = RKLSCP$$

Wischmeir[6] found that the best predictor of the rate of erosion (R) was the energy content of rainfall (E) multiplied by the maximum thirty minutes storm intensity:

$$R = \frac{EI_{30}}{100}$$

Division by 100 produces a normalized index.

Conversely, the erosive effect of steep slopes can be reduced by short slope lengths though the relationship is not linear.

Cover Factor (C) – This refers to the ability of a physical cover type to absorb the energy of the rainfall. Complete overall grass cover has a factor near 0 and unprotected bare ground has a factor of 1.0. Typical values may be obtained from Table 28.

TABLE 28

"C" Values for Mulching and Grasses

Mulch	
Very heavy 1500 to 2000#/acre	.10*
Heavy 1000 to 1500#/acre	.20
Moderate 500 to 1000#/acre	.40
Light 200 to 500#/acre	.60
Grasses	
Established	.01
New Seeding	
1st month	.60**
2nd month	.40
Remainder 1st year	.05

*Varies with type residue

**With seeding in mulched soil, use the lowest value of C for mulch or grass.

Source: "USDA-ARS Agriculture Handbook", No. 282, 1965, and "Soil Loss Prediction for Kentucky", USDA-SCS, 1964.

[6] W. H. Wischmeir, "A Rainfall Erosion Index for a Universal Soil Loss Equation", Proc. Soil Soc. of America, 23 (3), 246–249, 1959.

The maximum thirty minutes storm intensity (I_{30}) can be obtained from the U.S. Weather Bureau's Rainfall/Frequency Atlas.

The energy content (E) can be obtained using the following formula:

$$E = 916 + 313 \log_{10} I$$

where I is the average intensity of the rainfall under consideration (in this case I_{30}).

Wischmeir calculated values for R for the Eastern and Midwestern States as shown in Figure 44.

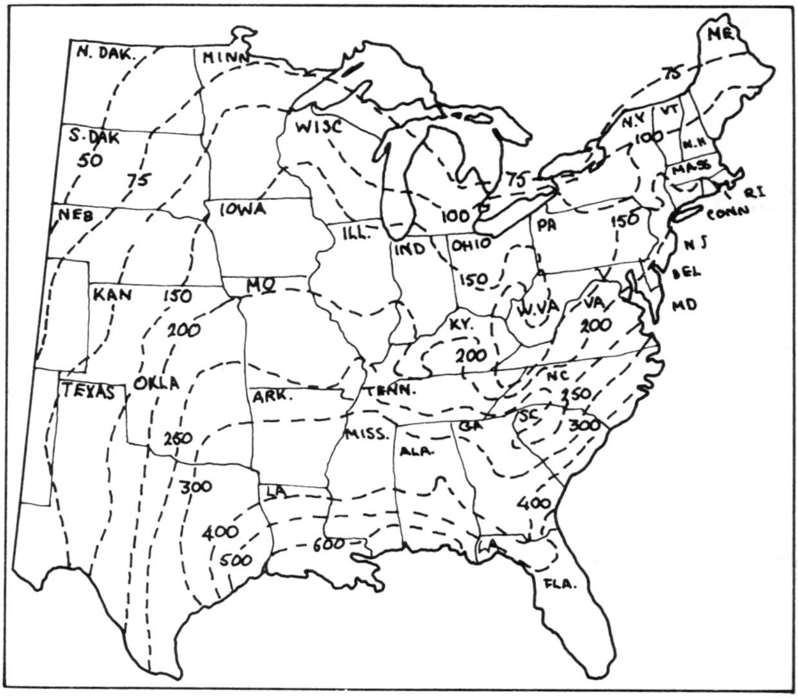

Figure 44. Average annual rainfall index.

Source: W. H. Wischmeir, ARS-SWC, Purdue University, February 1, 1971.

Values for the rainfall index (R) in the Western and Mountain States may be calculated using the above formula.

Soil Erodibility Factor (K) – This factor depends upon the physical and chemical characteristics of the soil.

The nomograph shown in Figure 45 can be used to estimate K factors quite accurately if they are not readily available. It should be noted that K values for the

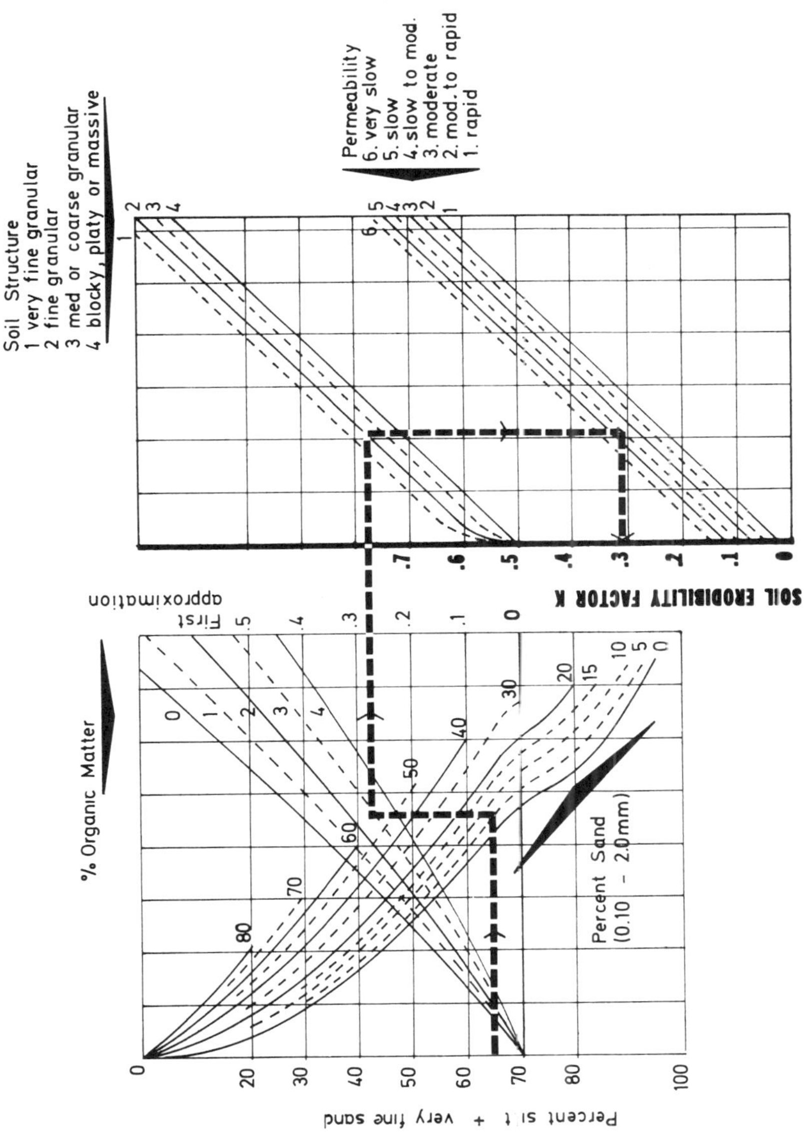

Figure 45. Soil erodibility nomograph.

Source: W. H. Wischmeir, ARS-SWC, Purdue University, February 1, 1971. First published in Journal of Soil and Water Conservation, with acknowledgments to the USDA Agricultural Research Service.

B or C soil horizons are more appropriate here since the A horizon is substantially disturbed during preliminary excavation and it is commonly the lower horizons which are used as final cover.

Length/Slope Factor (LS) – The quantity of soil eroded from a given surface by a specific rainfall increases rapidly with slope and length of slope. The LS factor can be obtained from Figure 46. Basically, steep slopes and long slope lengths cause higher losses.

Practice Factor (P) – This refers to such practices as contour plowing and terracing. Areas where little or no special care has been taken to reduce runoff velocity should have a practice factor of 1.0. Commonly, the practice factor for landfill is taken as 1.0 unless terracing is practiced for some special reason.

This particular factor has more specific value in an agricultural and mining context than in landfill practice.

We now have enough information to work out the soil loss from any area of land of given slope and soil type in the country.

The following example will illustrate how this information can be obtained. *Example:* Calculate the average annual soil loss from a ½ acre of land in Northern Kentucky 200 feet long with a 14 percent slope. The soil is 65 percent silt, 5 percent sand of fine granular structure with slow to moderate permeability, the B horizon is exposed and the soil is bare of vegetation.

1. R = 175 Northern Kentucky – Midway between 150 and 200 on Figure 44.
2. K = 0.30 See Figure 45.
3. LS = 3.3 See Figure 46.
4. C = 1.0 Table 28.
5. P = 1.0 (All construction sites)

Gross erosion A = RKLSCP

A = 175 × 0.30 × 3.3 × 1.0 × 1.0 = 173.25 tons/acre – year

Total Loss = 173.25 × 0.5 = 87 tons per year.

Consider the effect of seeding the area. Even after the first year, the C value is reduced to 0.05. The soil loss would be reduced to 4.35 tons per year at this value of C, a negligible amount.

Now consider the effect of increasing the slope to 20 percent with the same slope length. LS from Figure 46 = 6.0. Soil Loss = 7.9 tons per year. This represents a greater than 80 percent increase in soil loss though the slope was only increased by 6 percent.

As a general rule, soil loss should be held to low annual values. Losses of less than 5 tons/acre – year are acceptable. Anything over that, may require some additional controls either by changed practices or by the addition of retention ponds to settle out the sediment.

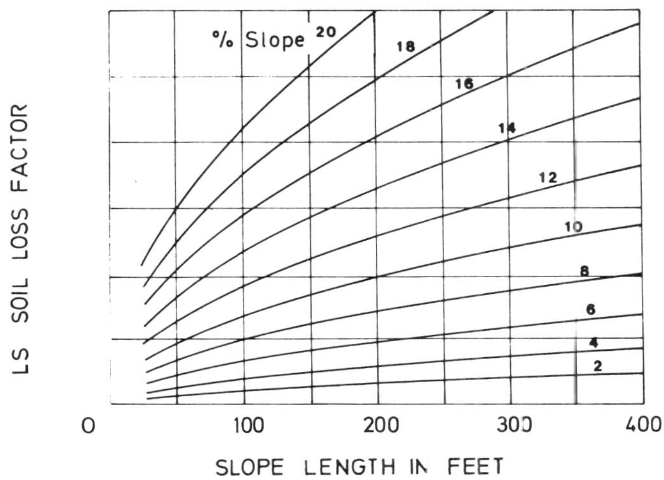

Figure 46. Curves for adjusting plot soil loss to length and steepness of slope.

Source: W. H. Wischmeir, ARS-SWC, Purdue University, February 1, 1971.

TABLE 29

Comparative Soil Loss

Loss (tons/acre/year)	Comments
0–5	Acceptable
5–20	Moderate (consider retention)
20–100	Poor (retention mandatory, consider changes in control practices)
100 and greater	Bad (radical changes in practice necessary)

Field Capacity

It may be recalled from the discussion of fundamentals that the period of time to achieve saturation in the landfill contents is called field capacity. The estimation of this period (in years) is important because it generally provides an indication of the onset of leaching.

Careful landfill design is directed towards extending this period as much as possible by limiting infiltration through the refuse. In certain areas of the country where rainfall is minimal and where drainage design is assisted by high rates of evapotranspiration, the field capacity can be infinitely extended. Indeed, a predicted moisture deficit may then be taken up by adding a liquid fraction to the solid waste load such as sewage sludge.

In other areas of the country where there is a net infiltration, some moisture movement through the fill can be predicted and in these areas estimates of field capacity can provide a warning of the probable onset of leaching. The frequency of sampling leachate monitoring wells should be increased around the period when the onset of leaching is predicted.

A study performed under a Federal solid waste disposal demonstration grant and performed under the direction of Ralph Stone provides some valuable data which can be used to predict field capacity.[7]

Field capacity is chiefly determined by infiltration rate and absorptive capacity of the landfill contents. We have already learned how to project the rate of infiltration. Now we must learn how to predict the absorptive capacity of a landfill.

By careful weighing, drying and saturation testing, the above referenced study developed Table 30 which contains the absorption capability for the common solid waste components of a landfill.

TABLE 30

Water Absorption Ranges for Solid Waste Components

Component	% Water Absorption Capacity		
	Maximum	Average	Minimum
Newsprint		290	
Cardboard (solid and corrugated)			
Other Miscellaneous	400		100
Lawn clippings (grass and leaves)	200		60
Shrubbery, tree prunings	100		10
Food Waste (kitchen garbage)	100		0
Textiles (cloth of all types, rope)	300		100
Wood, plastic, glass, metal (all inorganics)		0	

Source: "Disposal of Sewage Sludge into a Sanitary Landfill", U.S.E.P.A., 1974.

The figures can be used to estimate field absorptive capacity if the average composition of incoming refuse is known.

Clearly such figures are only approximate since the initial moisture content of each component will vary somewhat from one location to another. However, for the purposes of approximating field capacities of general areas of landfill, they are quite adequate.

[7] "Disposal of Sewage Sludge Into a Sanitary Landfill", U.S.E.P.A., 1974.

TABLE 31

Field Absorptive Capacity of Landfill Contents

Component	Average Composition (%)	Field Absorptive Capacity lbs of Water per 100 lbs Refuse Maximum	Average	Minimum
Newsprint	7		20	
Cardboard	8		14	
Miscellaneous Paper	23	92	57	23
Leaves and Grass	4	8	5	2
Prunings	6	6	3.5	1
Garbage	9	9	4.5	0
Textiles	2	6	4.5	3
Non-Absorbents	41		0	
TOTAL	100		109	

Source: "Disposal of Sewage Sludge Into a Sanitary Landfill", U.S.E.P.A., 1974.

In the example shown above, assuming average capacities for all components produces an average of 109 lbs. of water per 100 lbs. of refuse. This may be expressed as 1.75 cubic feet of water per 100 lbs. of refuse (1 pound of water = 0.016 cubic feet).

Assuming we are interested in a 5 acre area of fill with two eight feet lifts at an average density of 1000 lbs. per cubic yard this represents a total weight of refuse of:

Average (yards)^2		Depth of Fill (yards)		Density (lbs/yards^3)		Total Weight (lbs)
5 X 4840	X	$\frac{2 \times 8}{3}$	X	1000	=	129,066,000

Absorptive Capacity at 1.75 cubic feet per 100 lbs. of water.

$$\frac{1.75 \times 129{,}066{,}000}{100} = 2{,}258{,}655 \text{ cubic feet of water}$$

Returning to an earlier example where total infiltration was 6 inches per year. Over a 5 acre area, this would be equal to 108,900 cubic feet of water per second. If that rate of infiltration referred to this landfill, the field capacity would be derived as follows:

$$\frac{2{,}258{,}655}{108{,}900} = 20 \text{ years}$$

It should be noted that there are at least two additional factors which would mitigate against leaching actually occurring at this number of years hence. They may be considered safety factors in this context:

- moisture loss due to biochemical action.
- additional absorptive capacity of cover material.

As noted in earlier chapters, water is required in many of the chemical and bacteriological activities which occur in the course of the natural breakdown of landfill contents. This moisture is supplied by infiltration and used up in the process of promoting refuse breakdown.

The soil laid down as daily cover also has a certain absorptive capacity which will increase any estimate of field capacity. If known, the absorptive capacity of the soil could be included in the earlier calculation. However, the day to day changes in soil moisture are extremely variable and it is perhaps more appropriate to leave it out of consideration entirely assuming that it merely adds a certain safety margin to the computations.

Of course, it is also true that the soil interface between the landfill and the water table has a certain absorptive capacity and this has not been factored into the equation. Here again, the reason is largely due to lack of data. However, the absorptive capacity of the undisturbed soil interface is unlikely to be significant since this unweathered and heavily consolidated zone is commonly low in absorptive capacity.

Chapter 6

THE PLAN

The final plan must be all things to all men. A number of separate and distinct groups of people will be concerned about the documentation. Each group has a different perspective and a different need for information. Three groups may be identified:

- the regulatory agency
- the general public
- the owner and operator

The consultants chief obligation is to the owner of course, but part of that obligation is to assist him in ensuring that the other interested parties get the information in a form which satisfies them. The net result of this division of purpose often results in a plan which satisfies the regulatory agency, but no one else. As a result of the general concern to please the enforcers of the guidelines, the public and the owner often finish up with less than a satisfactory product. Considering the crucial nature of public acceptance, this little discussed fact is a serious fault in the development of many plans. Further, considering the fact that it is in the final analysis, the owner and operator who must carry out the plan it is perhaps an equal fault that such plans rarely address his needs to have a thoroughly practical and workable document with which to plan his daily activity.

The Regulatory Agency

The regulatory agency will no doubt expect detailed engineering drawings showing each phase of operations, plus a detailed report showing drilling records, soil analysis, calculations and everything else used to develop the plan.

Many agencies have developed report forms which lay out in great detail what information is required and how it shall be presented. Given the fact that the same state agency must review hundreds of such applications, it is understandable that some standard format should be required.

The state requirements should be fulfilled in the plan at a minimum. However, that does not necessarily mean that a separate presentation should not be prepared for the general public. Indeed, my guess is that a simplified, straightforward illustrated version of the plan would also be greatly appreciated by the state.

The General Public

The graphic requirements for any presentation to the general public have already been discussed at length. The presentation of the plan is no exception. Illustrations of the final use are an absolute necessity. Plans for cooperation with local recycle operations would probably also be appreciated. Above all, it is necessary to show realistically what the operation will look like both in its formative stages and upon completion. If the public do not feel they are denied access to this information, then they are more likely to treat a new operation as a neighbor or at least a known quantity. That is all good insurance in the long haul.

The Owner and Operator

In truth, the plan is really developed *for* the owner and operator. It is one of the paradoxes of increased regulatory agency involvement in this area that the plans required by the agencies are less and less meaningful and useful to the owner and operator.

The agency is looking for evidence of good design. As the state of the art improves, the agency requirements become more and more technically oriented and packed with theoretical data. This is hardly the kind of plan the dozer driver might be expected to study in order to plan his activity. Yet in the final analysis, he is the one person who should have a thorough grasp of what the plan is all about.

In states where the regulations are rigidly enforced, a separate operating plan should be drawn up for the purpose of planning activity on the site.

Operating Plan

I believe a thoughtful and responsible agency should see the advantages of a simplified operating plan.

This is particularly important when considering the graphic presentation of the plan. Sanitary landfill design is unusual in that it will often be constructed without engineering management over a number of years. It always seems extraordinary, therefore, that so many such plans are drawn up as though they were intended for immediate development by a skilled engineer.

Very often, the contours of each lift will be shown at two or even one foot intervals. Often, the exact location of temporary drainage ditches will be shown with great precision as though they would be placed by highly skilled surveying teams. It goes without saying that the final contour plans are generally shown in great detail as though the final cover will be rolled out with a giant hand according to every ripple in the final contour plan.

All this is well and good for the review agency if they like that sort of thing, but generally lost upon those responsible for achieving the final result. This is not intended to belittle operators in any way. The fact is that most people in their position simply require a general concept of the overall plan. The detailed final contour drawings are not highly memorable or useful items in this regard.

With the exception of the sites receiving large quantities of waste each day which experience dramatic daily changes in contours, I believe *concept sketches* should be used to convey the essence of the landfill design.

Plans should be very simple and straightforward, presenting simple wedges of fill where appropriate. Good weather areas should be easily identified compared to bad weather areas. All areas of fill should be represented by a series of sections showing the development of the site area by area.

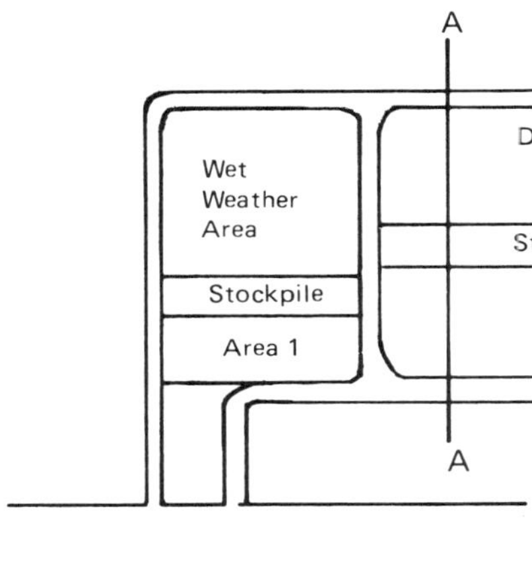

Figure 47. Concept sketch of landfill plan — Phase 1.

The basic information for each area of site development should be an area drawing showing size and location in the overall plan and two sections generally at right angles to show the finished contours.

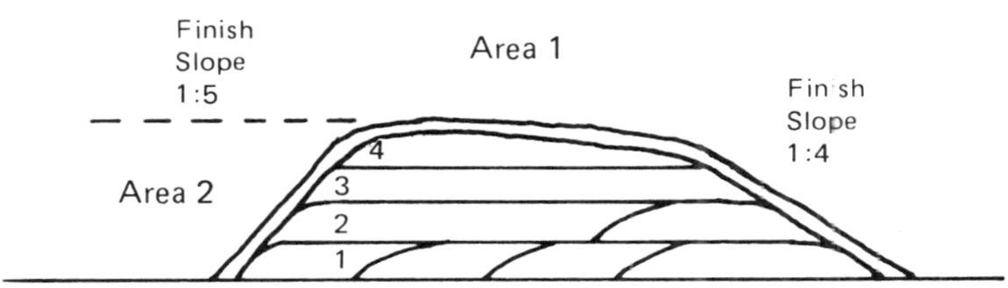

Figure 48. Sectional sketch showing section A — A.

By this means, the operator can easily visualize the overall concept and work towards it. The section should show the number of lifts planned in each area together with an indication of the general direction of the slope on each daily fill. It is pointless to attempt to show the boundaries of each days fill. The quantities of incoming waste will vary from day to day and to demand a certain fill volume before applying six inches of cover is to loose sight of the whole reason for applying daily cover.

The great value of concept sketches of the cross sections is that large versions of these sketches can be reproduced on milar or as photo-stats. These should be pasted on thick card and hung up in the site office where daily records can be kept of the area filled and the date of filling. In this way, detailed records of progress can be maintained whilst the site crew become familiar with the overall plan.

For a large site requiring placement of temporary drainage ditches, their line and slope also should be shown on the drawing together with a brief explanation of the purpose.

Some of the graphics prepared for the public can also be usefully put to work in the site office. The finished site drawing particularly should probably be hung up as a general reminder of the final plan. It could serve to give a conscientous crew something to strive for and it certainly would not do any harm to the public image (the site crew are members of the public too)!

Operational Sequence

Before the plan can be put into effect, each distinct operational area must be related to the next in terms of the sequence in which development will occur. In most cases, the operational sequence for each area will contain a description of the following items:

- Area – specific location of each distinct area on the site.
- Elevation of landfill base – depth of excavation for cover material.
- Proposed final elevation of the completed area.
- Location of cover material – cover material for this area may come from other locations on the site. The sequence of cover material requirement should be prescribed lift by lift.
- Surface water control – step-by-step surface water control should be prescribed.

A map is needed to visually identify each area and as previously noted it may be useful to show cross-sectional graphic representation of each area as it might look upon completion.

However, for a complex site which comprises several working areas, it may be found that a number of different cross-sectional diagrams can become very complex to work with. A sequential diagram may solve the problem. This particular technique has the added advantage that areas of cover material can also be shown

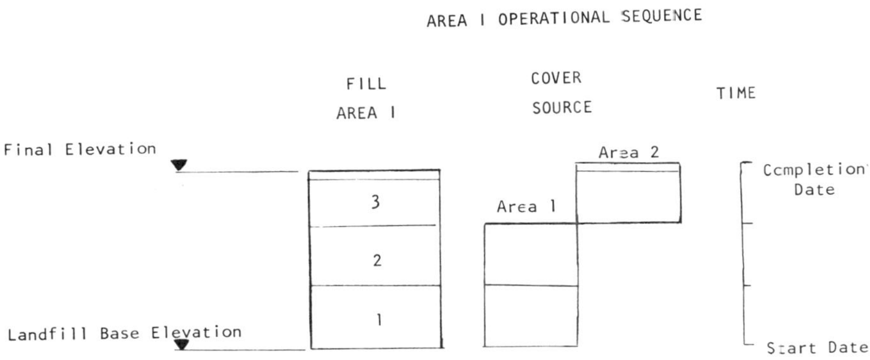

Figure 49. Operational sequence diagram.

together with a "real time" indication of progress on the site. The technique merely requires a box diagram portrayal of the site development. Figure 49 shows the general arrangement.

Column 1 shows the fill sequence for Area 1 rising from the base elevation to the final elevation in three lifts followed by final cover. The base and final elevations can both be shown on the chart. Upon completion of a lift (including daily cover, of course) the block can be filled-in with chalk or better still a felt pen. The time chart can be filled-in as each lift is completed.

The cover source column shows the location of cover material in relation to the fill area for which cover material is required. For example, in the above diagram cover material is only obtained from Area 1 for the first two lifts, thereafter and for final cover, the operator must go to Area 2.

Charts of this type can also show the sequential nature of landfill development area by area by the simple device of treating the bottom line as the starting date and showing each area above the other on the overall chart. For example in the following figure Area 1 must be completed according to the chart before Area 2 is begun.

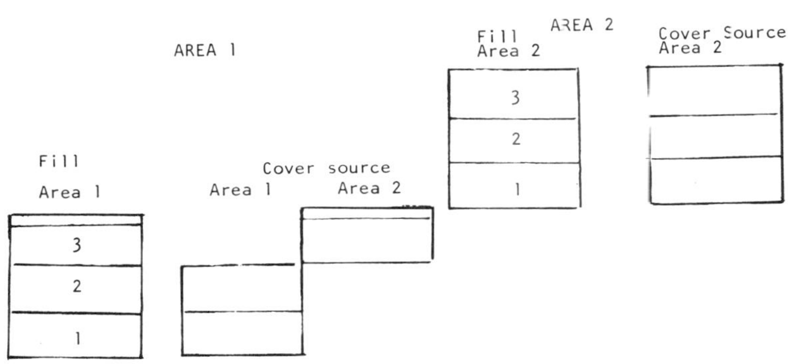

Figure 50. Operational sequence diagram for more than one area.

Charts of this kind can also be useful tools in educating site staff to the overall plan and in obtaining some degree of cooperation in reporting procedures. If the chart is kept up to date by the operator, it will provide a simple means of assessing progress and remaining life. In the long-term, the data can be used to plan future sites if it is known to be accurate and reliable.

Record Keeping

The information reported in the operational sequence chart is intended to provide a good general update of progress on the site. A monthly review of this information would provide an owner with a good general overview of progress on the site.

However, there are many other reasons why a sanitary landfill operation requires good data and in many cases, the operational sequence chart does not provide the necessary level of detail.

On most sites, data is needed for the following purposes:

- billing
- site operations planning
- long-range planning

Billing — Ideally, the users of sanitary landfill facilities should probably be billed according to the specific use which they make of the facility. However, in most cases, the difficulties (and, therefore, the cost) involved in instituting such a system outweigh the apparent advantages. I am speaking here of the individual customers as the users of the site, not specifically those who operate the collection vehicles.

In practice, of course, the collection vehicle operations pay a user fee and collect from the real users by a variety of different methods. The problem for the landfill operator is to institute a fee basis which is equitable as between the various different users of the landfill.

Commonly, the users fall into the following categories:

- municipal haulers
- private haulers
- industrial users
- members of the public

These categories describe those who specifically "come to the site" rather than the waste classification. Clearly, municipal and private haulers may bring solid waste to the site from any number of sources.

The latter point may be relevant to a billing procedure which is based upon waste classification, but it also accentuates the problems of this technique since it is obviously not possible to detect with any accuracy what each load contains.

Much has been written regarding the value of weigh scales at sanitary landfills. Some states go so far as to specifically mandate scales for sites above a certain size.

I feel sure there must be some isolated circumstances where scales are justified and useful but for most normal situations, I can see no major advantages. There are probably only two important characteristics of solid waste from the site operator's point of view:

- volume
- need for special handling

These are the two parameters which determine the quantity of work which must be performed on the waste in order to dispose of it in an acceptable manner.

Volume – Volume defines not only the quantity of work for the dozer driver, but also the amount of space taken up by the compacted load. It seems to me that payment on a volume basis rightly rewards the hauler for efficient use of compaction equipment. When well-compacted refuse arrives on site (in a rear loader for example) some degree of compaction may be an asset for the landfill operator. The load may retain a measure of compaction and in this event the dozer is necessary only to spread and finally place the partially compacted load. Payment on a weight basis would not provide any incentive to use heavy compaction equipment since the hauler would be charged the same regardless of the method of collection and transportation. Users of open trucks on the other hand would be penalized by the volume charging system since little or no compaction accompanies the loading operation. At the landfill, waste from open trucks requires substantially more work in order to place and compact the load. It seems reasonable that this requirement for more work should be recompensed. It is sometimes argued that this variation in the density of incoming waste makes it difficult to predict the rate of landfill usage. However, if the type of vehicle is recorded (as it usually is) along with the volume of incoming waste, I see little reason why this information should not be of greater value to the site planner than total weight. It would be a simple matter to convert open truck uncompacted volume to compacted volume in order to get a feel for efficiency of site utilization.

Need for Special Handling – There are three categories of waste which by their nature require special handling prior to burial in the landfill:

- Waste with high moisture content (generally greater than 80 percent)
- Bulky waste
- Hazardous waste

Techniques to achieve safe disposal of these categories of waste will be discussed in the chapter on special handling. However, it must be obvious at this point that each of the above categories requires some extra expenditure of energy and manpower in order to promote safe disposal. It, therefore seems appropriate that some additional surcharge system should be established which will impose an additional fee upon items in these special categories. Clearly, the surcharge will vary from site to site dependent upon the special techniques adopted to handle the special problems involved.

In summary, as far as data for billing is concerned, a detailed record of volume and special handling needs is all that is necessary to establish a balanced and equitable billing procedure.

Site Operations Planning – For planning site operations, some knowledge of in-place volume is necessary. Thus, in addition to in-coming solid waste volume a record should be retained showing the type of vehicle used. If the vehicle is an open truck, an allowance can be made for inplace density to arrive at the volume actually taken up in the landfill. For a rear loader where substantial compaction has already occurred a different correction factor will be required in order to arrive at an estimated in-place volume.

Long Range Planning – Data collected through the life of the landfill can greatly improve the efficiency of operations. The same data can also prove an excellent source of information for long range planning. Knowledge of incoming volume, seasonal and daily variations in quantity, utilization of land area and cover material will all greatly assist the planning of the next landfill and insure that it will be an even more efficient operation.

Chapter 7

SITE PREPARATION

Assuming at this point that site selection, final design and plan are completed the first steps toward implementing the plan also needs to be carefully planned. A list of site preparation needs is as follows in Table 32.

TABLE 32

Site Preparation Needs

- Public Relations
- Enclosure
- Clearance and Clean up
- Access Road
- Employee Facilities
- Equipment Facilities
- Monitoring

Public Relations

In spite of the earlier described extensive public relations concerns, which are probably necessary to establish any site, the relationship with the public does not end there. Even during the design process some local contacts should be maintained to be sure that no rearguard action is underway.

Even though the initial battle over the site was won, it is too much to hope that this action transformed dissenters into accomplices. The task at this point is to make friends once again with the immediate neighbors of the new operation. The public needs reassurance that the operation will be run with care and concern for the immediate environment.

Under some circumstances, it may be worthwhile to hold another public meeting but this time inviting only the immediate neighbors. The design and the operational plans for the site can be displayed at such a meeting.

A physical model of the final site is an ideal exhibit for this meeting. A model gives substance to a proposed plan and if there are waiverers who need to be convinced that the final plan is no mere caprice, the investment may be worthwhile. Drawings of the final site may be an adequate substitute. The main theme should be to show that early plans have come to fruition and that the promises made earlier will be honored. It is not enough just to do it. The message will be more widely taken-in if it is constantly repeated. It is sometimes necessary to say what will be done, then *do* the job and then *say* that it was done. The purpose of this final meeting is to *say* what will be done. (It might also be good for business!)

Enclosure

Even before site clearing begins, it is advisable to put up a fence. There are many reasons for this:

- safety and security
- privacy
- noise (psychological control)

The safety feature is obvious. Children should be discouraged from playing in the area. Ultimately when the operation is in progress scavengers and vandals can also be more easily excluded with a good stout fence.

The privacy factor really has an impact upon the extent of community acceptance of the operation. In other words, if the operation is visually obstrusive this single fact gives rise to many other complaints.

The genuine noise attenuation attributable to fencing is extremely minimal of course, but it is surprising the noise that may be tolerated if the origin of the noise cannot be seen. Natural vegetation plays the same role.

From the point of view of maintenance, security and repairability, cyclone fencing is usually the best option. Where visual isolation is required, several manufacturers make available a type with interwoven strips of plastic between the standard wire mesh. Where special effects are required such as at the entrance some of the common woven wood fencing types are suggested. This type of wooden fencing could easily be "mounted" directly on the cyclone fencing. At the approach to the site, it would be esthetically more attractive than the more sterile wire mesh.

Clearance and Clean-up — Before the landfill operation begins, it is worthwhile to clear the site of brush and undergrowth in addition to any debris which may have accumulated. Obviously, all this material will be the first to be buried when the operation starts.

Great care should be exercised in clearing the periphery of the site. Where possible, any existing mature of even semi-mature trees should be left standing. They form a more varied and attractive screen than a fence and they help considerably in improving relations with neighbors.

If noise berms are planned, special precautions will be necessary in order to preserve existing tree growth since most trees are sensitive to changes in ground

Figure 50a. Woven screening. With acknowledgments to Anchor Post Products Inc., Baltimore, Maryland.

elevation. Semi-mature trees are probably not appropriate candidates for special treatment since they can be more readily removed and replanted in the finished berm.

Large mature trees which would be very costly to move may be easily incorporated into the berm by constructing a suitable "sleeve" around the trunk up to the height of the berm. As shown in Figure 51, the "sleeve" may be constructed of split pipe, bricks or stone. At least 2 feet should separate the trunk from the stone edge so that adequate ventilation can take place and so that there is no danger of anyone getting accidently trapped in the hole.

Site clearance should also include setting out markers to indicate the boundaries of each fill area and the layout of the access road.

Access Roads – Access roads on the site represent the major arteries which keep the operation running smoothly. They should be planned with care in the manner described in the design section. They should also be carefully designed to take the projected traffic load.

In my experience, the design of the site road rarely receives the attention it deserves. Commonly, the road becomes a quagmire in wet weather and in dry weather site roads are often so pitted with potholes that they are all but impassable. But nothing is for nothing. Ultimately, such sites will be neglected by collectors in favor of more distant sites as the vehicle maintenance costs take their toll.

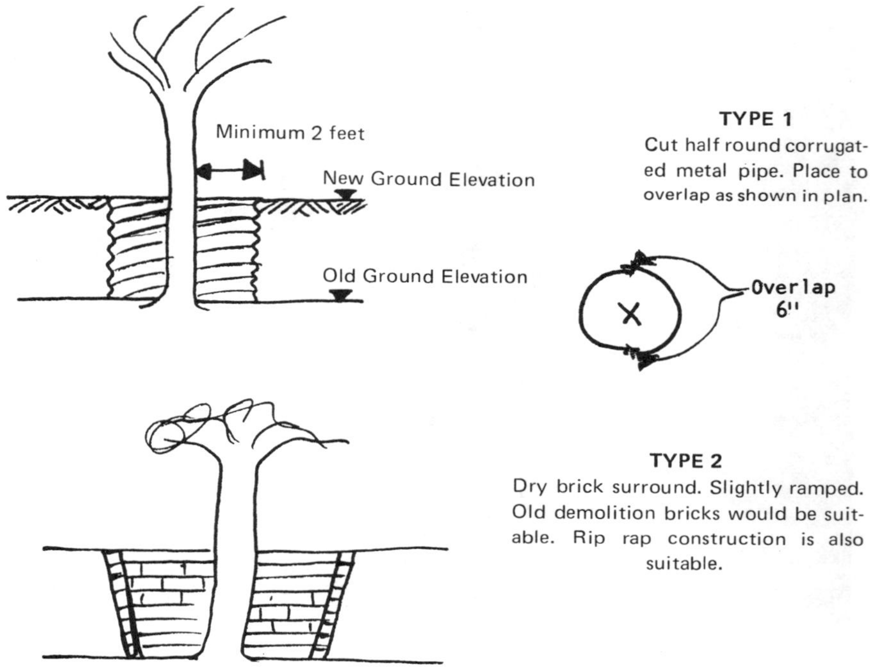

Figure 51. Protection of mature trees in raised berms.

In addition, the trail of mud leaving a badly designed site is a poor advertisement for the landfill business (both in general and in particular).

Wherever possible, I would strongly recommend an intermediate type — bituminous road which will stand the test of time and minimize maintenance costs.

Where bituminous roads are simply out of the question, the gravely soils will provide the best alternative. They present a good wearing surface and have excellent drainage characteristics. Table 33 presents a complete synopsis of the roadway characteristics of different soil types.

Equipment Facilities

Where possible, covered storage should be provided for all major items of landfill equipment. In northern states, the storage facility should provide adequate protection against sub-zero weather.

Whether storage facilities in excess of mere weather protection are provided depends largely upon whether maintenance will be carried out on the landfill or off the site in a service center or at facilities provided by the equipment supplier. Even in this latter case, it would seem advantageous to have an enclosed area on site where standard maintenance and small running repairs could be carried out.

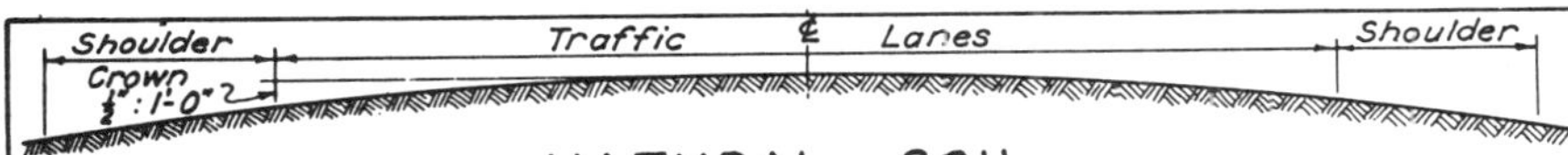

NATURAL SOIL

CLASS OF ROAD	TYPE OF TREATMENT
"F" and "G"	Graded and drained. Crowned ½" per foot and drainage ditches provided.
"E", "F" and "G"	Surface treated with road oil, bitum. material or calcium chloride.
"E", "F" and "G"	1" to 2" of gravel, shells, cinders or stone worked into surface by traffic.
"E", "F" and "G" "SAND CLAY"	Sandy or gravelly soils combined with silty or clay soils and stabilized mechanically.
"E", "F" and "G"	Stabilized with cement, chlorides, asphalts, tars or vinsol resin.

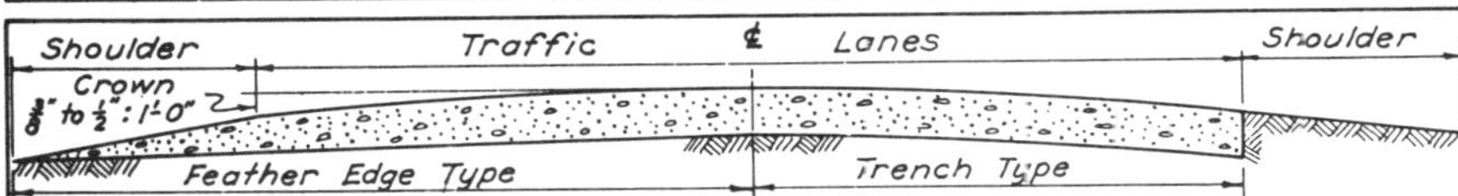

LOW TYPE

CLASS OF ROAD	SURFACE	SURFACE TREATMENT
"C", "D", "E" or "F"	6" of selected soil stabilized with Portland cement, bituminous material or calcium chloride.	Bituminous.
"C", "D", "E" or "F" "SAND CLAY"	6" to 10" of gravel, sand-clay, topsoil, lime-rock, caliche, scoria, shells, chert, shale, slag or cinders.	Road oil, calcium chloride or bituminous.
"C" or "D"	4" to 10" of traffic-bound macadam. 4" to 6" of water-bound macadam.	Bituminous.

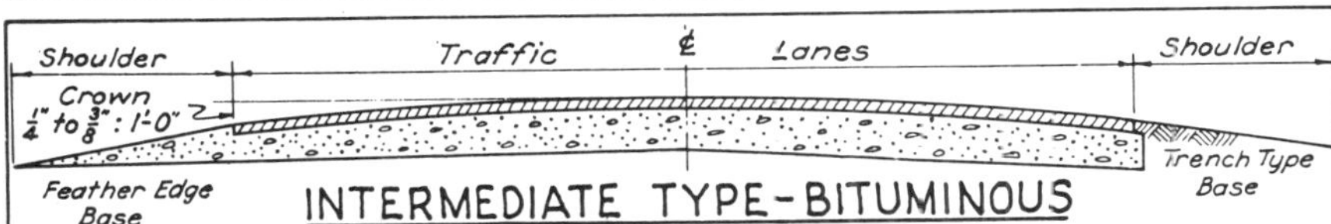

INTERMEDIATE TYPE-BITUMINOUS

CLASS OF ROAD	SURFACE	BASE
"C", "D" and "E"	2½" mixed-in-place (Road-Mix).	6" or more of compacted gravel.
"C" and "D"	2½" low cost plant-mix or pre-mix.	6" crushed gravel or equal.
"B", "C" and "D"	4" to 6" of sand-asphalt.	6" sand-clay or equal.
"B" and "C"	3" bitum. penetration macadam.	3" to 6" crushed stone.
"C" and "D"	1½" bituminous pavement.	6" of stabilized selected soil.
"B" and "C"	1½" to 2½" bituminous pavement.	6" cement-treated base.
"B" and "C"	1½" natural rock asphalt.	6" to 8" water-bound macadam.
"B" and "C"	2½" bitum. penetration macadam.	8" Telford, broken stone, field stone.

References: Highway Design & Construction by A.G. Bruce.
American Highway Practice, Vol. I & II by L.I. Hewes.

Figure 52. Roads — typical sections — 1.

Source: "Data Book for Civil Engineers — Design", Second Edition, Elwyn E. Seelye, John Wiley and Sons, Inc., New York, p. 3–44.

CLASS OF ROAD DAILY TRAFFIC	B 750 TO 4,000	C 300 TO 750	D UNDER 300	E, F & G E = 200 MAX. F = 100 " G = 50 "
Surfaced Width.	2-12' Lanes	2-11' or 12' Lanes	2-10' to 12' Lanes	20' Min. 2 Lanes 10' to 14'-1 Lane
Shoulders.	8' to 10' Stabilized	6' Minimum 8' Preferred		4' Minimum 8' Preferred.
Minimum Right of Way.	100'	100'	100'	66'
Recommended Design Speed.	50 to 75 M.P.H.		30 to 60 M.P.H.	
Maximum Horizontal Curve.	4°	5°	6°	8° to 24°
Minim. Sight Dist.	800'	800'	600'	500'

Figure 52a. Road class criteria.

The extent of additional area will depend upon the amount of maintenance contemplated. In general, an area at least equal to twice the plan area of the equipment is required to provide all-around access. Floor design should, of course, reflect the weight range of the vehicle not forgetting the concentrated load on a hydraulic jack.

The nature of the tools necessary will also reflect the extent of maintenance contemplated. An inventory for a small maintenance shop which would also provide tools for jobs around the site might be as shown in Table 34.

In addition, a small repair shop might be equipped with the equipment shown in Table 35.

At small sites, it may be more economical to have maintenance work carried out at a nearby service center. Alternatively, many of the tools mentioned above may be rented.

Employee Facilities — Most landfill site operators work out of doors all day at manual jobs requiring hard physical exertion. It is the landfill owner's responsibility to provide the men with clean and decent facilities for washing, eating and changing clothes.

The recent history of the development of the sanitary landfill from the bad old days of the burning dump has left a legacy of low-cost development in the landfill business which continues to smear the image of good landfill practice. The following paragraphs may be viewed as idealistic by the "old hands", but the day cannot be far off when proper facilities are provided as a matter of course on good sanitary landfill operations.

Most workers will accept that some jobs involve getting dirty if they have an adequately equipped and comfortable place where they can wash. Ideally shower

TABLE 33

Roadway Soil Characteristics

Major Divisions		Name	Value as Foundation When Not Subject to Frost Action	Value as Base Directly Under Wearing Surface	Potential Frost Action	Compressibility and Expansion	Drainage Characteristics	Compaction Equipment	Unit Dry Weight Lb. Per Cu. Ft.
COARSE GRAINED SOILS	GRAVEL AND GRAVELLY SOILS	Gravel or sandy gravel, well-graded	Excellent	Good	None to very slight	Almost none	Excellent	Crawler type tractor, rubber tired equipment, steel wheeled roller	125-140
		Gravel or sandy gravel, poorly-graded	Good to excellent	Poor to fair	None to very slight	Almost none	Excellent	Crawler type tractor, rubber tired equipment, steel wheeled roller	120-130
		Gravel or sandy gravel, uniformly graded	Good	Good	None to very slight	Almost none	Excellent	Crawler type tractor, rubber tired equipment	115-125
		Silty gravel or silty sandy gravel	Good to excellent	Fair to good	Slight to medium	Very slight	Fair to poor	Rubber tired equipment, sheepsfoot roller, close control of moisture	130-145
		Clayey gravel or clayey sandy gravel	Good	Poor	Slight to medium	Slight	Poor to practically impervious	Rubber tired equipment, sheepsfoot roller	120-140
	SAND AND SANDY SOILS	Sand or gravelly sand, well-graded	Good	Poor	None to very slight	Almost none	Excellent	Crawler type tractor, rubber tired equipment	110-130
		Sand or gravelly sand, poorly-graded	Fair to good	Poor to not suitable	None to very slight	Almost none	Excellent	Crawler type tractor, rubber tired equipment	105-120
		Sand or gravelly sand, uniformly graded	Fair to good	Not suitable	None to very slight	Almost none	Excellent	Crawler type tractor, rubber tired equipment	100-115
		Silty sand or silty gravelly sand	Good	Poor	Slight to high	Very slight	Fair to poor	Rubber tired equipment, sheepsfoot roller; close control of moisture	120-135
		Clayey sand or clayey gravelly sand	Fair to good	Not suitable	Slight to high	Slight to medium	Poor to practically impervious	Rubber tired equipment, sheepsfoot roller	105-130
FINE GRAINED SOILS	LOW COMPRESSIBILITY LL < 50	Silts, sandy silts, gravelly silts, or diatomaceous soils	Fair to poor	Not suitable	Medium to very high	Slight to medium	Fair to poor	Rubber tired equipment, sheepsfoot roller; close control of moisture	100-125
		Lean clays, sandy clays, or gravelly clays	Fair to poor	Not suitable	Medium to high	Medium	Practically impervious	Rubber tired equipment, sheepsfoot roller	100-125
		Organic silts or lean organic clays	Poor	Not suitable	Medium to high	Medium to high	Poor	Rubber tired equipment, sheepsfoot roller	90-105
	HIGH COMPRESSIBILITY LL > 50	Micaceous clays or diatomaceous soils	Poor	Not suitable	Medium to very high	High	Fair to poor	Rubber tired equipment, sheepsfoot roller	80-100
		Fat clays	Poor to very poor	Not suitable	Medium	High	Practically impervious	Rubber tired equipment, sheepsfoot roller	90-110
		Fat organic clays	Poor to very poor	Not suitable	Medium	High	Practically impervious	Rubber tired equipment, sheepsfoot roller	80-105
PEAT AND OTHER FIBROUS ORGANIC SOILS		Peat, humus, and other	Not suitable	Not suitable	Slight	Very high	Fair to poor	Compaction not practical	—

Source: Adapted from U.S. Army Corps of Engineers, Engineering Manual, Part XII, Chapter 2, Appendix "A".

TABLE 34

Tools for a Small Maintenance Shop

- Complete set of small hand tools (as often provided by the manufacturer)
- Two 15 ton screw jacks, one 20 ton hydraulic jack
- One ½ ton, one 1-ton and one 2 ton chain hoist
- An assortment of wire rope slings and shackles
- One 200 amp portable arc welder, one 300 amp stationary arc welder, or a gas combination welding and cutting outfit

TABLE 35

Equipment for a Small Repair Shop

- One bench vise and one large floorstand vice
- An assortment of drop lights and extension cords
- One bench-type drill press and assorted drills
- One bench-type grinder
- A good assortment of test equipment for electrical repair work

and washroom facilities should be provided preferably close to toilet facilities and a locker room. Obviously, appropriate storage space for stocking soap, towels, and toilet paper should also be provided.

Ideally, two lockers should be provided for each employee, one locker for street clothes and one for work clothes. In addition, the locker room (or an adjacent room in a large operation) should be furnished with a sturdy table and chairs or benches where personnel can relax and eat lunch. On a large site, this room might also contain a drinking fountain, a small gas or electric stove, a refrigerator, a sink and a storage cabinet.

For small sites with limited life, a minimum of portable sanitary facilities, safe drinking water and a place for eating lunch and changing clothes should be provided. The use of mobile trailors can solve the problem of the short-term site. For sites with over ten years life and a substantial annual waste input of greater than say 200,000 tons, permanent structures should be considered, particularly where these buildings may have continued utility as an important element of the final plan.

Chapter 8

EQUIPMENT SELECTION

Equipment, Plan and Personnel

A good engineering design is crucial to the success of any landfill operation. However, the value of the design can only be fully exploited with the right equipment and personnel. The secret of a good site is an effective working relationship between plan, equipment and personnel.

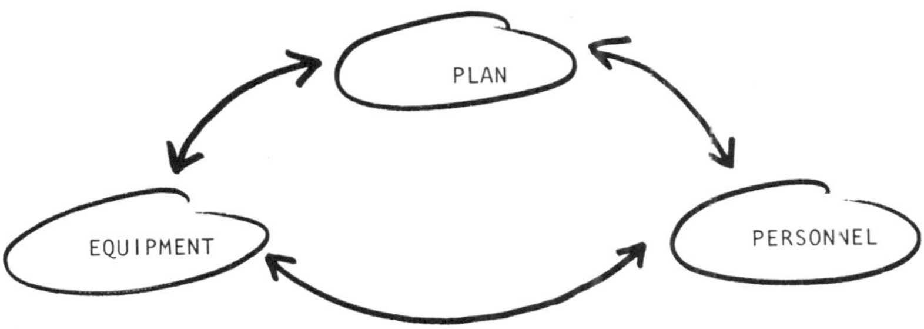

Figure 53. Interrelationship of plan, equipment and personnel.

Source: "The Selection of Sanitary Landfill Equipment", by Richard Eldredge, P.E., Eldredge Engineering Associates, Inc.

A defficiency or a "lack of fit" in any one of these three basic elements will generate problems and inefficiencies.

Of these three basic elements equipment is commonly the least flexible. Once a specific purchase has been made, the operation of the landfill is "required" to conform to the effective use of this equipment. Thus, if the equipment is badly chosen for the job, there is a tendency to "bend" both plan and personnel to cover the problem. This is obviously a poor beginning for a new operation and one which is likely to run into labor or operational difficulties sooner or later.

Strictly, the equipment, particularly major items, should be chosen to fit the plan. If the plan projects a very rapid growth in incoming waste, some allowance should be made for the fact that the total equipment capacity will need to be greater than that initially encountered.

Most landfill plans can be split up into a series of major work elements, each of which requires a different equipment characteristic. Commonly some of these required characteristics can be combined in order to reduce equipment costs.

Table 36 identifies equipment related work elements.

TABLE 36

Equipment Related Work Elements

Purpose	Work Element
Solid Waste Handling	Spreading Compaction
Cover Material Handling	Excavation Hauling Spreading Compaction

Clearly the quantity of work represented by each of these work elements will be different for every site. The amount of solid waste spreading and compaction will be a function of the following variables:

- total solid waste input
- characteristics of waste
- frequency of arrival of waste

The amount of cover material handling will be a function of the following variables:

- total solid waste input (since quantity of cover material will be directly proportional to waste quantity)
- type of cover material
- distance of source of cover material
- topography of site

Clearly different types of equipment are (or can be) engaged in each of these work activities. However, on a small site considerable efforts should be made to achieve the best compromise since the selection of specialized equipment for each work task would simply produce a great amount of idle equipment for part of each working day.

For larger sites multiple equipment requirements are common, since total waste input is usually more than one piece of equipment can handle. Waste input quantities in excess of 30,000 tons per year commonly require more than one piece of equipment. Thereafter, as the waste input increases, the equipment inventory can be increased according to:

- cost
- operating efficiency

Cost – The addition of each major piece of equipment has a direct impact upon disposal cost. Figure 54 shows very dramatically how each decision on equipment inventory is reflected in total cost. It also shows how additional utilization of equipment through overtime or second shifts markedly decreases disposal costs.

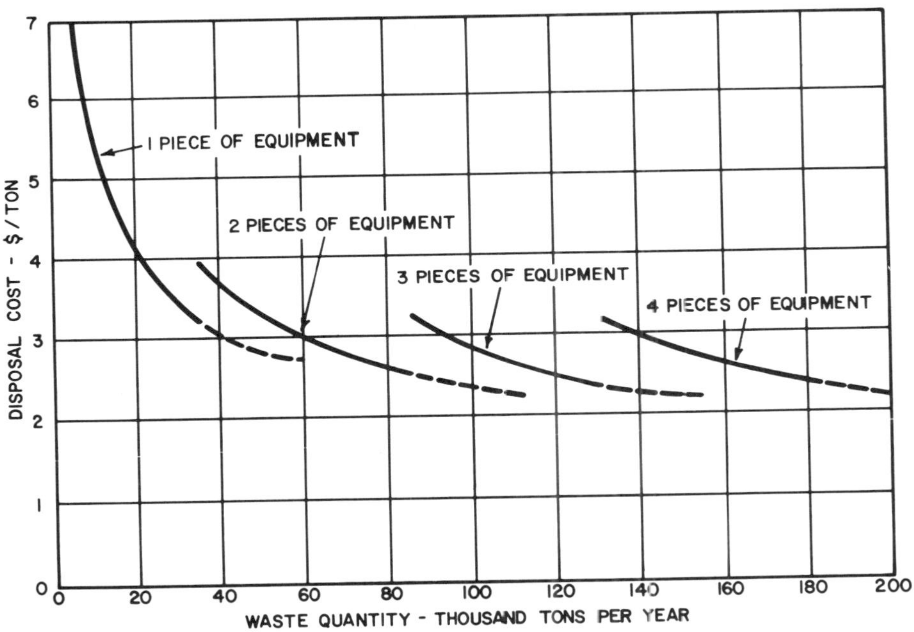

Figure 54. Sanitary landfill costs. This graph was developed by Malcolm Pirnie, Inc. Detailed information on considerations in developing the graph is given in the following notes on Landfill Costs. The dashed portions of the curve indicate overtime or second shifts allowing the site to be operated without purchasing additional equipment.

Source: Reproduced with permission of the New York State Department of Environmental Conservation.

Notes on Landfill Costs[1] – Landfill costs in the curve in Figure 54 are based on the following considerations:

1. Adequate cover material readily available at the site.

[1] From "Sanitary Landfill – Planning, Design, Operation, Maintenance", New York State Department of Environmental Conservation, Bureau of Solid Wastes Engineering, 1971. Reproduced with permission of the New York State Department of Environmental Conservation.

2. No special drainage facilities required.
3. No severe topographic restrictions.
4. A landfill consisting of 2 lifts, each containing 8 feet of refuse with one foot of intermediate cover and 2 feet of final cover.
5. A density of waste after compaction of 750 lbs/cubic yard.
6. Financing at the rate of 6 percent for 20 years for land and buildings and 6 percent for 5 years for equipment.
7. Land cost of $500 per acre.
8. A complete perimeter fence of chain link.
9. An all-weather, hard-surfaced road to the site, maintained all year around.
10. A building to house equipment, maintenance and office facilities. The building should have adequate utilities and a telephone.
11. Salary levels, including fringe benefits, of $7,500 per year for a heavy equipment operator, $7,200 for a truck driver and $6,000 for a laborer. The foreman's salary would be $8,500.
12. A site operated 8 hours a day with no second shifts or overtime.
13. Cost figures for upstate New York – 1970.

Other factors considered in estimating costs were:

1. Contractor overhead and profit or operating surplus of 20 percent of wages and operating costs.
2. Income taxes, in the case of a private contractor, and payroll taxes.
3. Engineering and other professional services including a site development plan, construction plans, perimeter and topographic surveys, soil borings and subsurface explorations, and administrative and clerical services.
4. Site redevelopment by planting with grass and tree seedlings.
5. Miscellaneous costs such as portable fencing to control blowing of paper, hand tools for maintenance, etc.

It is very important that *all applicable costs* be considered in estimating the true cost of a current or projected landfilling operation. Frequently some costs are overlooked if they are commonly shown under other headings in a municipal budget, resulting in an inaccurate accounting of current versus projected costs.

The specific costs of each item of heavy equipment vary somewhat from manufacturer to manufacturer. As in any purchase, it pays to shop around and to remember that "you get what you pay for". In few places is this dictum so true as in the heavy equipment business.

Operating Efficiency – Table 37 provides an excellent guide to the general capabilities of major items of landfill equipment.

TABLE 37

Equipment Selection Guide for Multiple Unit Sites

Purpose		Loader	Dozer	Compactor	Track Scraper	Rubber Tired Tractor Drawn Scraper	Dragline	Backhoe	Truck	Motor Grader
Solid Waste Handling	Spreading	A	A	A	O	O	O	O	O	O
	Compaction	A	A	A	O	O	O	O	O	O
Cover Material Handling	Excavate Cover	A	A	O	A*	A*	A	A	O	O
	Spreading	A	A	A	B	B	O	O	O	B
	Compaction	A	A	A	O	O	O	O	O	O
	Shaping	B	B	B	B	B	O	O	O	A
	Hauling									
	300′ or less	A	A	B	A	O	C	C	C	O
	300′–1000′	O	O	O	A	B	C	C	C	O
	More than 1000	O	O	O	O	A	C	C	C	O

A = Excellent Choice
B = Secondary Choice
C = "In-Combination Only" Choice
O = Not Applicable or Poor Choice
* = Scrapers may require loading assistance in tough soils and adverse weather conditions.

Source: "The Selection of Sanitary Landfill Equipment", by Richard W. Eldredge, P.E., Eldredge Engineering Associates Inc., Waste Age, Jan/Feb 1975.

"The selection of equipment for use in sites to be operated by multiple units is provided in Table 37. The relationships shown are general and, of course, there may be special instances where choices may be modified. Where refuse volumes and operating conditions warrant, equipment with "A" ratings should be provided for each of the steps. Where none but "A" ratings are selected, compromise is eliminated and the very best utilization of equipment may be obtained."[2]

Table 37 can be used to make a preliminary judgment of equipment suitability. This must be followed by a certain amount of "fine tuning" based upon known site conditions.

The already quoted paper by Eldredge goes on to present Table 38.

"To assist in understanding some of the physical limitations of track mounted, rubber tired and compactor equipment, Table 38 provides a relative evaluation. These relationships are considered generally applicable by the author; however, the specific definitions of "compaction," "flotation" and "abrasive" etc., are not provided and specific limitations are not implied by virtue of their use. These relationships are a current reflection of available literature, personal conversations, and received comments.

In some instances, other activities or interests affect the equipment selection. The need to construct roads on-site, assist in municipal maintenance roles, maintain multiple sites at separated locations, or support winter time snow removal activities will require specialized consideration."

"Rubber tired loaders working in refuse may be equipped with foam filled or steel breaker protection to prevent sudden tire failures.

Belly pan and radiator guards are desirable options. Exhaust pipe locations which will not ignite refuse or asphyxiate operators are a must. Reversible fan blades may be obtained to keep radiators clear of obstructions for reliable and continuous operation. (However side panels must then be screened to prevent litter being pushed through the radiator or impeding air flow.)

Landfill blade width can increase refuse handling capabilities. Special heaters and air conditioning will make the operator more comfortable and, therefore, more efficient.

In summary, the selection of "the" equipment or combinations of equipment for operation of landfills must consider the specific site requirements as well as equipment abilities. It is more desirable to select the equipment as may be required for optimum site operation than modify the operation to conform to the limitations of equipment."[3]

Monitoring

Regardless of any regulatory agency requirements, it is important that some provision should be made for ground water quality monitoring. Many agencies now

[2] From "The Selection of Sanitary Landfill Equipment", by Richard W. Eldredge, P.E., Eldredge Engineering Associates Inc., Waste Age, Jan/Feb 1975.

[3] Ibid.

TABLE 38

Physical Limitations of Major Landfill Equipment

Equipment Type	Compaction	Flotation	Resistance to Abrasive Materials	Earth Movement 300′ or less	Up to 600′	Traction Under Wet and Slippery Conditions	Excavation of Soils Granular	Tough Cohesive	Weathered Rock	Blade Width and Loading Capability	Ability to Provide On-Site Construction Assistance	Ability to Easily Provide Off-Site Assistance	Ability to Resist Puncture of Tires
Tracked Dozer	3	1	3	2	2	1	1	1	1	1	1	3	1
Loader	3	1	3	2	2	1	1	1	1	2	1	3	1
Rubber-tired Dozer	2	2	1	1	2	3	2	2	2	1	1	1	3
Loader	2	2	1	1	1	3	2	2	2	2	1	1	3
Steel-wheeled Compactor	1	3	2	2	3	1	3	3	3	1	3	3	1

1. Highly Rated for the Item Listed
2. Rated as Medium Abilities for Listed Item
3. Rated Low in Ability for the Identified Item

Note: These ratings are relative. A "three" rating does not indicate inability as much as greater problems in attaining the desired result. This greater effort may include the preparation of site or equipment, additional passes of equipment to obtain good results, or the operation of equipment at extremely low production rates.

Figure 55. The Balderson U-blade design is perfectly adapted for landfill use. The wide angled blade helps to crowd material to the blade center to limit side spill. Expanded net top sections allow the operator to obtain maximum efficiency of use.

Courtesy: Balderson, Wamego, Kansas

require at least one monitoring piezometer, but even where this is not mandated, the owner would be well advised to have one installed.

Monitoring protects the landfill owner from charges of groundwater pollution when third parties may be responsible. A good monitoring system will also provide an early warning of potential problems in sufficient time that evasive action may be taken to avoid irreversible impairment of groundwater quality and a lot of headaches for the landfill owner.

Piezometers

The monitoring device in most common use is the piezometer. A piezometer is no more than a test well drilled into the previously identified aquifer. A nonmetallic sleeve capped with a #10 screened well point is inserted into the hole and capped at the end with a screwed, vandal-proof top. It is common practice to place graded gravel or sand in the test hole to prevent soil accumulation around the screen blocking water movement.

Periodically, a pump is used to withdraw a sample of the groundwater for testing. Prior to sampling, it is important to evacuate water which has accumulated in the piezometer since this "standing water" might give a false reading of water quality. Piezometer diameters of 1½ or 2 inches are sometimes quoted as required standards. For water depths less than 34 feet, diameters of 2 inches may be workable, but beyond 34 feet depth to water table a 4 inch diameter would be more suitable since this will allow a submersible pump to be lowered into the well.

Samples should be taken for analysis at least once a quarter. The results should be carefully recorded and compared with earlier analysis.

Where several discrete water bearing strata can be identified, it is important that each aquifer should be sampled. In some states, the practice of "nesting" several piezometers in one test hole is accepted by regulatory agencies as an adequate placement technique in these circumstances.

While this shows a healthy regard for cost savings, it should generally be avoided since there appears to be no reliable means to prevent the aquifers becoming interconnected and giving false readings as a result. The location of the piezometer or piezometers is of vital interest to the regulatory agency and the owner of the landfill. It is for this reason that proposals showing test well locations are often required to be submitted to the agency for prior approval. It is in the interest of both parties that test wells should be located "down-gradient" of the site in the expected leachate path. However, there is no valid reason why they should be located directly on the landfill site itself since any short-circuiting down the sides of the well sleeve will automatically give false results. In addition, the full extent of the natural leachate renovation potential of which the site is capable should be allowed to operate prior to collection of samples for testing. "Down-gradient" piezometer(s) should, therefore, be located as far off the landfill disposal area as the site boundaries will allow.

Background Data

In some states where the importance of groundwater quality has been recognized, there may be extensive existing records already available which will show a history of groundwater quality in the area of interest over many years. In these circumstances, the fluctuations in groundwater quality readings which are typical of many sites may be more readily understood.

In the absence of this historical information, additional piezometers may be necessary in order to provide background readings of groundwater quality. Where previous site investigation has successfully established patterns of groundwater flow, it may be a simple matter to identify a suitable location for an "up-gradient" piezometer. Readings from this location will give an indication of background water quality in the area and through the life of the site the "down-gradient" quality should always be compared first with the "up-gradient" quality.

Analysis

Many states have differing requirements for sample analysis. The State of Illinois requires 32 different samples before an operating permit will be granted.

TABLE 39

Piezometric Analysis Required by the Illinois EPA

1.	Alkalinity, as $CaCO_3$	17.	Iron (Fe)
2.	Aluminum (Al)	18.	Lead (Pb)
3.	Arsenic (As)	19.	Magnesium (Mg)
4.	Bicarbonate (HCO_3^-)	20.	Manganese (Mn)
5.	Boron (B)	21.	Mercury (Hg)
6.	Bromides (B_r^-)	22.	Nickel (Ni)
7.	Cadmium (Cd)	23.	Nitrate (NO_3)
8.	Calcium (Ca)	24.	pH
9.	Carbonate ($CO_3^=$)	25.	Phenol
10.	Chloride (Cl)	26.	Phosphate (PO_4)
11.	Chromium — Trivalent (Cr^{+3})	27.	Potassium (K)
	Hexavalent (Cr^{+6})	28.	Sodium (Na)
12.	COD	29.	Specific Conductance*
13.	Copper (Cu)	30.	Sulfate (SO_4)
14.	Cyanide (CN)	31.	Total Dissolved Solids
15.	Floride (F^-)	32.	Zinc (Zn)
16.	Hardness, as $CaCO_3$	33.	Percent Error

As a starting point, it is probably wise to perform an extensive analysis of this nature in order to establish a comprehensive set of base readings. However, once the landfill is operating there should be little need for such an extensive analytical program every quarter. In general, the rule should be to test for what may be a problem. For a landfill handling only residential waste, there is little value in testing for a variety of exotic metals. On the other hand for a landfill handling sewage sludge, some biological indicators may be more appropriate.

Chapter 9

SPECIAL PRECAUTIONS

In the course of attempting to merely bury refuse on a daily basis, it is astonishing the problems an unsuspecting operator can get himself into. Basically, a good landfill operation can be compared to a well-organized earth moving contract and yet the contractor of the latter operation does not face one tenth of the abuse and complaints that many landfill operators receive.

Clearly, there may be additional need for special precautions because refuse and not earth is being moved around, but in addition to that fact, I feel sure the generally unpopular nature of the activity tends to make the public at large much more critical.

In the course of planning the landfill operation, it is, therefore, well worth drawing up some special precautions to assist the operator in both staying out of trouble and more efficiently operating the site.

The following problems are likely to occur on most sites at some time or another and a suitable control program should be available to handle each contingency:

- dust
- litter
- mud
- erosion
- rats
- fire
- odor
- frost
- accidents
- gas
- problem wastes

Dust

As a result of any earthmoving project, dust will inevitably develop as a problem particularly in an urban or suburban area.

The points of origin of the dust are as follows:

- cover material stripping
- refuse compaction
- access road

The stripping of cover material during dry weather can give rise to local dusty conditions though problems in this area occur infrequently.

A simple but potentially unrealistic solution is not to excavate in these conditions. After periods of rainfall, the top surface of the earth will be sufficiently wet that dust will not develop.

Alternative favorable conditions for stripping cover material occur during dry periods when the wind is low or when the wind direction is away from inhabited areas.

The value of good tree cover can be appreciated in the context of dust control. Even a single line of native trees can reduce wind speed by as much as 40–50 percent.

Minor dusting sometimes occurs during refuse compaction. Slight wetting of refuse at the dumping face can help here. (Industrial refuse high in dust or powder should be handled with great care and checked for occupational hazard at the earliest opportunity.) Fortunately, normal domestic refuse is not excessively dusty and the dust problems which arise in this area are usually minor.

Dust from the access road is commonly the major source of dust complaint. Mud is carried off the site by each truck, it dries on the access road and is then kicked up by the subsequent passage of vehicles often to be entrained in the wind and carried (inevitably) in the direction of the nearest residential area. Some organized means of dust control is absolutely essential to handle dust from this source. There are three common expedients:

- water
- calcium chloride
- waste oil

The first of these is preferred. A simple spray truck with a spray bar at the back can be put to work quickly and easily before any alternative could be organized.

In very cold, dry conditions calcium chloride presents an interesting alternative. The crystals are broadcast by machine or by hand and by hygroscopic action moisture is taken from the atmosphere thus wetting the dust at the same time and reducing potential air entrainment.

Waste oil is still in common use as a dust suppressor in some areas and it can be an effective means of control. Great care must be taken that the resultant runoff does not cause problems.

Litter

Most of the previous statements concerning dust, particularly those with reference to wind screens, apply to litter control. There is generally only one point of origin, however, and that is at the point of discharge of the refuse load onto the working face. (A second source may exist if open trucks are used to carry refuse, but this problem would be more appropriately corrected by covering the contents of the truck.)

At the working face, the refuse should be dropped at the toe of the cell slope at all times if possible.

Moveable litter fences are a must on landfills subject to high winds and these should be in place before operations begin. It would be false economy to wait for the first complaint. A considerable amount of good public relations effort can be lost very quickly if word gets around that the operation is causing litter problems. Just one or two early episodes could be enough to lose a reputation which might take years of care to reinstate.

There are basically three types of litter fence. Most of them are homemade efforts engineered for the particular site:

- moveable screen
- fixed screen
- suspended net

Moveable Screen – The moveable screen appears to be the most common type in use today. It may be moved about reasonably easily and positioned *downwind* of the working face in order to catch airborne trash. In high winds, the screen may need to be staked down so it should be equipped with rope tie-lines for that event. It may be constructed of wood or steel and the net may be wire or rope. The screen must be cleared of litter periodically of course otherwise it will be unsightly and easily upset in high winds.

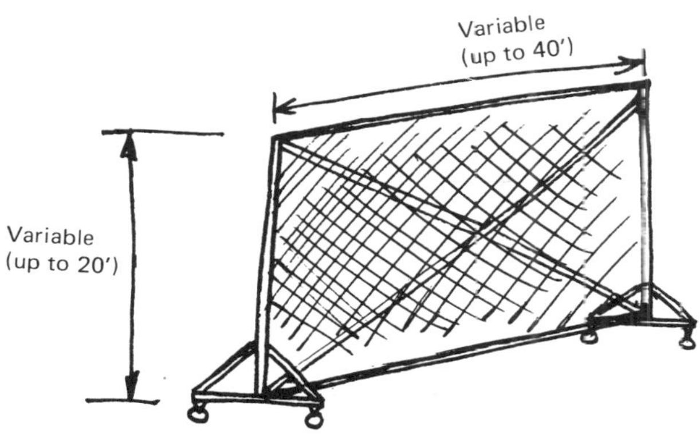

Figure 56. Moveable screen.

The only serious shortcoming is the fact that one screen does not cover a wide area and a gusting wind of variable direction can easily blow trash beyond its range. It may therefore be necessary to have a set of 3 or 4 available.

Fixed Screen – The fixed screen can be staked out to cover a much larger area. However, it must be placed with care since semi-permanent location must serve for several wind directions and several working faces. The main vertical posts can be driven by machine or by hand, but in either event, the maximum height is usually

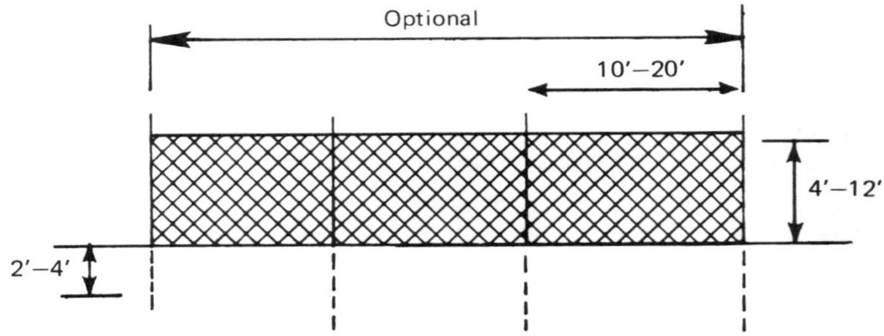

Figure 57. Fixed screen.

rather more limited than the fixed screen. Snow fences make an excellent and inexpensive fixed screen.

Where littering is a severe problem, this type of screen may serve as a back-up to the more versatile moveable screen.

Suspended Net – Some enterprizing landfill operators have attempted to exploit the advantages of both previously described systems by suspending a rope net from steel lines drawn across the site.

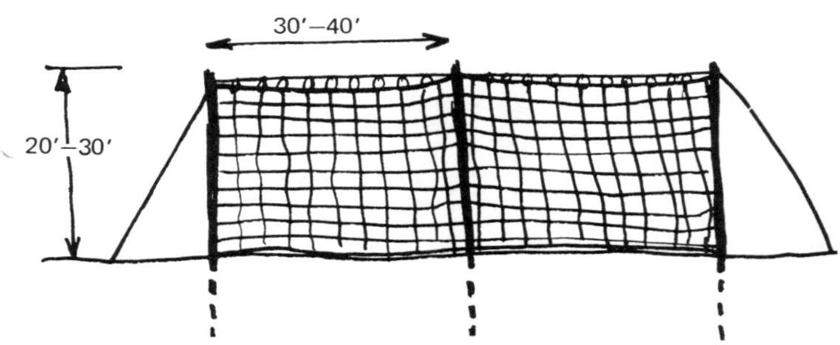

Figure 58. Suspended net.

With careful planning, the positions of the main poles can be determined prior to landfilling. All postholes should then be drilled and poles set in one operation – this would minimize the cost of what might otherwise amount to a prohibitively costly series of drillings to set poles each time the operation moves to another area. For a large site, with limited tree screening, the suspended net may represent some definite advantages. After the initial investment, the system can be relatively trouble free. The area of screen coverage exceeds most alternatives unless several moveable screens are used side by side. Compared to this latter arrangement the maintenance and operational costs are minimal.

Litter Maintenance – Whatever system is used, some additional maintenance will undoubtedly be necessary. There are days on most sites when the direction of the wind is so variable that blown litter will miss any screens. In situations like this, a surrounding screen of trees and fencing provides valuable support by preventing litter escaping onto nearby property. Litter should be picked up from fencing or from adjacent areas on at least a weekly basis. This kind of housekeeping should not be overlooked since the image of the operation is almost as important as the reality and the site may be the subject of visual inspection at a most any time.

Mud

No site can completely avoid mud even on the best regulated sites. Muddy conditions should be expected to occur at the fo lowing locations:

- at the working face
- on the site roads
- on the approach roads

There are a number of precautions which may be taken to reduce the problems. Precautions may be listed as follows:

- wet weather areas
- mixing of sand
- site road cleaning
- tire cleaning

Wet Weather Areas – On many sites, the areas where particularly muddy conditions can be expected to develop in wet weather may be easily identifiable. Areas of clay, clay loam, silty clay and sandy clay will be particularly difficult in wet weather.

There may also be areas of the site relatively free of adverse wet weather conditions.

Wherever possible, areas free of mud in wet weather should be designed as wet weather areas and used only on those days when difficult conditions prevail on the remainder of the site. If possible, the site access road should be designed so that the wet weather area is closest to the main entrance. This reduces on-site travel distance to a minimum at a time when road conditions on site even in special areas will not be ideal. The wet weather area designation also helps to reduce the amount of mud carried onto the site roads by vehicles leaving the working face.

Admixtures – The presence of sandy soils on the site, which would normally be inappropriate for landfill due to their high permeability, may be turned to good advantage by mixing with soils heavy in clay. This admixing may be achieved on site with several passes of a dozer over the wet clay area while dozing a dry sand load into the mixture.

In wet weather, the mixture will be much less "fatty" and less inclined to stick to truck tires. Demolition debris can also be used with the same effect.

Site Road Cleaning – However well the site is operated, mud will be tracked off the disposal area and onto the site roads from time to time. Muddy site roads are a safety hazard. They also cause delays in turn-around at the site and unsightly conditions on the approach roads. It is bad practice and poor public relations to allow these conditions to continue uncontrolled.

The best form of control is to patrol the roads with a truck mounted snow shovel or in heavy mud conditions a rubber-tired tractor with scraper blade may be necessary.

As fast as the mud builds up on the site roads, it should be shoveled off to the side as in a snow clearing operation. No additional equipment is necessary and this reduces mud on the road to a minimum. Obviously, well engineered site roads are required for this operation to be a success. Roads constructed of mud in the first place will always give trouble in wet weather conditions. In the sparsely populated or outlying areas, this may be acceptable, but in urban/suburban regions, I believe the need for cleanliness and good public relations practically mandates good site roads capable of mud clearing and snow clearing without undue damage to the road bed.

Silty soils which slurry in wet conditions cannot be treated in this way though the build up on the site road may be such that cleaning is still necessary. A small battery powered sweeper may be necessary to take up this kind of liquid mud. The same type of equipment may be used to keep the roads clean in dry weather conditions to assist in reducing emissions of fugitive dust from the site.

Figure 59. A truck mounted snow blade.

Source: Reproduced by permission of the Glenhill Road Machinery Company, Galion, Ohio.

Tire Cleaning – One of the most significant causes of poor relations between landfill and general public is the tendency for site vehicles to muddy public roads. For the private site, this problem can also cause bad feelings with the local municipal officials. For the municipal site, muddy roads are definitely not conducive to good public relations or good relations with elected officials.

Usually, the problem is caused by mud sticking to the tires of disposal vehicles. Some simple means of cleaning the mud off the tires before the vehicles reach the public highway is required.

Unfortunately, this is a fairly specialized area, and I am not aware of any proprietary device specifically designed to handle such a problem. However, there is no reason why a resourceful landfill operator could not develop an appropriate cleaning system for the equipment which is generally available. The following ideas are offered in the hope that even if they may not be directly applicable, they may establish the germ of a workable system.

Ideas — I would judge that the chief design criterion should be that any tire cleaning system should be as trouble free as possible. I do not personally see the advantages of a tire cleaning system if it takes up considerable time for maintenance and possibly also manual operation.

Brushes — Vertically mounted brushes probably present the best option. Stiff bristled circular brushes could be mounted on either side of a small by-pass off the main exit road. In wet weather when mud begins to adhere to the wheels, the driver would simply drive between the brushes. A simple trip switch would rotate the brushes against the direction of travel of the vehicle while a scraper blade to the rear of the brushes would continually prevent mud building up in the brush. A pit below the brushes would collect mud falling from the scraper blade. Discarded municipal street cleaning brushes might be adequate as a prototype.

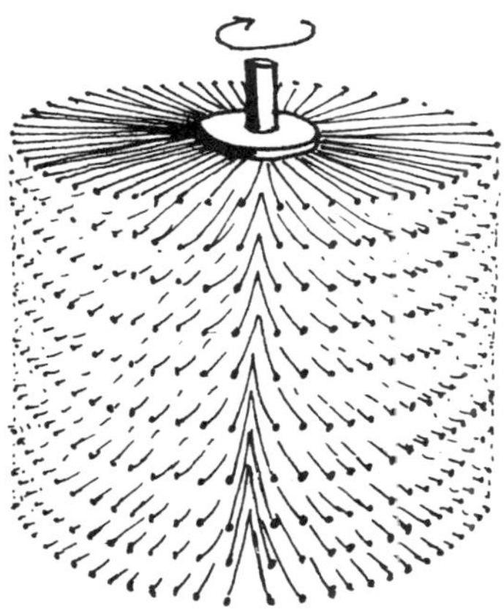

Figure 60. Rotating brush tire cleaners.

Obviously, the system could be made more sophisticated by adding a water spray and spring mounted brushes rather like those in an automated car cleaning operation.

A further sophistication might include "inside" brushes to handle mud adhering to the inner walls of the wheels.

Split Tires — Another idea which might possibly be tried with material available on the site utilizes car tires. Car tires on rims could be mounted vertically one above the other to resemble the vertical brushes previously described. If the tires are then cut across the thread and about 3–4 inches into the side wall, they resemble a cylindrical rubber brush. (Obviously, steel belted tires should be avoided.) Additional flexibility could be achieved by cutting each tire along the face of the tread all the way around.

Mounted on a vertical axis these tire "brushes" could be rotated against the tires to clean the outside edge. They may even be more hard wearing than brushes and they are certainly easily replaceable.

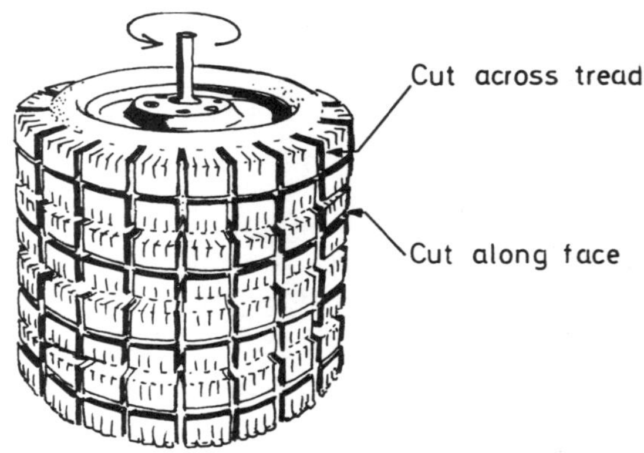

Figure 61. Split type cleaner.

Water Spray – Perhaps the only truly effective treatment to clean tires of mud is the portable water spray. Equipment of this type should be readily available on site for dozer cleaning and maintenance, of course, but it is rarely pressed into service for cleaning trucks leaving the site. I have mixed feelings about the use of such equipment here because additional labor would obviously be required to conduct the cleaning operation. However, this method may represent the most efficient technique for thorough cleaning. In addition, there can be no doubt that it is certainly flexible and could be adapted to clean any type and size of vehicle.

Sophistications include the option of hot water or steam cleaning. These latter may not be particularly appropriate for mud removal, but they are important for equipment cleaning and maintenance.

Fixed Washers – If portable spray units require too much manpower to operate, fixed spray units in a special cleaning shed could do the job very quickly and efficiently. The initial outlay may be high, but for a site with a significant mud problem and a large daily truck count, it may be the most economical solution.

As a general rule, of course, landfill operators do not supply labor, equipment or facilities to work on contractor's trucks. No doubt there are problems of liability. For the benefit of both parties, these inherent difficulties should be overcome because the problems created by the lack of cooperation create conditions from which no one can benefit.

Figure 62. High pressure washer.

Source: With acknowledgment to Citation Manufacturing Company, Siloam Springs, Arkansas.

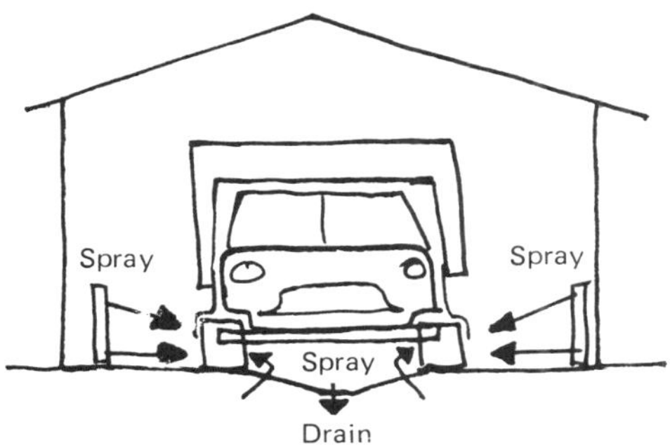

Figure 63. Spray cleaning shed.

In the event that no appropriate cleaning system can be developed, excessively long site roads may be an effective solution. If the roads are continually maintained a fast run down the site road might assist in shaking most of the mud loose.

Erosion

In spite of the best laid plans for erosion control planting, as already described, contingencies may occur on the site which unattended could quickly develop into serious erosion problems. Finished areas should be patrolled at least once a quarter for the first 3–4 years after completion to check that erosion is being held in check. One particular problem is local settlement.

It has been previously observed that large crushed objects such as cars, drums or refrigerators should be placed at the toe of the working face. This serves two purposes. Primarily the toe of the slope provides a wedge against which the dozer can compact the object into the landfill. Additionally, any future corrosion of the object itself is less likely to cause local settlement. The effect of the void caused by the corrosion of this large object may thus be spread over a large area and the surface of the landfill may settle reasonably evenly. However, some local settlement does occur from time to time and it is vitally important that these depressions should be filled and reseeded before they cause problems. In some sites where depressions remain unattended often for years they form a local reservoir. Rainfall gradually infiltrates the surface and taking the line of least resistance may easily emerge at the side of the fill where gully erosion is promoted very rapidly.

Cover material should always be used to fill these local depressions as there is an urgent need to revegetate and this presents a better growth medium than refuse. New plantings should be set immediately and the area carefully tended back to the same condition as the remainder of the finished area.

Rats

The sanitary landfill method of garbage disposal which requires six inches of compacted cover daily should normally eliminate the risk of rats developing on the site. However, one improperly covered cell is sometimes enough to present an open invitation to a storehouse of food and before very long a rat population can develop that represents a real threat to the continued good public relations of an otherwise well run site. It is important, therefore, that the landfill operator should be an expert at recognizing the first signs of rats.

Droppings — Fresh droppings are usually moist, soft shiny and dark, but in a few days, they become dry and hard. Old droppings are dull and grayish and crumble when pressed with a stick. In a building or storage area droppings are an ideal clue but on the landfill area itself they look surprisingly like everything else. Usually only snow cover provides the necessary setting for this clue to be identified.

Burrows — The Norway rat prefers burrows for nesting and harborage whereas the roof rat burrows only occasionally. Burrows may be found in earth banks, under rubbish, in old refrigerators and car hulks, even in a pile of old tires. Frequent body contact leaves a dark greasy smear and these fresh rub marks at the opening of a suspected harborage will often indicate a well established burrow.

Norway Rat

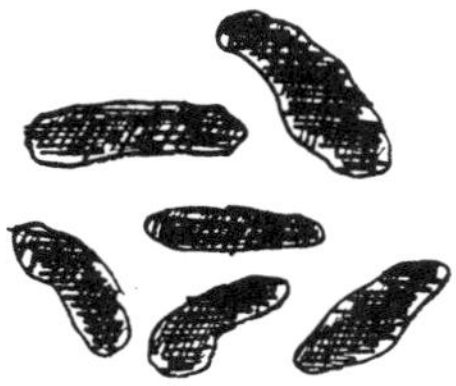

- Blunt
- 1/2" – 3/4 long
- 1/4" thick

Roof Rat

- Pointed
- 1/3" – 1/2" long
- 3/16" thick

Figure 64. Rat recognition.

Visual Identification – The Norway rat is predominantly a burrowing rodent. The adult may grow to as much as 18 inches to 20 inches from head to tail. Common names are brown rat, barn rat, sewer rat or wharf rat.

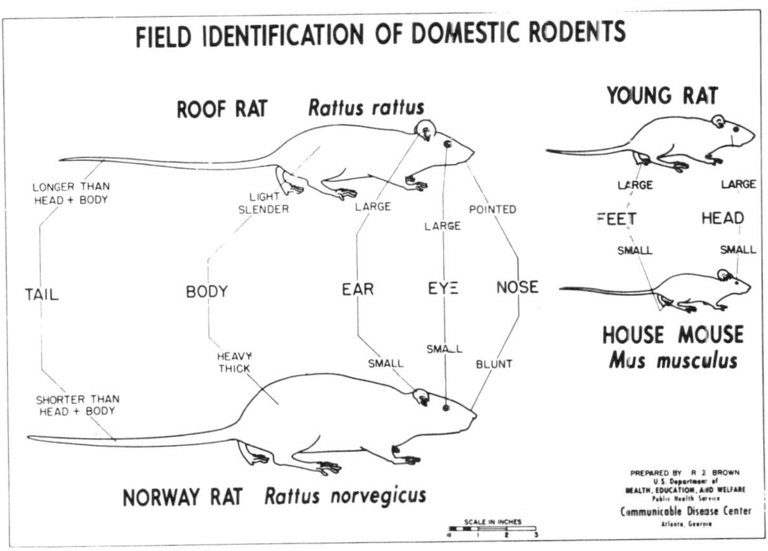

Figure 65. Field identification of domestic rodents.

Source: "Control of Domestic Rats and Mice", by Bjornson, Pratt and Littig, U.S. Department of Health, Education and Welfare. Reprinted by the Bureau of Solid Waste Management, 1970.

The Roof rat is somewhat smaller than the Norway and a more agile climber. In the United States, its range is confined largely to the South, to the Pacific Coast and Hawaii.

Rat-Borne Disease — Rats are responsible for the spread of a number of diseases, either directly or by food contamination. Rats can cause rat bite fever, leptospirosis, salmonellosis, trichinosis, murine typhus fever, plague and rickettsial pox. A number of other diseases are also associated with rats.

There should be no question that immediately upon identification measures should be taken to control the rat population.

Rat Control — This is not a job for amateurs. The function of the landfill operator is to identify the problem, but it cannot be overstressed that a rat control program should be established by a professional.

Rat control is not just a question of killing rats. In fact, killing rats without careful safeguards actually insures a very rapid increase in population since the young survive easily when there is a limited competition. Also, an amateur killing program is almost guaranteed to produce bait shyness or immunity. For this reason, available poisons and baits are not discussed here — where necessary, this information can be supplied by the professional appointed to plan the control program.

Nothing should be attempted by the landfill operator before a control program begins. There is little to be gained from closing burrows except to make the job of the exterminator much more difficult.

Ideally, all the signs which prompted his appointment should be left intact.

When the control program does begin, it should also include measures to eliminate harborage and promote systematic daily cover in order to eliminate food supply.

Handled with intelligence and care there is no reason why the rat population should not be contained on the site and eliminated on the site. (It may be appropriate to add that a thorough rat extermination program is a vitally important preliminary to closing open dumps.)

Fire

Fires may be a hazard both in the landfill and on the equipment. In the landfill one of the virtues of the daily cell construction is that the six inches of compacted cover can act as a fire wall. In most conditions where an underground fire does breakout, it can be contained within the cell structure.

However, where the integrity of the earthen cell is broken, an underground fire can spread and smoulder for a considerable period.

Typically, it is much easier to explain what should not be done than it is to provide positive solutions. Underground fires in landfills can be a serious problem for active working operations and often the necessary manpower to correct the problem may not be available. In consequence, short cut solutions are sometimes attempted which create undesirable side effects. Complete saturation of the offending area may or may not put out the fire, but for reasons that will be obvious from

TABLE 40

Comparison of the Norway Rat and the Roof Rat

	Norway Rat	Roof Rat
General Description	The Norway rat (Rattus norvegicus), predominantly a burrowing rodent, is the most common and the largest of the domestic rats. It is distributed generally throughout the temperate regions of the world, including the United States. Common names for the species are the brown rat, the house rat, the barn rat, the sewer rat and the wharf rat.	The roof rat (Rattus rattus) is somewhat smaller than the Norway rat and is a more agile climber. In the United States its range is confined largely to the South and to the Pacific coast and Hawaii. Worldwide, it is found most abundantly in the tropical or temperate regions but is rare or absent in the colder portions of the world.
Adult Weight	16 or more ounces; adults average about a pound.	8 to 12 ounces
Fur	Coarse, generally reddish brown to grayish brown, with many variations.	Three color phases in the United States; the black rat (Rattus rattus rattus), black to slate-gray colored; the Alexandrine rat (Rattus rattus alexandrinus), tawny above and grayish-white below; the fruit rat (Rattus rattus frugivorus), also tawny above, with white to lemon colored belly. All three sub-species have long prominent guard hairs and they interbreed.
Body	Heavy-set, and with muzzle blunt	Slender, muzzle pointed
Tail	Bicolored, shorter than body and head combined.	Single color, longer than body and head combined.
Ears	Small, close-set.	Large, prominent.
Droppings	Large (½ to ¾ inches long), capsule-shaped (See Figure, Page 7).	Medium-sized (½ to ½ inches long), spindle-shaped (See Figure, Page 7).
Sexual Maturity	Attained in 2 to 3 months	Attained in 2 to 3 months
Gestation Period	Averages 22 days	Averages 22 days
Young	Average 8 to 12 per litter	Average 6 to 8 per litter.
Number of Litter	Averages 4 to 7 per year.	Averages 4 to 6 per year.
Number Weaned	Averages about 20 young per year per female.	Averages about 20 young per female.
Length of Life	Averages about 1 year.	Averages about 1 year.
Harborage	Outdoors-in burrows in the ground and under foundations of buildings and in rubbish dumps; indoors-between floors and walls, in enclosed spaces of cabinets, shelving and appliances, in piles of rubbish and in any other space concealed from view.	Above ground level; indoors-in attics, between walls, and in enclosed spaces of cabinets and shelving; outdoors-in trees and dense vine growth, and in some California cities, in sewers.
Range	Frequently 100–150 feet.	Frequently 100–150 feet.
Food and Water	Omnivorous; garbage, meat, fish and cereal baits are well accepted daily requirement is ¾ to 1 ounce of dry food, ½ to 1 ounce of water.	Omnivorous; vegetables, fruits and cereal grains preferred; daily requirement is ½ to 1 ounce of dry food and up to 1 ounce of water.

Source: "Control of Domestic Rats and Mice", by Bjornson, Pratt and Littig U.S. Department of Health, Education and Welfare. Reprinted by Bureau of Solid Waste Management, 1970.

the design considerations, it can have nothing but damaging consequences. In any event, the point of smoke emergence is not necessarily the source of the fire itself since the smoke is merely escaping along the line of greatest permeability.

Opinions vary over suitable control techniques. In general, if the underground fire appears to be fairly limited in scope an additional layer of heavily compacted impermeable cover material may be an adequate method of control. The fire will not continue to smoulder long without oxygen and the additional cover material may be adequate to complete the seal. However, if the fire appears widespread and continues to smoke for more than 2–3 days, it will be necessary to dig out the whole area and systematically roll-out the smouldering contents. The contents should then be thoroughly wetted and the cell rebuilt with ample quantities of cover material.

Fires on equipment are rather more easily handled, though the first time is all too often the occasion when the operator discovers he is *not* carrying extinguishers. Most reputable manufacturers will recommend appropriate fire extinguishers for their heavy equipment, these should be carried at all times and periodically checked to assure that they will work in an emergency. The equipment storage and maintenance area should also be equipped with suitable fire extinguishers.

Odor

Daily application of six inches of compacted cover should be adequate to eliminate any odors which may be caused by particularly noxious wastes. However, complaints do sometimes occur and all too often they are stored mentally by the operator. Subsequently, they may be followed up by making some change in operating procedure, but the detailed information is often forgotten. The first sign of a well run landfill operation is effective, systematic record keeping and records of odor complaints are no exception. Indeed, record keeping in this area could actually help in solving the problem. The minimum information necessary when logging an odor complaint is as follows:

- Date and time of odor episode
- Geographical location of complaint
- Wind velocity and direction
- Temperature

When the complaint is episodal in nature, there are only three possible explanations. The complaint must be episodal in relation to:

- unrelated activities in the general area
- wind direction
- waste load

If an examination of the surrounding area reveals no suspect unrelated activities a comparison of the odor complaint data and the daily inventory of waste should reveal a correlation supporting one or other of the latter two relationships.

If the complaints is directional, i.e., a certain group of people only complain when the wind is in a particular direction, then any treatment necessary can be reserved only for application during that particular wind direction. Furthermore, some additional planting should be planned between the source and the complaint. A tree screen will assist by reducing the wind velocity and increasing the turbulence of air from the site. The turbulence will increase the mixing with the atmosphere and aid in diluting and dispersing any remaining odors. Also, the tree screen is as psychologically beneficial where odors are concerned as it is when dealing with noise problems. If the screen eliminates the direct view of the landfill operations, it should reduce the extent to which the complaint is due to visual identification (i.e. landfills *should* be odorous).

The most common and effective form of treatment for odors is to spray the offending area with potassium permanganate solution (1–2 percent KMO). Most odor causing compounds are rapidly oxidized by potassium permanganate to form generally odorless compounds.

If the complaint can be correlated with waste from a particular industry then some alternative courses of action are open. The most obvious solution may be to decline to take the waste since the problems caused by the odor may begin to impact upon the good name of the site generally. Alternatively, special precautions may be established which will go into effect whenever malodorous waste from this particular client requires disposal.

- The most obvious precaution is to avoid windy days, particularly when the wind is blowing in the direction of residential areas near the site.
- The load may also be sprayed with an appropriate odor reducing agent before it leaves for the landfill.
- At the site, earth cover can be immediately available to cover the offending waste.

Remembering the earlier discussion of a fair basis for user changes, it should also be obvious that the foregoing special handling requires reimbursement. It seems to me to be entirely equitable that the extra precautions necessary to handle this waste properly should be directly chargable to the client whose waste might otherwise cause problems on the site.

Frost

Frost is a problem at landfill sites chiefly because of the difficulty it causes in penetrating stockpiles of cover material. In areas of the country where frost penetration may be as much as 2–3 feet stockpiles can often be rendered useless unless adequate precautions are taken.

Leaves — There are few precautions as simple and inexpensive as covering the stockpile with leaves. When they arrive at the site in plastic bags they are ideally suited for covering the stockpile. Alternatively loose leaves may be used and held down by logs or scrap lumber.

Hay – Hay represents another simple and flexible precaution. In the winter, when the material in the stockpile is needed, the hay can be easily pulled aside and replaced again after use.

An alternative to loose hay bundles is hay bales. These are obviously more expensive, but they are more durable and have substantially better insulating properties. On a large site where substantial volumes of cover material need to stockpiled for use during the winter months, a "stockade" can be built of hay bales and the cover material dumped inside it. Upon completion, the top can be covered with bales to provide complete protection. To avoid problems in removal and to increase insulation, the top should then be covered with a waterproof covering.

Polystyrene – In areas where hay is not available, operators should keep a sharp eye open for large polystyrene sheets. Roughly tied together several of these sheets provide excellent insulation.

A large sheet of polystyrene has the advantage over leaves and hay that it is also waterproof. In addition if used with care it can provide adequate protection for several years.

Calcium Chloride – A careful operator could save all the time involved in previously described covering and uncovering by methodically building up the stockpile layered with generous applications of calcium chloride. Provided each load of stockpiled material is adequately treated, this simple treatment should prevent cold penetration from forming hard ice packs and in the winter, the dozer should be capable of ripping out cover without difficulty.

Accidents

The accident rate for landfill personnel is not known. It is thought to be quite high based on information regarding solid waste collection workers who have the highest injury rate of all occupations except logging. According to the National Safety Council, solid waste collectors have an injury rate which is nine times higher than industrial workers.

Hazards at sanitary landfill sites are very similar to those encountered by collection personnel with the possible exception of risks associated with lifting heavy containers.

At disposal sites, hazards to personnel include:

- direct injury from explosion or fire
- inhalation of contaminants and dust
- asphyxia from smoke
- falls from vehicles
- accidents associated with the operation of heavy earth moving equipment
- accidents from attempting to repair equipment while engine is operating
- exposure to extremes, frost and heat
- traffic accidents at or near the site

Quite apart from the human cost, accidents are expensive. The indirect costs include:[1]

- lost time of injured employee
- lost time of employees assisting the injured
- lost time of supervisors helping the employee, determining and reporting the cause of the accident, and arranging for a replacement
- damage to equipment
- interference with schedules and production
- increased compensation, insurance rates and settlements

Accident Prevention – The chief protection against accidents is careful and thoughtful personnel. An accident prevention program for all supervisors and employees is a must in order to instil habits of care and safety on the site.

Important elements of an accident prevention program include:

- safety training
- safety records
- protective clothing and equipment
- first aid

Safety Training – In some respects safety training begins with personnel selection. Those potential employees who show an intelligent awareness of the hazards on a disposal site are inclined to be much more responsible employees and much more responsive to on-site training.

According to a publication[2] of the U.S. EPA, the National Disposal Contractors, Inc., assigns its department heads and supervisors the following safety responsibilities:

TABLE 41

Safety Responsibilities

1. providing employees with thorough job instruction and training;
2. setting a good example;
3. talking to employees daily about safety;
4. enforcing all safety rules and policies;
5. maintaining good housekeeping;
6. promptly investigating, classifying, and recording all accidents;
7. insisting on participation in the safety program by all employees;
8. distributing safety literature; and
9. observing collection vehicles in normal operation.

[1] From "Guidelines for Local Governments on Solid Waste Management", U.S. EPA, 1971.
[2] Ibid.

Safety is undoubtedly the responsibility of labor and management. It is the responsibility of management to assign safety standards and the responsibility of labor to carry them out. The development of an attitude of care in the exercise of safety standards is a legitimate area for a reward or incentive program. However, a good incentive program is often constrained by the lack of good data available. Accurate and dependable records are very important not only to maintain an incentive program, but also to provide an indication of the effectiveness of safety programs.

Safety Records — The requirement for good safety records is a national one as well as a concern of each site.

For many years, the National Safety Council (NSC) has been attempting to collect statistics on the injury rate in the solid waste industry. There are standard forms available to make the reporting process as straightforward as possible, and yet the response is usually very low.

It has been suggested[3] that there may be some confusion surrounding the reporting methods and the reasons why specific questions are asked. (The NSC assures the confidentiality of all responses so there should be no concerns about lack of privacy.)

Accident rates are required, of course, rather than numbers of accidents. The latter may be easier to report, but it is not a useful statistic if the number of hours of exposure on the job are all different. Clearly, a man working 60 hours a week on the site has a longer period during which he may have an accident than one working 20 hours a week. Therefore, in any week, the probability of his having an accident is greater. Conversely, an accident free record for the 60 hour a week man is better than a similar record for the 20 hour a week man.

Another source of confusion is the fact that there are two standard definitions for accident rate in use nationally. The older standard from the American National Standards Institute (ANSI) only includes accidents which result in death, permanent disability or lost time. ANSI also uses a normalizing factor of 1,000,000 hours of exposure to calculate a standard. (This is merely a device to reduce the order of magnitude of the numbers expected.)

The newer method of computing injury rate is that developed by the Occupational Safety and Health Administration (OSHA). OSHA accident rates include many less severe accidents than ANSI such as medical treatment cases. However, neither rate includes first aid cases.

In the March, 1975 edition of Waste Age, Kelly King provided an example of a procedure which could be used to calculate both the ANSI and the OSHA rate.

The example may be found in Appendix F. It is provided here as a service to the National Safety Council in the hope that it may be copied and mailed to them in order to improve national records for this industry. An appropriate Accident Report Form is also appended. Obviously, if the site records are immediately

[3]"How to Calculate Injury Rates", by Kelley King, Waste Age Management, March, 1975.

compatible with ANSI and OSHA annual reporting to these two bodies could be simply a matter of copying site records.

It is important to stress that the battle to improve the solid waste disposal industries appalling safety record can be fought on many different levels. At the national level, it is important to develop an overall statistical picture of what are the most accident prone activities on the site. New accident trends can also be spotted at a national level. This knowledge can help to keep the industry informed of accident "soft spots" and hopefully direct attention to improving these areas.

At the local level, safety standards should be scrupulously enforced and safety training should be constantly reiterated. Safety habits should become second nature.

Protective Clothing and Equipment – This feature of accident prevention is often overlooked. On small sites, the temporary nature of the operation and the low level of investment on facilities mitigates strangely against expenditures in this area. However, some attention to the protection which normal work clothing provides site workers should be a must.

Clothing on site should provide both comfort and protection. Too often, it provides only comfort. Employees on a sanitary landfill who work in all weathers should have adequate clothing to provide protection against rain, snow, heat and extreme cold.

In addition, protective equipment should be available for use when conditions at the site warrant extra precautions.

Table 42 provides a checklist of the equipment which should be available to ensure safety under a variety of conditions.

First Aid – In spite of precautions, incentives and protective equipment, accidents do sometimes happen and the landfill site should be properly equipped to handle any eventuality. It cannot be overstressed that at least one member of the site staff should have attended a comprehensive first aid course. For municipal operations such courses are often available for municipal workers in other branches of local government service and every effort should be made to secure attendance for landfill personnel. For private operations, local evening courses often provide first aid instruction. For one employee to be proficient in first aid is a small insurance against minor accidents on the site developing into major problems through lack of proper help.

Dependent upon the first aid skills available and the proximity of a local hospital, there are two approaches to supplying first aid equipment:

- the minimum to sustain life awaiting assistance.
- all necessary support equipment.

It must be stressed that neither approach is possible without someone trained in first aid, but in the former case, the emphasis would be upon maintaining warmth and comfort until an ambulance arrived on the scene.

TABLE 42

Equipment Used to Protect Employees at Refuse Disposal Facilities

Head (including hair and ears)	Skull guard (hard hat)
Face and eyes	Hood
	Goggles
	Face shield
	Welding helmet
Respiratory system	Oxygen breathing apparatus
	Air respirator
	Canister and cartridge respirator
	Filter respirator
Hands and feet	Gloves and hand leathers
	Safety shoes
	Foot guards
Body	Chemical resistant clothing
	Heat resistant clothing
	Cold resistant clothing
	Water resistant clothing
	Radioactive resistant clothing
	Special purpose equipment:
	Life line
	Life ring
	Reflector belt

Source: "Municipal Refuse Disposal", 1966. Prepared by the American Public Works Association. Copyright by the Public Administration Service.

In cases of accidents where fractures are suspected, the emphasis should be upon not moving the victim and on maintaining physical comfort.

Where amputations or severe cuts are involved, it is no longer considered good practice to use a tournequet. Pressure should be applied with a large sterile cotton wad directly at the injury until help arrives. In case of poisons or overdoses, there are some simple antidotes which may be applied pending the arrival of help. A useful list provided by the American Druggist may be found in Appendix G.

Where a highly competent staff skilled in first aid is reliably available, a supplementary kit (Table 44) is suggested to provide them with the necessary equipment to prepare a victim for transport to a hospital for complete medical attention.

Finally, the nearby hospital should be alerted to the development of the sanitary landfill facility. It may be helpful to discuss the nature of the business so that the hospital may be aware of the type of emergencies they may be called upon to handle.

A list of emergency phone numbers should be compiled and placed near the site telephone as shown in Table 45.

TABLE 43

Suggested Minimum First Aid Kit

1. Sterile gauze pads
2. Roll of gauze bandage
3. Band-aids, both regular and Telfa*
4. Adhesive tape
5. Bandage scissors
6. Sterile cotton balls
7. Roll of sterile absorbent cotton
8. Safety razors
9. Triangular bandages
10. Mouth gag
11. Safety pins
12. Cotton-tipped applicators
13. Tweezers
14. Plastic measuring cup
15. Ampoule of Spirits of Ammonia**
16. Aspirin tablets 300 mg in strength (5 grains)
17. Darvon capsules***, this is a non-narcotic pain relieving medication, 32 milligrams (1½ ounce). These tablets are stronger than aspirin and may be combined with aspirin.
18. Hydrogen peroxide
19. Antibiotic ointment (Neosporin, Bacimycin or Furacin)
20. Soap, A cake of plain white soap
21. Blankets and pillows
22. Activated charcoal tablets. Used for poisoning cases. Can be obtained at any pharmacy.
23. Syrup of Ipecac, 15 cc (4 teaspoons). To stimulate vomiting in poisoning cases.
24. Sterile razor of scalpel (for snake bites requiring incision at point of penetration of the fangs). (In an emergency, the use of the non-sterile instruments would be permitted.)

*Telfa-Trade name for non-adherent dressing. Supplied by Kendall Co., Bauer & Black Division. There are other good non-adherent dressing supplied by other manufacurers.

**Vaporole – Aromatic Ammonia, Burroughs, Wellcome Company.

***Ely Lily, Indianapolis, Indiana 46206.

Source: "New Ways in First Aid", by Joel Hartley, M.D., F.A.C.S. Hart Publishing Company, Inc., New York. This list, in addition to Table 44, was adapted from the above source with the assistance of Dr. Derek Stubbs and Dr. Joe Bruno.

TABLE 44

Supplementary First Aid Kit

1. Gauze or universal dressing. 25 cms by 22.5 cms (10 inches by 9 inches). The universal dressing comes packed and ready to use.
2. Alcohol wipettes
3. Plastic air splints
4. Padded wooden splints
5. Sterile cotton applicators
6. Sterile eye pads
7. Eyecup
8. Paper cups
9. Pencil and pad
10. Flashlight
11. Sterile tongue depressors
12. Short board splint for neck injuries
13. Long board splint for spinal injuries
14. Collapsable stretcher

Source: "New Ways in First Aid", by Joel Hartley, M.D., F.A.C.S. Hart Publishing Company, Inc., New York.

TABLE 45

Emergency Telephone Numbers

Local		
Poison Control Center		
Pharmacy		
Police		
Hospital		
Ambulance Service		
Fire Station		

Gas Production

It will be remembered that anaerobic decomposition occurs in a landfill in the absence of oxygen and in the presence of excess moisture. The biproducts of anaerobic activity are gases, the major components of which are methane, carbon dioxide and nitrogen. Oxygen and hydrogen sulfide may also be present in small quantities.

Over the years, various attempts have been made to establish the relationship between landfill life and gas production, but it is still too early to say with accuracy what that relationship may be. It appears likely that the onset of gas production may occur around the time that field capacity is reached since anaerobic conditions occur at that time. However, it is also possible that local areas of gas production may occur prior to field capacity where isolated cells may become saturated by virtue of the cells contents.

It also appears likely that gas production declines with age though the specific relationship probably varies from site to site. Obviously, the characteristics of the refuse also play an important role in the generation of landfill gases. In general, solid waste containing a large percentage of biodegradable material will tend to produce a greater volume of gas compared to inert solid waste.

The EPA quotes[4] a rate of gas production of 40 cubic feet of gas per cubic yard of solid waste, but it is important to note the particular contents of the experimental fill.

TABLE 46

Contents of Experimental Fill and Gas Production

Gas Production = 40 feet^3 per yard^3 of Waste	
Contents	**Percent**
Paper	42.7
Grass & Garden Clippings	38.0
Plastic	3.0
Glass	5.0
Metal	7.0
Dirt	5.0

Source: Merz and Stone, Special Studies of a Sanitary Landfill, U.S.P.H.S., 1970.

It is probable that a landfill containing food waste would produce a somewhat greater volume of gas per cubic yard of waste.

Gas production is a problem in landfills chiefly because the gas contains methane

[4]Sanitary Landfill Design Criteria, EPA, 1971, Page 15.

which can be explosive. Methane is colorless and odorless, and highly explosive in concentrations of 5 to 15 percent in the presence of oxygen.

In the absence of controls, the gas may migrate laterally from a site and accumulate in nearby basement structures where it could easily achieve critical concentrations. Control techniques are designed to vent the gas to the atmosphere where natural mixing reduces the concentration to safe levels.

The technique for gas control depends upon the degree of permeability of the final cover material and as noted earlier, cover soil should be as impermeable as possible in order to reduce infiltration.

By reducing infiltration, the onset of gas production is delayed significantly since excess moisture is needed to promote anaerobic decomposition. However, in most temperate zones, where there is an excess of rainfall after allowing for evapotranspiration and runoff, field capacity will ultimately be reached and gas production will start.

One school of thought treats the production of gas as a contingency item thus the chief mode of handling the problem is to vent the gas wherever it happens to form.

This solution first involves identifying the area of local gas production. A hole is excavated at the point of gas emission to a depth of 6–8 feet and a vertical standpipe is placed in the hole. The pipe (usually 4–6 inches in diameter) is then held vertical while the hole is filled with coarse gravel or crushed stone. As in the figure, the final two feet should consist of replaced cover material and the general ground level around the standpipe should be raised at least one foot above the surrounding area in order to prevent runoff accumulating in the area and passing directly to the cells below through the gravel. The vent pipe itself should consist of a straight section with a 180° bend at the top to prevent rainfall passing directly down the vent to the cells beneath the cover material.

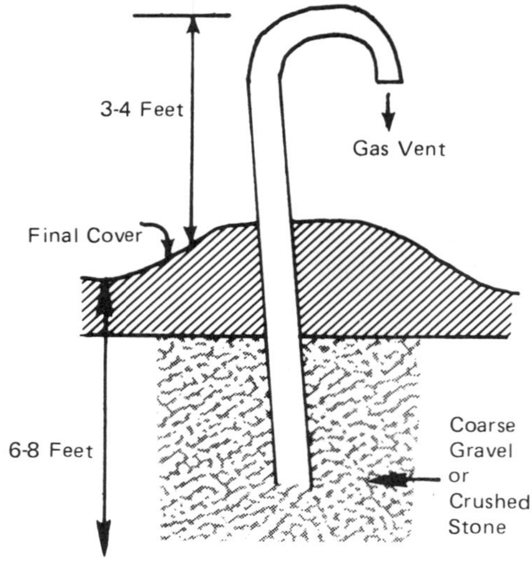

Figure 66. Landfill gas vent.

This technique is by far the most common and appears to work well in most circumstances. Early recognition of gas venting clearly reduces the risk of migration since establishing a vent at source allows no opportunity for horizontal movement of the gas.

However, the technique depends upon the supposition that the final cover material must be impermeable in some areas. While that may be true, I am not convinced that it will always be true. Several additional techniques are suggested where the earlier described method proves unsuccessful.

Gas Control Measures for Landfills with Impermeable Final Cover – Clearly, the requisite design feature in these circumstances should vent the gas which might otherwise migrate from the landfill area. Venting must be carefully established in such a manner that it does not impair the ability of the final cover to reduce infiltration. The safest method, in my opinion, with this latter concern still in mind, is to construct a deep gravel trench around the periphery of the site as shown in Figure 67.

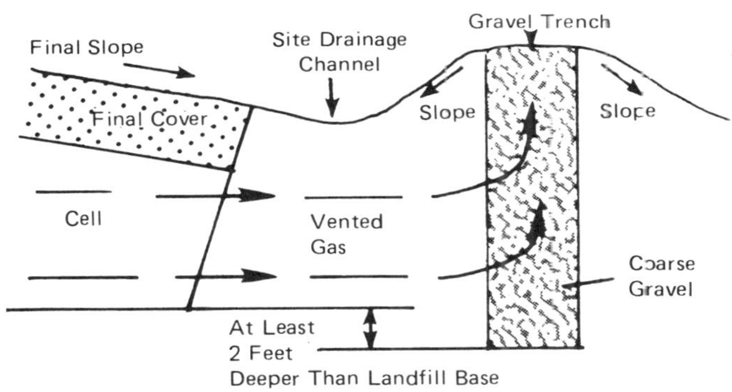

Figure 67. Gas vent – Peripheral Method.

The slopes above the gravel trench should be carefully raised above the site drainage channel. There are two reasons for this important precaution:

- Runoff will be prevented from flowing into the gravel trench and defeating the purpose of the impermeable cover.
- Fine silt eroded from the surface will not be carried into the gravel trench where it might block the gas diffusion channels.

Some texts describe a similar gravel trench arrangement, but oriented across the landfill rather than around the periphery. It is possible that this may be necessitated by the geometry of a site, but I think it should be understood that this arrangement is much less satisfactory from the point of view of providing assurance of no gas migration. Gas movement in the landfill is fundamentally a function of soil permeability and if the peripheral soils will allow gas diffusion, I see no reason why a gas vent across the center of the site would prevent out-migration.

It should be noted that landfills rarely experience a rate of gas production (in addition to odor causing gases) which requires anything other than venting to the atmosphere. Should the rate of production reach a nuisance level, the peripheral trench method will serve well as the basis of a gas collection system.

Perforated field tiles can be laid in the gravel trench (about 2–3 feet below the surface) and gas burners placed at intervals around the closed loop of the collection system.

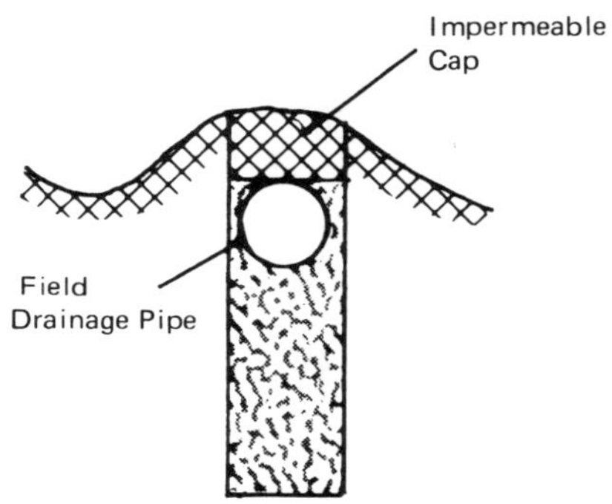

Figure 68. Gravel trench and gas collection system.

It is also feasible to lay a drainage tile in a gravel trench across the surface of the landfill. However, the same limitation holds true, as before, there is no guarantee that all potentially migrating gas will be collected by this method. Differential settlement also makes this method somewhat impractical. Furthermore, it has the added disadvantage that the trench and the gravel locally increase permeability of the final cover and contribute to an increase in infiltration.

Gas Control Measures for Landfills with Semi-Permeable Cover Material – In the semi arid regions of the country where there is little or no net infiltration, it may be permissible to use cover material which allows some moisture movement to occur. In these circumstances, gas control measures are generally unnecessary since gas production is usually so low.

In isolated cases where gas control is required, it should only be necessary to ensure that the periphery of the landfill is impermeable to gas movement. This may be achieved by filling a peripheral trench with a good consistent local clay.

By this means, the design will ensure that gas will only escape through the cover materials.

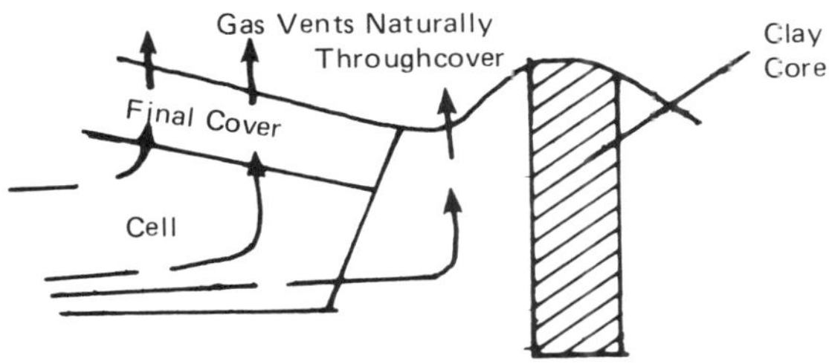

Figure 69. Gas control for landfill with semi-permeable cover.

It should be noted that it is still worthwhile to ramp up the side slopes of the clay surround in order to reduce the risk of infiltration between the sides of the core and the trench.

Problem Wastes

Contingency plans should be made for the disposal of problem wastes, particularly since special handling and preparation is involved in most cases. Problem wastes which require special attention are as follows:

- Waste with high moisture
- Bulky waste
- Hazardous waste

High Moisture Content – This category may include either liquids or sludges. In general, anything with a moisture content greater than 80 percent will create problems on almost any site unless special care is taken in the disposal.

Most sludges with a moisture content greater than 80 percent will flow under gravity and cause instability in the fill which may lead to isolated areas of settlement later in the life of the landfill. Where possible, the excess moisture should be taken up by the addition of a suitable absorbent before sludges are incorporated into a sanitary landfill. The most plentiful absorbent may be the refuse itself though whether or not this is suitable depends upon absorptive characteristics. As observed previously, however, the absorptive properties of a landfill are a valuable safeguard against early leaching and every drop of moisture added to the landfill prior to final cover hastens the day when field capacity is reached.

An alternative absorbent may be surplus dry cover material though it is a fortunate operator who has enough of this commodity to spare for mixing with sludges. Ideally, dry sand should be imported into the site in order to absorb surplus moisture. This may seem like an extravagant gesture merely to assist in landfilling sludges but it must be remembered that the price of good practice today is insurance against bigger problems in the future.

Unfortunately, the legacy of poor practice persists even today in the general expectation that sanitary landfill disposal of sludges is cheap. Sanitary landfill disposal of sludges is never cheap. Even if the price to the customer is low and, therefore, the special attention is minimal, the price is ultimately paid by the operator in frequent maintenance and early leaching.

I do not believe the problem of sanitary landfill disposal of liquid sludges has received the attention the subject deserves. Unless the sludge is subject to special treatment, the presence of excess moisture will hasten the day when leaching begins. This fact alone should change the current attitudes toward the problem.

But instead, *there appears to be a conspiracy to ignore the difficulties since the alternatives are expensive.*

Bulky Waste – Compared to liquid sludges, the problems associated with bulky waste are a welcome relief. However, the problem of traditionally low price expectations for this type of waste also persists for bulky wastes.

In my opinion not nearly enough special handling costs are passed on to the client for bulky waste disposal and yet of all waste disposal problems probably only hazardous waste requires more special care.

Most bulky items present a multiplicity of problems:

- vermin harborage
- visual blight
- physical hazard to operators
- lure to scavengers
- danger to children

Burial is clearly preferable. Cars, appliances and 55 gallon drums are partially crushable, particularly with heavy machines (in the 30–50 ton range) and some landfill volume may be conserved. Placed at the base of the lift, they will be less likely to cause differential settlement and they will be more effectively crushed with each pass of the dozer. Obviously, most bulky items (particularly those insulated with plastic filler) do not degrade rapidly, if at all, and consequently care should be taken to surround such objects with solid waste of similar character such as demolition debris or similar crushed items. In this way future settlement can be "managed" by careful operations planning and future maintenance can be kept to a minimum.

Construction and demolition debris; particularly, brick, concrete, asphalt and plaster should be laid down as a landfill base wherever possible. As noted earlier, this strategy will assist in raising the pH of the leachate and thereby enhance the renovative potential of the underlying soils. Excess quantities of construction and demolition debris can be used to form excellent sub-base material for on site road construction.

Tree stumps and timbers present the greatest problem to the landfill operator. Where space and volume permits, they should be buried and covered, preferably in a separate area where no biodegradable material is buried. In this way, the risk of differential settlement is kept to a minimum. Where the volume of this type of waste is significant compared to available landfill space and where air laws will permit, the air curtain destructor should be considered as an attractive alternative. The volume taken up by tree stump disposal can be truly prodigious and the operator must assess whether the service is worth the effort compared to servicing regular clients with "conventional" disposal problems.

In many instances, tree stump disposal problems, in particular, are actually created by short-sighted legislation in other areas of environmental control. Typically, building regulations require that all tree stumps be excavated and removed from the site. Also typically, the legislation omits to describe the location of their final disposal! I do not see why municipal officials and law makers should not comprehend the fact that a problem does not go away simply as a result of moving the problem around.

The energy and productivity wasted in moving tree stumps around must be truly prodigious. Considering that the disposal problems are identical whether the stump is buried on the site or in the landfill, it seems asinine to me that they are still such mobile items and such regular landfill problems.

Hazardous Waste — This is truly an area for which a very careful contingency plan should be available. I think it is horrifying to realize that the number of landfill sites in this country with a workable contingency plan for hazardous waste disposal probably amounts to less than 0.1 percent[5] of all landfill disposal facilities. Due to the relative difficulty of obtaining information on the number and type of occupational accidents on landfills, it is not known the extent to which the absence of such plans causes death and injury due to careless disposal of hazardous wastes.

Not only is there a general lack of information at working landfills, but few if any[6], state agencies boast a workable set of rules and regulations for specific forms of hazardous waste disposal. In the face of such a huge lack of forsight, the landfill operator is faced with imposing some rudimentary guidelines of his own. In my opinion, it is too much to expect a landfill operator to impose a complex set of analytical constraints upon a customer. In those few states where some rudimentary efforts are being made to establish some controls, it is the customer who is required to complete an application form giving all the analytical details necessary to establish an appropriate level of treatment. The range of alternative hazards and appropriate treatments is too great to expect this identification/classification procedure to be imposed at any other level than the state or federal regulatory agency.

All the landfill operator can do is exercise rudimentary caution particularly where he may be dealing with clients who are not exactly forthcoming as to the risks involved in handling their waste. The landfill operator really needs a simple warning or beware list so that he can ask the right questions. My suggested list is shown in Table 47.

For those who want to really play it safe, I would say do not touch any of them. At very least, the operator should begin to ask questions if these items begin to turn

---

[5] Personal estimate.

[6] At the time of writing (1975) only California and Minnesota have developed comprehensive legislation for hazardous waste control.

TABLE 47

Beware List for Landfill Operators

- Fibers, dust or powders
- Liquids and sludges
- Gases
- Radioactive material
- Pathological waste

up in normal loads of refuse from industrial/commercial clients. The first item is the only one that might be considered marginal. Some fibers, dust or powders may be acceptable but the landfill operator should obtain a written statement from the client describing the contents before such material is landfilled. Even where this material is relatively innocuous and appears to contain no occupational hazard such material has a tendency to blow around easily and create nuisance conditions in the surrounding area. The operator should weigh very carefully whether or not the problems (short and long term) justify the business.

In my own opinion, the other items on the list should be banned from all sanitary landfill sites. However, all too often the operator will accept them, at an inflated price (because the need to dispose of such materials is very great). However, at this point, very little of the excess profit is plowed back into financing special handling for the problem waste and yet compared to the risks involved no dollar value can be set for disposal that would be adequate to pay subsequent compensation if the worst should happen.

Even where wastes can be neutralized, liquids and sludges still contribute toward the total moisture budget in the landfill and as observed previously hasten the day when field capacity is reached. Gases, whether in normal or pressurized containers should never be disposed of in a landfill site. Radioactive waste, by law, must be disposed of under the strict control of the Atomic Energy Control Commission.

Pathological waste from hospitals, research establishments and drug manufacturers should be disposed of in an appropriate pathological incinerator. The problem is compounded for the landfill operator by the fact that pathological waste forms a very small proportion of hospital solid waste and may be easily discarded along with everything else thus escaping detection.

I do not wish to sound alarmist on this subject, but contrary to the belief that hospitals are scrupulous protectors of public health, this is rarely true in the area of solid waste disposal. My advise to the operator who receives waste from hospitals would be that he should specifically see the pathological incinerator and assure himself that it is in use for the purpose intended. For more information on this complex subject, the reader is referred to "Hospital Solid Waste Management" by Frank L. Cross and George Noble.

To summarize this section on hazardous wastes, I believe the operator must be more than the simple receiver of solid waste which he may have been in the past.

With the increasing complexity of industrial development and the increasing pressure upon industry to abstain from on-site disposal, landfills are themselves under pressure to receive many materials which they are not strictly capable of handling. In the absence of any practical guidelines, the operator must be vigilant. It pays to know the type of industry using the landfills services. It pays to know the type of waste that industry generates. It may even pay to know a little about the manufacturing process where there is suspicion that the whole story is not being told. If business is lost as a result of asking too many questions, the conscientious operator can be reasonably assured that he is better off without the particular waste in question. No customer is worth the risks inherent in inadequate disposal or the risks of injury or death of valued employees.

Some insurance against serious risk may be obtained by keeping a useful reference available so that once the hazardous material is identified the risks can be reasonably easily determined. The following list is woefully limited but these four works provide, as far as I am aware, the only comprehensive source on this subject:

1. "Dangerous Properties of Industrial Materials" by Sachs
2. "Toxic Substances List", U.S. Dept. of HEW
3. "Laboratory Waste Disposal Practices" by the Manufacturing Chemists Association
4. "Handling Guide for Potentially Hazardous Materials" by M_2S – Management and Safety Inc., Niles, Illinois.

Of these four only the last item is specifically written with waste handling problems in mind.

Chapter 10

LANDSCAPING

Careful planning of the final vegetation is critically important for the future well-being of the site, and perhaps the sanitary landfill business in general.

It should be clear at this stage that there are several natural phenomena which will rob the site of its value as a safe and effective disposal area unless each phase of the site development is carefully engineered and executed. The final vegetation plan is no exception.

Careful planting will assist in preserving the following important features of the final site:

- binding surface soils to prevent erosion
- increase evapotranspiration
- preserve attractiveness for final use

Considering the vital importance of vegetation to landfill technology, it is somewhat surprising that so little research has been conducted in this field. Only in the field of landscaping reclaimed strip mining areas has extensive research been carried out and though this work may be applicable in some circumstances, it specifically addresses the problems of establishing vegetation on mined spoil material. In such circumstances acidity and stoniness are of overriding concern whereas choosing vegetation for landfill cover material can be an exercise rather similar to choosing material for landscape development of almost any site. Perhaps the only general constraint to growth which is typical of a sanitary landfill is the risk that methane gas generation will accumulate around root structures and literally asphyxiate the plant. This problem may be avoided by the choice of vegetation having lateral root systems which in any event would be natural choice to prevent erosion.

Selection of appropriate vegetation for the site is based upon the following:

1. Identification of plant material adaptable to the climate and soil conditions of the site.
2. Identification of climate and soil conditions of the site.
3. Refining the list of plant material to select trees, shrubs, or groundcovers with characteristics which are compatible with the final use of the site and necessary to serve special purposes such as erosion control on steep banks or screening for visual control.

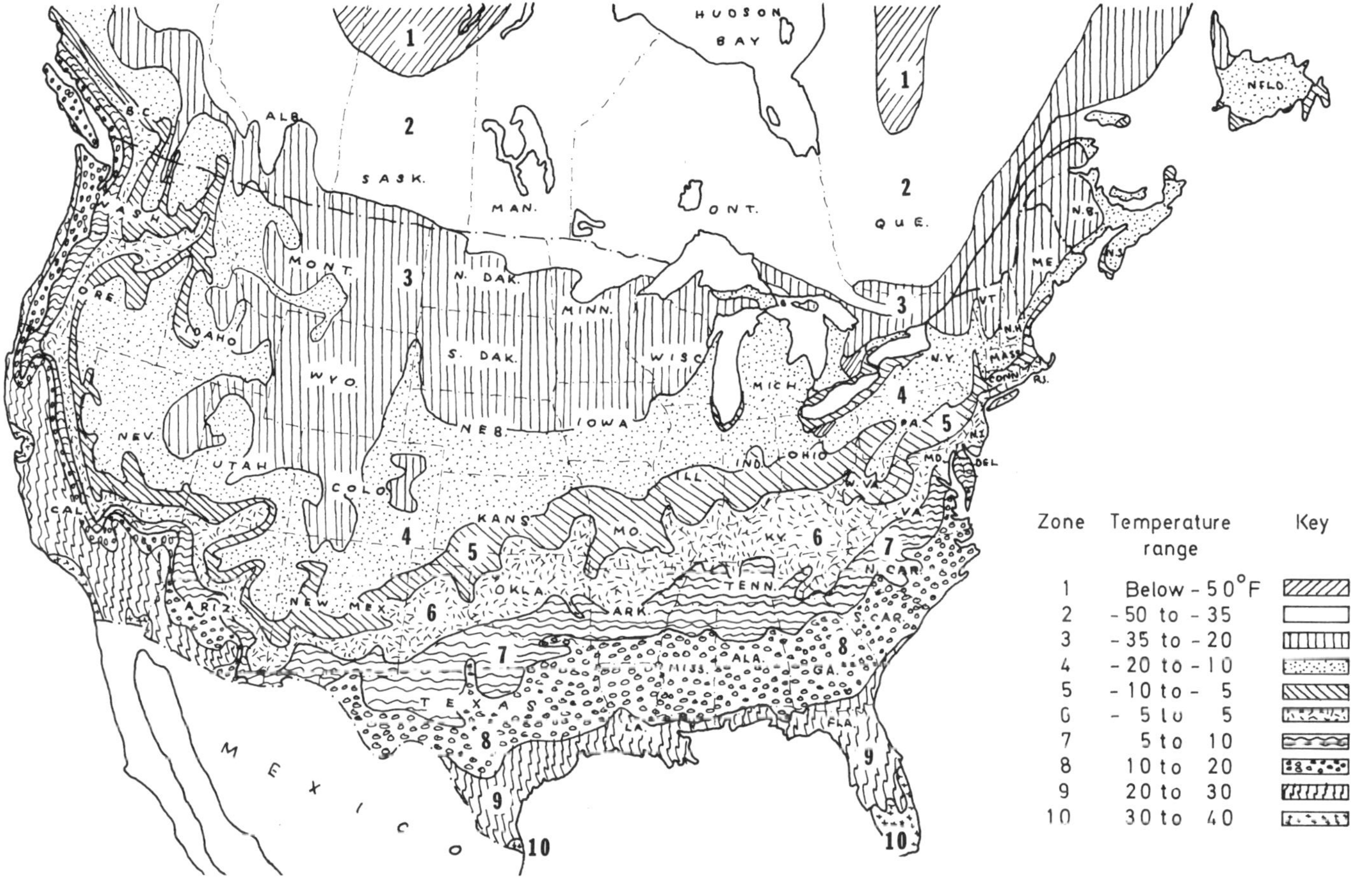

Zone	Temperature range	Key
1	Below - 50°F	
2	- 50 to - 35	
3	- 35 to - 20	
4	- 20 to - 10	
5	- 10 to - 5	
6	- 5 to 5	
7	5 to 10	
8	10 to 20	
9	20 to 30	
10	30 to 40	

Compiled by the Arnold Arboretum, Harvard University
Jamaica Plain, Mass., May 1, 1967.

From "Shrubs and Vines for American Gardens" by
Donald Wyman, The Macmillan Company.
Copyright © 1969 by Donald Wyman

Figure 70. Hardiness zones of the United States and Canada.

Hardiness

The appropriateness or suitability of plant material to a given location is based upon a number of factors which when considered together are considered to be measures of "hardiness". The major concern is that the plant material selected should be "hardy" in the specific climatic zone in which the landfill is located. The overriding consideration is correct *selection* of plant material. It is necessary to select plant material which is suited to the climate in which it will be growing. It is desirable too to select plant material which is suitable for the particular soils of the site as they exist. However, some preliminary soil preparation is commonly necessary to establish vegetation on a sanitary landfill. It is, at the same time, clearly more desirable that soil maintenance be minimized or eliminated once vegetation is established.

Hardiness is directly associated with three things:

- soil
- rainfall
- temperature

Of these three factors, temperature is the most critical. For this reason, simplified hardiness maps can be constructed based only on average annual temperatures. Soil and rainfall conditions ultimately play an important role in final selection, but a simplified hardiness map is useful in establishing a specific range of plant materials appropriate to the climatic or hardiness zone. The Hardiness Zone map shown in Figure 70 is based on data collected by the U.S. Weather Bureau for a period of over forty years. The United States is divided into nine zones. Zones are based on 5, 10 or 15 degree differences in annual minimum temperatures.

To supplement the use of the Hardiness Map, the hardiness zone of the landfill site in question may also be identified by the presence of plants, growing in the area, which serve as indicators of the types of plants typical of that zone.

The following plants are listed in the coldest zone where they will grow normally. They may be taken as reliable indicators of the type of plant material that will grow in each zone. (See Table 48)

Most plant handbooks use the hardiness zone methodology and later when we begin to consider other constraints, it is important to remember that the hardiness zone provides an important geographical yardstick to suitability. In other words, when vegetation is selected later on the basis of soil suitability and erosion control, it is with the implicit understanding that it generally complies with the hardiness zone in which the landfill is located. Much valuable information can be obtained locally from landscape architects at work in the area or from Agricultural Extension Agents in the country.

Soil Testing

Once the general hardiness zone has been established, it is important that the soil

TABLE 48

Indicators of Hardiness Zone Vegetation

Zone	Botanical Name	Common Name
2	Caragana arborescens	Siberian Pea Tree
	Cornus alba "Sibirica"	Coral Dogwood
	Prunus tomentosa	Nankin Cherry
	Rosa rugosa	Rugosa Rose
	Viburnum dentatum	Arrowwood Viburnum
3	Euonymus alatus	Winged Euonymus
	Ligustrum amurense	Amur Privet
	Lonicera tatarica	Tatarian Honeysuckle
	Rhus typhina	Staghorn Sumac
	Syringa vulgaris	Common Lilac
4	Cornus mas	Cornelian Cherry
	Ligustrum vulgare	Common Privet
	Lonicera morrowii	Morrow Honeysuckle
	Rosa setigera	Prairie Rose
	Taxus cuspidata	Spreading Japanese Yew
5	Abelia grandiflora	Glossy Abelia
	Berberis triacanthophora	Threespine Barberry
	Forsythia suspensa	Weeping Forsythia
	Pieris japonica	Japanese Andromeda
	Weigela florida	Old Fashioned Weigela
6	Hydrangea macrophylla	House Hydrangea
	Ilex crenata	Japanese Holly
	Myrica cerifera	Wax-myrtle
	Rhododendron fortunei	Fortune Rhododendron
	Taxus baccata	English Yew
7	Berberis darwinii	Darwin Barberry
	Camellia japonica	Common Camellia
	Lonicera nitida	Box Honeysuckle
	Prunus laurocerasus	Cherry Laurel
8	Campsis grandiflora	
	Euonymus japonicus	Evergreen Burning Bush
	Myrtus communis	Myrtle
	Nerium oleander	Oleander
	Pittosporum tobira	Japanese Pittosporum
9	Ficus pumila	Creeping Fig
	Hibiscus rosa-sinensis	Chinese Hibiscus
	Lonicera hildebrandiana	Hildebrand Honeysuckle
	Sabal minor	Dwarf Palmetto
	Viburnum odoratissimum	Sweet Viburnum

Source: "Shrubs and Vines for American Gardens", by Donald Wyman. Published by Macmillan. Copyright © 1969 by Donald Wyman.

to be used as final cover should be tested in order to establish a basis for plant selection and to determine the additional needs of the soil to support plant growth.

It is important that care should be exercised in obtaining the soil sample to be tested since vegetation selection will be based upon results of a small statistical sample. Factors of importance in obtaining a soil sample are:

- Time
- Location
- Method
- Preparation

Time – Wherever possible, soil samples should be obtained in the early spring or the late fall either before or after the active growing period of any existing vegetation. Clearly, because existing vegetation obtains nutrients from the soil most actively during the spring and summer, any soil tests taken during this period would tend to indicate low nutrient levels. A fertilizer program based upon these supposed low levels would add excessive nutrients to the system. Previous heavy rainfall also impairs the value of soil sampling since nutrients tend to be leached from the soil during this period. Nutrients such as nitrates and ammonia are particularly vulnerable to loss from leaching. (An advantage of fall sampling is that fertilizer prices are discounted around this time of year.)

Location – If the area appears to be quite uniform, it may be appropriate to take samples from several locations and mix them together to obtain a composite sample. Professional assistance is commonly necessary to make this judgment, but in a preliminary way, it is possible to obtain some clue to variability by watching for variations in soil color, slope, texture and the degree to which the soil supports different existing vegetation. Variations in these factors indicate that separate samples should be taken and clearly marked to show where (and when) they were taken.

Avoid sampling in the following areas[1] as the results will give false indications of soil characteristics:

- Dead furrows or back furrows
- Lime or manure piles
- Animal droppings
- Fences or roads
- Rows where fertilizer has been banded
- Eroded knolls
- Low spots

Method – A number of references recommend the use of a special sampling probe or soil auger to obtain samples. Where these may be unavailable a trowel or spade may be used to make a V-shaped hole, 6 or 7 inches deep. The loose dirt

[1] From: "Sampling Soils for Testing", by Schulte, Walsh and Pionke, Circular 664 University Extension, University of Wisconsin.

should be removed from the hole and a thin uniform slice should be cut off the straight side of the hole from top to bottom. This insures a uniform sample from the surface down to roughly plow depth.

If the sample is to be composited, at least 5 samples should be taken and mixed together. One composite sample for every five acres is about the right level of sampling though indications of general uniformity in the existing vegetation may allow a lower level of sample coverage (say one composite every 20 acres).

Preparation — After sampling and subsequent compositing (where applicable) the soil should be carefully spread out to dry on a clean surface. Great care should be taken to avoid contamination of the sample from dust, fumes, or chemicals of any kind.

When thoroughly dry, the soil should be packed for shipment in a canvas soil sample bag or a plastic jar.

Analysis

The soil samples and a complete fact sheet showing location of samples and geographical details may be sent to the County Agricultural Extension Agent if there is one in the area. Frequently, he can make arrangements to have tests conducted by a competent laboratory and any lime and fertilizer recommendations can be returned within a few weeks.

Alternatively, it is advisable to obtain recommendations before shipping samples to a commercial soil testing laboratory.

The Results

The results of the soil analysis and their importance to soil preparation can hardly be better stated than by an extract from the Soil Test Report Sheet by the Cooperative Extension Program of the University of Wisconsin.

pH — In general, most herbaceous plants and the great majority of common trees and shrubs thrive best in soil which is very nearly neutral (pH 7.0). Some plants appear to be almost indifferent to pH and will thrive over a wide range (pH 4.0–10.0). It is best to maintain the cover soil as nearly neutral as possible since this provides the widest range of vegetation to choose from and the ideal conditions for growth.

Reducing Soil pH — In a highly alkaline soil (high pH) it is necessary to take steps to reduce the pH. It is comparatively simple and inexpensive to lower soil pH or at least bring it to a neutral level. It is commonly well worth doing early in the establishment of the final cover since additives can easily be disked into the cover material at this time.

The ideal method to reduce soil pH is to incorporate additional organic matter into the soil. The following material would be suitable and may even be found among the normal refuse load arriving at the site:

- Leaf mold
- Rotting bark and wood from old logs and stumps of hemlock or oak

EXPLANATION OF SOIL TEST REPORT

The results of the laboratory analyses are listed and interpreted on the top half of the report. Lime and fertilizer recommendations are given on the bottom half. Information received with the sample is printed on the bottom of the report. If more than one sample was submitted from the same area, the average results are used in the visual interpretation and in making lime and fertilizer recommendations.

pH

Soil pH is important because it affects (1) the availability of most of the essential plant nutrient elements, (2) the activity of soil microorganisms, and (3) the ability of the soil to hold nutrients. The optimum pH for most plants and soil microorganisms is between 6.0 and 7.0. However, some plants such as blueberries, cranberries, strawberries, and others, tolerate or prefer more acidic conditions.

ORGANIC MATTER

Organic matter serves many useful functions, the most notable of which are: (1) improves soil structure, and, thereby, water infiltration, drainage and soil aeration, (2) serves as a reservoir of available plant nutrients, and (3) increases water holding capacity of sandy soil.

The organic matter content of a soil depends mainly on the vegetation under which the soil developed, soil texture, drainage, extent of erosion and average temperature. It is a basic property of the soil and cannot be changed easily. When soil organic matter is low, large amounts of peat, compost, crop residues or other organic amendments are required to change the overall organic matter content of the soil.

SMP BUFFER

The lime requirement is calculated from soil pH, organic matter content and the SMP buffer test. The soil pH measures the "active" acidity and indicates whether or not lime is needed. If lime is needed, the organic matter and SMP buffer tests are used to measure the "reserve" or potential acidity and to calculate the exact amount of lime that is needed for the plants to be grown.

SOLUBLE SALTS

This test is used mainly to indicate whether too much fertilizer or other salts have been applied. Since dissolved salts are able to conduct an electrical current, soluble salts are easily determined by measuring the conductivity of the soil sample in water. The results are reported in units of electrical conductivity, mhos $\times 10^{-5}$/cm. If too much fertilizer has been applied, the excess must be leached away before plant growth will return to normal.

PHYSICAL ANALYSIS

A physical analysis of soil refers to a determination of the weight percentages of sand, silt and clay particles present in the soil. The physical analysis gives information concerning soil texture, and from this property one can approximate the water holding capacity of the soil and the likelihood of drought, drainage, compaction and erosion problems that may occur. It is a constant for a particular soil; hence, it is not necessary to have a physical analysis every time the soil is tested.

Source: Soil Test Report Sheet, Cooperative Extension Program, University of Wisconsin.

- Sawdust from hemlock or oak
- Acid muck from swamps and stream banks

It is a good idea to conduct additional pH tests after mixing this material with the cover soil in order to determine whether the pH has been brought to the desired level.

If the previously described organic material is not available it may be necessary to resort to a chemical additive though this will be a more expensive solution. Sulfur or aluminum sulfate are among the most satisfactory chemical additives which may be used to increase soil acidity. The rate of application is shown in Table 49.

TABLE 49

Table for Increasing Soil Acidity

Change from pH	Sulphur, lbs. per 100 sq ft	Aluminum Sulfate, lbs. per 100 sq ft
8.0–7.0	2.0	4.5
8.0–6.5	3.0	7.0
8.0–6.0	4.0	10.0
8.0–5.5	5.5	13.5
8.0–5.0	7.0	17.5
7.5–7.0	1.75	3.5
7.5–6.5	2.0	5.0
7.5–6.0	3.5	7.5
7.5–5.5	5.0	11.5
7.5–5.0	6.5	15.5
7.0–6.5	1.5	2.5
7.0–6.0	2.0	5.5
7.0–5.5	3.5	9.0
7.0–5.0	5.0	13.0
6.5–6.0	1.5	3.0
6.5–5.5	2.5	6.5
6.5–5.0	4.0	10.5
6.0–5.5	1.5	3.5
6.0–5.0	3.0	7.5
5.5–5.0	1.5	4.0

Source: "Floriculture", by Laurie and Reis, 1942.

Increasing Soil pH – In a highly acid soil (low pH) it may be necessary to take steps to *increase* the pH to obtain neutral or alkaline conditions.

Lime is of particular value in increasing soil pH since it also hastens bacteriological action in the soil thus releasing plant nutrients and making them more readily available to support plant growth.

TABLE 50

Rate of Application of Lime[1]

SOIL ACIDITY	LIGHT SANDY SOIL				MEDIUM SANDY SOIL			
	HYDRATED LIME		GROUND LIMESTONE		HYDRATED LIME		GROUND LIMESTONE	
	per 1,000 sq. ft.	*per acre*	*per 1,000 sq. ft.*	*per acre*	*per 1,000 sq. ft.*	*per acre*	*per 1,000 sq. ft.*	*per acre*
pH 4.0	60 lbs.	2610 lbs.	90 lbs.	3915 lbs.	80 lbs.	3480 lbs.	120 lbs.	5220 lbs.
pH 4.5	55 lbs.	2392 lbs.	82 lbs.	3567 lbs.	75 lbs.	3263 lbs.	112 lbs.	4872 lbs.
pH 5.0	45 lbs.	1957 lbs.	67 lbs.	2914 lbs.	60 lbs.	2610 lbs.	90 lbs.	3915 lbs.
pH 5.5	35 lbs.	1522 lbs.	52 lbs.	2262 lbs.	45 lbs.	1957 lbs.	67 lbs.	2914 lbs.
pH 6.0	None	None	None	None	None	None	None	None

SOIL ACIDITY	LOAM AND SILT LOAM				CLAY LOAM			
	HYDRATED LIME		GROUND LIMESTONE		HYDRATED LIME		GROUND LIMESTONE	
	per 1,000 sq. ft.	*per acre*	*per 1,000 sq. ft.*	*per acre*	*per 1,000 sq. ft.*	*per acre*	*per 1,000 sq. ft.*	*per acre*
pH 4.0	115 lbs.	5002 lbs.	172 lbs.	7482 lbs.	145 lbs.	6307 lbs.	217 lbs.	9439 lbs.
pH 4.5	105 lbs.	4567 lbs.	157 lbs.	6827 lbs.	135 lbs.	5872 lbs.	202 lbs.	8787 lbs.
pH 5.0	85 lbs.	3697 lbs.	127 lbs.	5524 lbs.	100 lbs.	4356 lbs.	150 lbs.	6525 lbs.
pH 5.5	65 lbs.	2827 lbs.	97 lbs.	4219 lbs.	80 lbs.	3480 lbs.	120 lbs.	5220 lbs.
pH 6.0	None	None	None	None	None	None	None	None

Note: A light application of lime at the rate of 25 lbs. per 1,000 sq ft has proved beneficial on certain soils, even though the soil has a pH value of 6.0.

[1] As it is not advisable to apply more than 50 lbs. of hydrated lime or 75 lbs. of ground limestone per 1,000 sq ft at any one time, it will be necessary to make several successive applications on strongly acid soils and raise the pH gradually. It is often detrimental to plant growth to make an excessively heavy application of lime at one time.

Source: "America's Garden Book", by James and Louise Bush-Brown. Published by Charles Scribner's Sons.

Lime is generally applied as ground limestone or hydrated lime. In general, hydrated lime acts much more quickly to increase pH, but it has the disadvantage that it is not as lasting in its effect.

Application should be by surface spreading and disking into the upper few inches of soil. Rates of application are dependent upon the form of lime used and the soil texture. See Table 50 for general guidance.

Soil Conditioners

The best soils for use as sanitary landfill cover are the clay loams. However, the high clay content so necessary to reduce infiltration presents a problem for plant growth. Heavy clay soils have limited pore space between the soil particles and low air capacity. Clay soils also lack the ability to absorb moisture readily thus making the application of soluble fertilizers very difficult and wasteful.

It is necessary to strike a balance between the vital need for impermeability and the equally necessary requirement for vegetation in order to reduce erosion.

Soil conditioners have been developed for the purpose of rendering clay soils more porous and loose in order to aid plant growth.

The best and cheapest conditioner is compost or decaying organic matter. The large quantities of leaves and grass clippings which find their way to the landfill provide a cheap and ample supply. They should be stock-piled in a separate area of the landfill and lightly disked into the cover material after it is placed. The chief advantage which synthetic conditioners have over organic varieties is that they are quicker acting. However, they tend to be costly and given the large quantities of suitable organic material at hand on the landfill, it appears unnecessary to look elsewhere for soil conditioners. Later when we consider soil erosion needs, it will become apparent that there may be some advantage in *growing* the soil conditioner first and then disking it into the cover material.

As a simple means of determining soil need for conditioner application, a simple test is recommended in Table 51.

Cover Crops

It is common agricultural practice to grow certain crops for the sole purpose of plowing under to improve soil texture, organic content and fertility. These crops are termed "cover crops" and the procedure represents one of the least expensive methods of improving poor soils.

From the standpoint of improving fertility, the best cover crops are the legumes such as soybeans, alfalfa, clover and the vetches, since they all possess the ability to make nitrogen available for plant growth.

> "A well-grown leguminous crop which is plowed under at the proper stage of growth will frequently add from 100 to 150 pounds of actual nitrogen to the soil per acre. This would practically equal 10 to 15 tons of good animal manure."

TABLE 51

Field Test to Determine Soil Need for Conditioner

TEST

1. Take a *small* handfull of soil.
2. Add four or five drops of water using a medicine dropper.
3. Mix the soil and water together thoroughly until every granule of soil crumbles.
4. Roll the soil between the palms of your hands into the form of a cigarette.

	RESULT	INTERPRETATION
A	Cigarette holds its shape.	The soil will BENEFIT from application of conditioner.
or B	Soil remains granular and cannot be formed into a cigarette.	Conditioner NOT NEEDED. The soil will derive no benefit from application.
or C	A "weak" cigarette is formed which flops over at the end when lifted.	Application of conditioner will be of enough benefit to warrant use.

Source: "America's Garden Book", by James and Louise Bush-Brown. Published by Charles Scribner's Sons.

Where cover crops are planned for soils of very low fertility, it is frequently necessary to add commercial fertilizer to aid the first growth.

Many of the cover crops have the advantage of rapid growth and on steep banks where erosion may be expected to be a problem without early vegetative support a quick growing cover crop may provide the answer. In real problem areas, the crop may be applied in a carefully designed mulch to hold the seed until germination. When the mulch begins to disintegrate, the root system takes over and provides early stability. Later, the cover crop may be plowed under to improve soil quality and a more appropriate vegetation planted consistent with whatever the final use may be. Table 52 presents desirable cover crop requirements.

Erosion Control

Regardless of the ultimate use intended for the landfill, the main aim of a vegetation plan should be to control erosion on sloping areas. This will prevent erosive forces from carrying away soil particles and carving gullies through final cover into the refuse cells.

TABLE 52

Desirable Cover Crops

Species	Time and Rate of Seeding	Soil Requirements	Time of Plowing Under*
LEGUMINOUS CROPS			
Alfalfa – A perennial crop and consequently requires a longer time to become established than the annual and biennial crops, but there are instances when its use as a cover crop is justified.	The seed may be sown in early spring or late summer at the rate of 10 to 12 pounds per acre and it should be inoculated. If the field is weedy late summer sowing is preferable.	A well-drained soil of reasonably good fertility with a pH of 6-5 or above.	Hay may be harvested for a number of years and the crop may be plowed under in the fall or in the spring when the ground is desired for some other purpose, such as a lawn area, garden or orchard. When plowed under, the tops and roots contain approximately 2.65 percent nitrogen and as much as 150 pounds per acre of actual nitrogen may be added to the soil.
Hairy Vetch – A winter annual and is one of the most valuable of all soil builders.	May be sown early in September at the rate of 3 to 4 pecks per acre. The seed should be inoculated.	It thrives best on a sandy, well-drained soil with a pH of 6.5 or slightly above.	Should be plowed under about the middle of May. If turned under at this season the top growth and roots will contain approximately 3.75 percent nitrogen.
Mammoth Red Clover – Being a biennial, it is necessary for red clover to occupy the ground for a considerable length of time and it is, therefore, not as well suited for a cover crop as are some of the other legumes, although it is rated as a splendid soil builder.	It should be sown very early in the spring, preferably with a small grain crop such as oats or on a wheat crop sown the previous fall. Rate of seeding, 8-12 pounds per acre. Innoculate seed.	A soil of at least moderate fertility with a pH above 5.5.	Should be plowed under just before full bloom the year after sowing.
Soybeans – Are one of the most valuable of all cover crops as they are a summer annual and make rapid growth.	Soybeans should be sown in May or early June at the rate of 2 bushels per acre.	If grown on very poor soil a complete fertilizer should be added. Soil should have a pH above 5.	It is very important that soybeans be plowed under before the vines have become too mature. The most generally accepted practice is to turn them under when the beans in the pods are about half grown. At this stage, the tops and roots will contain approximately 2.65 percent nitrogen.
Sweet Clover – Is a biennial which possesses tremendous soil building potentialities.	Sweet clover should be sown early in the spring either on, or with, a small grain crop. The usual rate of seeding is 10 to 12 pounds per acre.	Sweet clover is one of the few legumes which will grow well on very poor soil. It thrives reasonably well on washed clay land. Its one definite requirement is a soil high in lime content with a pH of 6.5 or above.	The crop may be plowed under in the spring following seeding. In order to secure good results, it is essential that it be turned under before it reaches maturity. If plowed under between April 15 and May 1 the maximum percentage of nitrogen will be secured, often running as high as 3.75 percent.
NON-LEGUMINOUS CROPS			
Rye – Is considered the most valuable of all the non-legumes as a cover crop. Although it does not increase the actual nitrogen content of the soil appreciably, it adds a considerable amount of organic matter.	Rye should be sown during early September, being seeded at the rate of 1½ to 2 bushels per acre.	Rye succeeds reasonably well on a soil of average fertility and is almost entirely indifferent to the matter of soil acidity, thriving well on decidedly acid soil.	Rye should be plowed under in the spring when it has reached a height of approximately 10 inches. At this time, the plant contains a greater percentage of nitrogen than at any other stage of its development, sometimes running as high as 1.75 percent.

*Dates recommended for plowing under are applicable to the latitude of Pennsylvania and Ohio. In any locality, cover crops should be turned under when slightly immature.

Source: "America's Garden Book", by James and Louise Bush-Brown. Published by Charles Scribner's Sons.

Where the slope length exceeds 200 feet, erosion should be controlled by the construction of diversion channels across the direction of slope. Diversion channels help to reduce the momentum of surface flow and direct runoff to a safe outlet.

As central channels for runoff, diversion channels are themselves vulnerable to erosion and need to be carefully planted to achieve maximum protection.

In some areas where the risk of severe erosion can be identified, a preliminary seeding is recommended where fast protection is needed. More permanent types of cover would then be seeded on those areas where additional protection is required.

Wind Erosion – The climatic factors which influence the susceptability of soils to wind erosion include:

- Wind direction
- Intensity
- Duration

It is commonly more difficult to establish vegetation on slopes facing the prevailing wind. Existing tree screening can reduce this problem to a limited extent.

Where no existing trees can provide protection, windbreaks may be needed to prevent severe erosion in isolated areas.

> "It has been found that trees moderate the effect of the wind on their leeward side for a distance on the ground equal to approximately twenty times their height."[2]

As mentioned previously windbreaks also have the advantage of reducing dust and litter problems.

A windbreak may be a single row of deciduous or evergreen trees. The "break" is more effective the greater the number of rows of planting.

Water Erosion – As previously discussed erosion due to rainfall action is chiefly a function of the velocity with which the raindrop hits the soil. The runoff which follows the accumulation of rainfall on the surface of the landfill carries away the eroded material. In the process of surface flow, more soil particles are carried forward and depending upon whether flow takes place over the surface of the land or in specific channels, the resultant erosion is termed:

- Sheet erosion

or

- Gully erosion

Plants can be used to control erosion from water in at least 3 ways:

- Leaves and branches reduce the velocity of falling rain drops.
- Fibrous roots hold soil particles
- Fallen leaves and the remains of dead plants on the soil help to increase organic content and reduce erodibility.

---

[2] From "Trees for American Gardens", by Donald Wyman, Published by the Macmillan Company, New York. Copyright © 1951, 1965 by Donald Wyman.

TABLE 53

Trees for Shelterbelts in the Great Plains

"These are recommended for the Great Plains area of the Midwest since they grow and have very tenacious roots even with comparatively little rainfall. Trees like the Hackberry, Honeylocust and Pinus ponderosa have roots known to go 10 to 20 feet deep in the soil."

Botanical Name	Common Name
Acer negundo	Box-Elder
A. saccharinum	Silver Maple
Caragana aborescens	Siberian Pea Tree
Celtis occidentalis	Hackberry
Elaegnus angustifolia	Russian Olive
Franxinus pennsylvanica lanceolata	Green Ash
Gleditsia triacanthos	Honey Locust
Juglans nigra	Black Walnut
Juniperus scopulorum	Western Red Cedar
J. virginiana	Eastern Red Cedar
Maclura pomifera	Osage-orange
Picea glauca	White Spruce
P. pungens	Colorado Spruce
Pinus ponderosa	Ponderosa Pine
Populus species	Poplar species
Prunus americana	American Wild Plum
P. virginiana	Choke Cherry
Pseudotsuga menziesii	Douglas Fir
Quercus macrocarpa	Bur Oak
Salix alba	White Willow
S. pentandra	Laurel Willow

Source: "Trees for American Gardens", by Donald Wyman. Published by the Macmillan Company, New York. Copyright © 1951, 1965 by Donald Wyman.

TABLE 54

Trees for Windbreaks

"These trees make good windbreaks in areas other than the Great Plains. The best windbreaks are of course the evergreens, but since they grow slowly and it takes some time for them to be effective, fast-growing deciduous windbreaks are frequently planted in front of the evergreen and then eventually cut out when the evergreens become large enough to prove effective."

Botanical Name	Common Name
Acer ginnala	Amur Maple
A. platanoides	Norway Maple
A. pseudoplatanus	Sycamore Maple
A. rubrum	Red Maple
A. saccharum	Sugar Maple
Carpinus betulus	European Hornbeam
Cornus mas	Cornelian Cherry
Crataegus phaenopyrum	Washington Hawthorn
Eucalyptus species	Gum
Fagus species	Beech
Franzinus americana	White Ash
Franxinus pennsylvanica lanceolata	Green Ash
Juniperus species	Juniper
Ligustrum lucidum	Glossy Privet
Maclura pomifera	Osage-orange
Malus baccata	Siberian Crab Apple
Picea abies	Norway Spruce
P. glauca	White Spruce
P. omorika	Serbian Spruce
Pinus nigra	Austrian Pine
P. resinosa	Red Pine
P. strobus	White Pine
Populus alba	White Poplar
P. berolinensis	Berlin Poplar
P. nigra italica	Lombardy Poplar
P. simonii	Simon Poplar
P. tremuloides	Quaking Aspen
Pseudotsuga menziesii	Douglas Fir
Quercus (many species)	Oak
Q. imbrica (one of best)	Shingle Oak
Q. phellos (one of best)	Willow Oak
Rhamnus davurica	Dahurian Buckthorn
Syringa amurensis japonica	Japanese Tree Lilac
Thuja species	Arborvitae
Tilia species	Linden
Tsuga caroliniana	Carolina Hemlock
Viburnum prunifolium	Blackhaw Viburnum

Source: "Trees for American Gardens", by Donald Wyman. Published by the Macmillan Company, New York. Copyright © 1951, 1965 by Donald Wyman.

Sheet and gully erosion are best controlled by vegetation which has shallow fibrous roots which spread through the soil in all directions to bind the soil particles against the flowing runoff. The vetches are particularly good in this capacity since they also grow rapidly and reduce the risk of erosive forces carrying away seedling growth.

Table 55 provides numerous examples of plant material suitable for bank protection.

TABLE 55

Shrubs and Ground Cover Plants for Bank Stabilization

Botanical Name	Common Name
Ajuga reptans	Carpet Bugle
Akebia quinata	—
Arabis alpina	Alpine Rockcress
Arctostaphylos uva-ursi	Bearberry
Arenaria montana	Mountain Sandwort
Berberis thunbergii	Japanese Barberry
Bignonia capreolata	Crossvine
Calluna vulgaris	Heather
Ceanothus americanus	New Jersey — Tea
Celastrus species	Bittersweet
Cerastium tomentosum	Snow-in-Summer
Clematis paniculata	Sweet Autumn Clematis
C. vitalba	Travelers Joy
C. Viticella	—
Clethra alnifolia	Sweet Pepperbush
Comptonia asplenifolia	Sweetfern
Comptonia perigrina	Sweetfern
Convallaria majalis	Lily-of-the-Valley
Cronus alba	Tatarian Dogwood
C. Amomum	Silky Dogwood
C. Stolonifera	Red Osier Dogwood
Cotoneaster Horizontalis	Rock Cotoneaster
Cytisus scoparius	Scotch Broom
Diervilla sessilifolia	Southern Bush-honeysuckle
Euomymus fortunei "Vegetus"	—
Eunymus radicans	Wintercreeper
Euonymus radicans coloratus	—
Euonymus radicans "Minimus"	—
Forsythia "Arnold Dwarf"	"Arnold Dwarf" Forsythia
F. suspensa sieboldii	Siebold forsythia
Hendera helix	English Ivy
Hemerocallis fulva	Tawny Daylily
Hypericum buckleyi	Blue Ridge St. Johnswort
H. Calycinum	Aaronsbeard St. Johnswort

(continued)

TABLE 55 *(continued)*

Botanical Name	Common Name
Indigofera incarnata alba	White Chinese Indigo
I. kirilowii	Kirilow Indigo
Jasminum nudiflorum	Winter Jasmine
Jiniperus chinesis sargentii	Sargent Juniper
Juniperus communis depressa	Spreading Juniper
Juniperus hosizontalis	Creeping Juniper
Juniperus Sabina	Savin Juniper
Leucothoe Fontanesiana	Drooping Leucothoe
Lonicera japonica Halliana	Hall's Honeysuckle
Lonicera henryi	Henry Honeysuckle
Lycium chinese	Chinese Matrimony Vine
Lysimachia nummularia	Moneywort
Lycium halimifolium	Matrimony Vine
Menispermum canadense	Common moonseed
Myrica species	Bayberry
Nepeta Missini	Nepeta
Pachysandra terminalis	Japanese Spurge
Parthenocissus henryana	Silver Vein Creeper
P. quinquefolia	Virginia Creeper
Phlox subulata	Moss Phlox
Physocarpus opulifoliua	Common Vinebark
Plumbago Larpentae	Leadwort
Pueraria lobata	Kudzu Vine
Rhus aromatica	Fragrant Sumac
Rhus canadensis	Sumac
R. copallina	Dwarf Sumac
Robinia hispida	Bristly locust
R. kelseyi	Keksey Locust
Rosa humilis	Pasture Rose
Rosa lucida	Virginia Rose
Rosa "Max Graf"	—
R. nitida	Bristly Rose
R. rugosa "repens"	—
R. Setigera	Prairie Rose
R. Virginiana	Virginia Rose
Rose Wichuraiana	Wichurian Rose
Salix tristis	Swarf Gray Willow
Spiraea billiardii	Billiard Spirea
S. salicifolia	—
Sedum acre	Stone Crop
Spiraea tomentosa	Hardhack
Stephanandra incisa "Crispa"	Cutleaf Stephanandra
Symphoricarpos species	Coralberry
Thymus Serpyllum	Thyme
Vaccinium angustifolium	Lowbush "Blueberry"
V. pallidum	Dryland Blueberry
Viburnum dentatum	Arrowwood Viburnum
V. lentago	Nannyberry Viburnum
V. trilobum	American Cranberry bush Viburnum
Vinca minor	Periwinkle
Vitis species	Grape
Xanthorhiza simplicissima	Yellowroot

Source: "Shrubs and Vines for American Gardens", by Donald Wyman. Published by the Macmillan Company. Copyright © 1949, 1969 by Donald Wyman.

Rapid growth is also important for erosion control. Table 56 provides an excellent summary of appropriate tree species which will grow rapidly to provide erosion protection.

TABLE 56

Trees of Rapid Growth

Deciduous	
Acer dasycarpum	Silver Maple
Negundo	Box Elder
platanoides	Norway Maple
rubrum	Red Maple
Ailanthus glandulosa	Tree of Heaven
Betula Maximowicziana	Monarch Birch
populifolia	Gray Birch
Catalpa speciosa	Western Catalpa
Fraxinus americana	White Ash
Ginkgo biloba	Ginkgo
Gleditsia triacanthos	Honey Locust
Gymnocladus dioica	Kentucky Coffeetree
Larix decidua	European Larch
Liriodendron tulipifera	Tulip-tree
Magnolia acuminata	Cucumber Tree
tripetala	Umbrella Magnolia
Paulownia tomentosa	Empress Tree
Platanus acerifolia	London Plane Tree
Populus alba	White Poplar
nigra italica	Lombardy Poplar
Prunus serotina	Black Cherry
Quercus palustris	Pin Oak
Robinia Pseudoacacia	Black Locust
Salix alba	White Willow
vitellina	Golden Willow
Sorbus aucuparia	European Mountain Ash
Syringa japonica	Japanese Tree Lilac
Tilia americana (T. glabra)	American Linden
Ulmus americana	American Elm
glabra	Scots Elm
Evergreen	
Picea excelsa	Norway Spruce
Pinus resinosa	Red Pine
rigida	Pitch Pine
strobus	White Pine
sylvestris	Scots Pine

Source: "America's Garden Book", by James and Louise Bush-Brown. Published by Charles Scribner's Sons. New York and London.

Final Use — The proposed use of the site also plays an important part in the final selection of plant material.

If the site area is being returned to natural vegetation, then the main concern is erosion control. Plants appropriate for bank planting would be used and wherever possible they would be native species known to be hardy in the particular climatic zone.

On the other hand, if the site is planned for recreation then a number of additional requirements may be identified:

- Groundcover able to withstand intensive foot traffic
- Good shade trees
- Thornless trees and shrubs to prevent injury
- Thorny trees and shrubs in certain areas to control movement
- Trees and shrubs for screening

Foot Traffic — Ability to withstand foot traffic is very important. The greater the anticipated use, the more important the selection of plant material which will be able to withstand intense foot traffic.

Table 57 indicates the order of plant material which can withstand the most intensive traffic.

Kentucky Bluegrass will withstand the most — Memorial Rose, the least.

TABLE 57

Plants Able to Withstand Foot Traffic

FLOORS of exterior spaces created by plant materials rated by their ability to withstand foot traffic and the ease with which they are able to be walked upon.

1. POA PRATENSIS KENTUCKY BLUEGRASS	6. CERASTIUM TOMENTOSUM SNOW-IN-SUMMER
2. THYMUS SERPYLLUM THYME	7. HEDERA HELIX ENGLISH IVY
3. ANENARIA VERNA CAESPITOSA MOSS SANDWORT	8. AJUGA REPTANS BUNGLEWEED
4. ANTHEMIS NOBILIS CAMOMILE	9. COTONEASTER ADPRESSA PRAECOX CREEPING COTONEASTER
5. MATRICARIA TCHIHATCHEWII TURFING DAISY	10. ROSA WICHURAIANA MEMORIAL ROSE

Source: "Plants/People/and Environmental Quality". A Study of Plants and Their Environmental Functions, by U.S. Department of Interior, Park Service.

Shade Trees — Canopy density can be important in determining the effective shade producing qualities of selected tree species. Table 58 goes from most dense to least dense.

Barrier Plants — In some circumstances, thorny plants are ideal for controlling circulation while in other areas such as play areas they should be scrupulously avoided.

TABLE 58

Shade Trees

CEILINGS representative plants indicating the relative degrees of visual density (spaced to give uniform ceiling density).

1. THUJA OCCIDENTALIS
AMERICAN ARBOR-VITAE
2. JUNIPERUS VIRGINIANA
EASTERN RED-CEDAR
3. PSEUDOTSUGA TAXIFOLIA
DOUGLAS-FIR
4. ACER PLATANOIDES
NORWAY MAPLE, SUMMER
5. LARIX DECIDUA
EUROPEAN LARCH, SUMMER
6. PICEA PUNGENS
COLORADO BLUE SPRUCE
7. QUERCUS ALBA
WHITE OAK, SUMMER
8. CELTIS OCCIDENTALIS
AMERICAN HACKBERRY, SUMMER
9. ULMUS AMERICANA
AMERICAN ELM, SUMMER
10. FRAXINUS PENNSYLVANICA LANCEOLATA
GREEN ASH, SUMMER
11. QUERCUS ALBA
WHITE OAK, WINTER
12. BETULA POPULIFOLIA
GREY BIRCH, WINTER
13. MACLURA POMIFERA
OSAGE-ORANGE, WINTER
14. QUERCUS PALUSTRIS
PIN OAK, WINTER
15. SORBUS AUCUPARIA
EUROPEAN MOUNTAIN ASH, WINTER
16. CLADRASTIS LUTEA
AMERICAN YELLOW WOOD, WINTER
17. GINKGO BILOBA
GINKGO, WINTER
18. GLEDITSIA TRIACANTHOS
COMMON HONEY-LOCUST, WINTER
19. RHUS TYPHINA
STAGHORN SUMAC, WINTER
20. ROBINIA PSEUDOACACIA
BLACK LOCUST, WINTER

Source: "Plants/People/and Environmental Quality".

Screening — Development of a vegetative screen could serve as a visual barrier to block undesirable views. As previously mentioned, screening can also be used for control of erosion due to wind. Thus, trees and shrubs can be selected which will serve both purposes. The degree of control can also be managed by the careful selection of trees and shrubs.

For effective visual screening, knowledge of the density of the plant material is important. Table 61 indicates screening from the densest screen — Thuja occidentalis (Eastern Arborvitae) to the least dense — Ulmus Americana (American Elm). Spacing is indicated along with special shearing and time of year during which the screen is most effective.

TABLE 59

Trees with Thorny Twigs (Barrier Plants When Restrained to Many Stems)

Botanical Name	Common Name
Aralia elata	Japanese Angelica Tree
Crataegus species	Hawthorn
Gleditsia triacanthos	Honey-locust
Hemiptelea davidii	David Hemiptelea
Hippophae rhamnoides	Common Sea-Buckthorn
Kalapanax pictus	Castor-aralia
Maclura pomifera	Osage-orange
Pyrus species	Pear
Parkinsonia aculeata	Jerusalem Thorn
Poncirus trifoliata	Hardy Orange
Prunus cerasifera	Sour Cherry
Rhamnus davurica	Dahurian Buckthorn
Robinia species	Locust
Xanthoceras species	Yellowhorn
Ziziphus species	Common Jujube

Source: "Trees for American Gardens", by Donald Wyman. Published by Macmillan. Copyright © 1951, 1965 by Donald Wyman.

TABLE 60

Shrubs and Vines with Thorny Stems (Chiefly of Value as Barrier Plants)

Botanical Name	Common Name
Acanthopanax species	–
Aralia species	Aralia
Berberis species	Barberry
Carissa grandiflora	Natal-plum
Celastrus flagellaris	Bittersweet
Chaenomeles species	Quince
Elaeagnus species	Eleagnus
Halimodendron halodendron	Salt Tree
Hippophae rhamnoides	Sea-Buckthorn
Lycium species	Matrimony-Vine
Paliurus spina-christi	Christ Thorn
Poncirus trifoliata	Hardy Orange
Prinsepia sinensis	Cherry Prinsepia
Pyracantha species	Firethorn
Rhamnus catherartica	Buckthorn
Robinia species	Locust
Rosa species	Rose
Severinia buxifolia	–
Smilax rotundifolia	Common Greenbrier
Sophora davidii	–

Source: "Shrubs and Vines for American Gardens", by Donald Wyman. Published by Macmillan. Copyright © 1949, 1969 by Donald Wyman.

TABLE 61

Representative Plants Indicating the Relative Degrees of Visual Density (at least 6′ high)

1.	THUJA OCCIDENTALIS	SHEARED 18″ O.C.
2.	TSUGA CANADENSIS	SHEARED 2′ O.C.
3.	PSEUDOTSUGA TAXIFOLIA	SHEARED 2′ O.C.
4.	TAXUS MEDIA HICKSI	SHEARED 18″ O.C.
5.	JUNIPERUS CHINENSIS HETZI	SHEARED 2′ O.C.
6.	BUXUS SEMPERVIRENS	SHEARED 9″ O.C.
7.	PICEA GLAUCA	SHEARED 2′ O.C.
8.	PINUS STROBUS	SHEARED 2′ O.C.
9.	JUNIPERUS VIRGINIANA	SHEARED 2′ O.C.
10.	RHAMNUS CATHARTICA, 'TALLHEDGE'	SHEARED 1′ O.C., SUMMER
11.	RIBES ALPINUN	SHEARED 9″ O.C., SUMMER
12.	BERBERIS MENTORENSIS	SHEARED 6″ O.C. DOUBLE ROW, SUMMER
13.	EUONYMUS ALATUS	SHEARED 18″ O.C., SUMMER
14.	LONICERA TATARICA	SHEARED 2′ O.C., SUMMER
15.	LIGUSTRUM AMURENSE	SHEARED 9″ O.C., SUMMER
16.	ROSA MULTIFLORA	SHEARED 6″ O.C., SUMMER
17.	ULMUS PUMILA	SHEARED 2′ O.C., SUMMER
18.	CRATAEGUS CRUSGALLI	SHEARED 2′ O.C., SUMMER
19.	PHYSOCARPUS APULIFOLIUS	SHEARED 9″ O.C., SUMMER
20.	SYRINGA VULGARIS	SHEARED 9″ O.C., SUMMER
21.	VIBURNUM LANTANA	SHEARED 18″ O.C., SUMMER
22.	CORNUS MAS	SHEARED 18″ O.C. WINTER
23.	BERBERIS KOREANA	SHEARED 9″ O.C., WINTER
24.	RHODOTYPOS SCANDENS	SHEARED 1′ O.C., WINTER
25.	MALUS SARGENTI	SHEARED 2′ O.C., WINTER
26.	COTONEASTER NITIUS	SHEARED 18″ O.C., WINTER
27.	QUERCUS IMBRICARIA	SHEARED 2′ O.C., WINTER
28.	POPULUS NIGRA ITALICA	8′ O.C. WINTER
29.	ELAEAGNUS ANGUSTIFOLIA	10′ O.C., WINTER
30.	PHILADELPHUS VIRGINALIS	5′ O.C., WINTER
31.	HIBISCUS SYRIACUS	6′ O.C., WINTER
32.	SALIX MATSUDANA TORTUOSA	8′ O.C., WINTER
33.	AMELANCHIER CANADENSIS	10′ O.C., WINTER
34.	CARPINUS BETULUS	10′ O.C., WINTER
35.	BETULA POPULIFOLIA	15′ O.C., WINTER
36.	POPULUS TREMULOIDES	20′ O.C., WINTER
37.	GLEDITSIA TRIACANTHOS	25′ O.C., WINTER
38.	TILIA CORDATA	30′ O.C., WINTER
39.	ACER SACCHARUM	35′ O.C., WINTER
40.	ULMUS AMERICANA	40′ O.C., WINTER

Source: "Plants/People/and Environmental Quality".

Woodland – It should be emphasized that recreation related final uses are not by any means the only feasible uses for a finished landfill. Recreation uses may be found most expedient from the point of view of public acceptance, but where this is not a problem, there may be commercial value in developing the land as woodland.

The experience of reclamation of woodland on strip mine spoil has been very successful. There seems little reason why the same success might not be achieved on a completed sanitary landfill.

To quote:

"Approximately 80 percent of all areas that have been strip-mined since 1945 have been planted to trees. There are several reasons for this. Many of the areas that have been stripped were originally covered by forests and much of the remaining coal lies under woodland. Trees can be planted on most stripped areas regardless of the type of quality of the backfilling job. The area does not have to be fitted for planting. Trees will survive and grow on a wider range of spoil conditions than other plants. The use of lime and fertilizer is usually not necessary. Thus, tree planting has proved to be an economical method for successfully revegetating strip-mined areas."

"Trees may be planted for a variety of uses. They may produce wood products such as pulpwood, poles, and sawtimber. Growing Christmas trees on spoil has become popular in recent years. Trees may be planted for their food and cover value in wildlife plantings, to protect the site where no other vegetation will grow, and for watershed protection and erosion control on steep slopes. Evergreen trees are especially suitable for roadside plantings to hide unsightly spoils along major highways."

The literature indicates that many species can survive in the difficult conditions typical of spoil recovery areas. In many respects, landfill cover conditions should be less difficult.

Most species of conifer have matured well to produce commercially valuable timber stands. Hardwood species have not matured so well in Pennsylvania with the exception of hybrid poplars, European alder and black locust.

In New Jersey, remarkable success has been achieved by Mr. Miles W. Fry by the use of the Hybrid Poplar planted for timber on the Fort Dix sanitary landfill. After only one growing season, the growth was about 3 feet with a survival rate of 90 percent. In four years, the trees stand about 10 feet in height. Thinning for pulp can be carried out in about 10 years with a full harvest of large timber of 20 years.

Figure 71. Hybrid Poplars for Reclaiming Sanitary Landfills.

Pictured above is the hard waste at Fort Dix, New Jersey. Tin cans, metal cartridge casings, metal, wood, trash and garbage of all sorts were pushed into a 10 to 15 foot trench 300 yards long. The waste was packed down with a crawler and covered with 3 to 4 feet of New Jersey sandy soil.

In April 1970, 50,000 Hybrid Poplar rooted cuttings were planted on 70 of these desolate acres of sanitary fill at Fort Dix. The above picture was taken in Sept. 1970 after one growing season. Survival rate was 90%. Growth, about 3 feet.

Pictured above is the same area in September 1971.

Pictured above is the same area in September 1972.

Pictured above is the same area in September 1973.

With grateful acknowledgments to Miles W. Fry & Sons, Frysville, Ephrata, Pennsylvania.

Chapter 11

LEGAL ASPECTS

Since the practice of dumping solid waste has been confused with sanitary landfilling for many years, it is not surprising that many communities have sought to limit the development of sites by all means at their disposal. The most obvious and powerful method has been through the zoning laws.

Commonly, there is no mention of suitable areas for sanitary landfill in zoning ordinances and usually a specific mandate against open dumps. Even in areas where sanitary landfills are badly needed they are frequently accepted only as a variance of an industrial zoning classification. This latter arrangement often defeats the values inherent in a final site utilization plan since recreation areas or similar final uses are rarely the most sought after commodity in industrial zones. Zoning restriction of this type is evidence of a general fear on the part of the public and elected officials that all attempts at refuse burial turn into open dumps. It is often an uphill struggle to turn this attitude around. It is clearly ostrich-like on the part of local elected officials since the problem does not "go away" by merely legislating against it. Sanitary landfill remains a viable means of ultimate disposal (perhaps the only one with a history of practical and efficient operation) and it behooves local authorities to consider writing zoning ordinances to permit this use.

Before we consider suitable criteria for an ordinance, it may be instructive to review some of the relevant court cases. There appear to be no real landmark cases, but a number of cases depict some general trends rather clearly.[1]

In the past, municipal governments (those who have recognized the need for landfills) have sought to avoid their own zoning restrictions by seeking immunity from them. Typically, municipal authorities have claimed that the operation of a sanitary landfill is a *governmental* function and not a *proprietary* function. If it could be clearly proven that the former is the case (i.e. municipal operation is a government function) then the zoning laws would not apply.

In the State of New York, there have been a number of cases which address this question. In Bressett v. Ogdensburg 180 NYS 2nd 908 (1958), it was the City of

[1] Planning for Solid Waste Management", by Mary McLean, published by ASPO identifies the important cases very well.

Agdensburg which sought to establish the proprietary right to run a landfill. The question was presented as to the right of the City to commit an act that would be denied any private individual. The neighborhood where the landfill was proposed had been zoned as residential. The City proposed to operate a sanitary landfill which in specific terms was excluded from the area by the ordinance.

It was held that the disposal of garbage and refuse constitutes a proprietary function and it is subject to the same standards and restraints which are applicable to a private individual or corporation in the conduct of the same or similar business. See:

- Morton v Mayor, etc. of New York
  140 N.Y. 207, 35 N.E. 490, 22 L.R.A. 241
- O'Brien v Town of Greenburgh
  239 App. Div. 555, 268 N.Y.S. 173

In the latter case, the court stated in 239 App. Div. at Page 559, 268, N.Y.S. at Page 177:

> "The general authority conferred on the defendants to establish disposal plants did not involve an express or implied authority to establish them, or any of them, in violation of zoning ordinances adopted pursuant to the same source of power — the Legislature, and under which the rights of individuals have become fixed."

In Sinn v Board of Selectmen of Acton 259 N.E., 2nd 557 (1970) the town of Acton possessed a zoning by-law provision that specifically *exempted* all municipal uses from use restrictions. This provision was upheld in this case.

Conversely, the City of Philadelphia has a zoning ordinance which specifically binds the City to *observe* its own zoning ordinance. Kelly v Philadelphia 115 A. 2nd 238 (1955) upheld this explicit provision.

Where public disposal sites are planned by county authorities, however, the courts have generally held that the county authority is not bound by the zoning ordinance of municipalities within the county.

In Appelbaum v St. Louis County 451 S.W. 2nd 107 (1970), it was held that the county could construct an incinerator and landfill in an area taken by the county under eminent domain without regard to the original zoning ordinance of the municipality in which the property lay.

The court quoted State ex rel. Heisel v Board of County Commissioners, Ohio 79 N.E. 2nd 698 704, which held that zoning ordinances of two municipalities could not prevent a county from acquiring and using property within such areas for a public airport: "To suppose that zoning ordinances may limit or prevent the public use for which land is taken is to invest municipalities with power to restrict the exercise of the power of eminent domain."

In Society of Divine Word v County of Cook 247 N.E. 2nd 21 (1969), the court held that the statute which prohibited *dumping* within one mile of the corporate limits of a municipality did not apply to *sanitary landfills*. It was observed that the nuisance statute was adopted prior to the "alleged development of the sanitary landfill method."

Where private disposal operations are concerned, there is no question that this is a proprietary function and therefore, subject to all the limitations and restrictions of the zoning ordinance. Even in those instances where municipalities may have built-in exemptions to their own ordinances for special purposes, it is clear that these exemptions do not apply to private individuals. In Kersheske v Township of Thomas 138 N.W. 2nd 509, it was claimed that the collection and disposal of refuse is an essential service and thereby excepted from the provisions of the ordinance. The relevant section of the ordinance provides that:

> "Essential services shall be permitted as authorized and regulated by law and other ordinances of the Township, it being the intention thereof to except essential services from the provisions of this Ordinance."

However, the Court of Appeals held that provision of zoning ordinance that "Essential services shall be permitted as authorized and regulated by law and other ordinances of the Township," was intended to exempt township from municipal services which it might operate and did not exempt collection and disposal of refuse as "essential service".

In a curious case, Dougherty v Messner et al. Supervisors of the Township of Union 172 A. 2nd 288 (1961), the Supreme Court of Pennsylvania even upheld the right of municipalities to impose zoning restrictions on growth of existing operations while also prohibiting development of new operations.

A man named Arnoni had dumped garbage in the township for nineteen years until he was brought up in court and as a result was dumping, in accordance with the latest approved methods, garbage brought from various neighboring townships. In January, 1960, defendants adopted the ordinance in question, prohibiting the expansion of Arnoni's dump and allowing him to continue, but also prohibiting the establishment of any other dump in Union Township. Plaintiffs who also ran a landfill operation alleged discrimination and unconstitutionality.

The court declared the problem to be less one of the need to dump garbage or even the competitive right to do so than it is the power of the local authorities to control it. Considered in such terms it was held that the ordinance was reasonable in relation to "public health, safety and general welfare."

In the above case, the disposal of domestic refuse was involved. However, in Wildlife Preserves, Inc. v Poole 201 A. 2nd 337 (1964) "dirt, stumps, trees, rock, concrete, brick and plaster" were being dumped in a wetland area zoned for residential development. No evidence of pollution was adduced. The trial judge expressed the opinion that the plaintiff had not "established here that (the operator) had created any unsanitary condition." In his summary of findings, he stated:

> "I am satisfied from the proofs that the operator is not conducting a commercial dump; he is not conducting a business on the nine acres I find that he is filling in the premises to prepare them for a use permitted under the zoning ordinance."

Keller v Farmington 99 N.W. 2nd 578 (1959) was concluded in the same fashion but in both cases the disputed filling was construction debris and the filling was merely designed to enhance the ability of the land to be used for the purpose zoned.

We may summarize the general standing of sanitary landfill with respect to the courts as follows:

- Zoning ordinances are commonly drafted to restrict the development of sanitary landfills.
- In many instances, the zoning authority itself must conform to its own ordinances unless it expressly exempts itself.
- County authorities may avoid zoning restrictions as a result of exercising rights of eminent domain.
- Private sites are subject to all the limitations of zoning ordinance.
- There are signs that the courts recognize the distinction between a dump and a sanitary landfill.

In many respects, zoning restrictions on sanitary landfills are a backward step unless they are imposed based upon the criteria discussed in the chapter on Site Selection.

Commonly, this is not the case. In most instances, rather nebulous and unsubstantiated concerns about public health are adduced to limit development of sanitary landfill. With the progressive development of State Solid Waste Management Laws which define standards of selection, operation and maintenance, it would seem appropriate that zoning ordinances should permit conforming sanitary landfills wherever existing and future land uses would allow.[2]

---

[2] In a recent Illinois case (H. Carlson v Village of Worth) it was found (and upheld by the State Supreme Court) that the village did not have the right to disallow a sanitary landfill on the grounds of no local zoning if the landfill had received a state permit. If this decision is upheld in further court action it may have momentous impact upon the whole question of zoning controls over sanitary landfills.

Chapter 12

HIRING A CONSULTANT

It will be clear from the foregoing that the site selection and design of a sanitary landfill capable of meeting emerging Federal Guidelines and State Regulations is not a simple business. From the emergence of the idea of a landfill to the finalization of an operating plan, the prospective owner must be served by many disciplines. The needs of a good plan run the gamut of modern environmental disciplines from soil chemist, hydrogeologist and public relations expert to professional engineer, hydrologist and landscape architect. Commonly, these disciplines can only be found in a so-called multi-disciplinary organization. Occasionally, a gifted professional engineer may have mastered sufficient of the peripheral fields to handle the whole job. Either way, the important thing is to find the right consultant. The steps to be taken are as follows:

- search
- evaluation
- selection
- compensation

Search

Dependent upon the size of the project and the deadlines, some decision should be made initially as regards the number of consultants to be included in the initial search. Commonly, a "field" of 5 to 10 may be invited in the initial search. Commonly, a "field" of 5 to 10 may be invited to submit qualifications, details of experience and expressions of interest. The problem is basically how to find the field.

There are few better testimonials than personal experience. Where direct or reliable second-hand experience reports of good work being achieved in a specific area, the consultant responsible should definitely be on the list. When Federal guidelines will govern the landfill, this defines a need for a consultant with some experience in that area.

The Federal Office of Solid Waste Management Programs may be in a position to supply a list of candidates with experience on Federal projects (they are strictly not permitted to make specific recommendations). Where State Regulations will be the controlling factor, the State regulatory authority may also have a similar state-wide list (usually the State NSPE list).

The trade publications in this field (in particular, Waste Age and Solid Wastes Journal) may also have articles by leading consultants in this field from time to time. Many certainly display calling cards.

Professional associations of consulting engineers also make available lists of qualified consultants. The following bodies in particular would no doubt assist in developing the field:

- Council of Engineering Consultants (CEC)
- National Society of Professional Engineers (NSPE)
- American Academy of Environmental Engineers (AAEE)

Thus, it should not be too difficult to put together a list of 5 to 10 options. Even in those cases, where an owner has knowledge of one particular consultant and has some appreciation for his work, I would recommend a brief comparison with others from time to time. This makes sense simply from the point of view of confirming that his service stays ahead of the field.

Once the list has been put together, the next step is to circulate a brief description of the project to those in the field, and invite (equally brief) statements of interest, qualifications and experience.

Important items to include in this initial letter are:

- project deadlines
- size of landfill operation desired
- desired life of site
- status of project

Important items to request in the reply are:

- availability
- name and biographical details of the project manager
- specific experience with state and local regulatory agencies

Evaluation

The purpose of an evaluation phase is to reduce the field to 2 or 3 who will be invited to an interview.

Some consultants will no doubt disqualify themselves due to other commitments. Of the others, it is merely a question of determining who best meets the test of experience and general all-around capability. The designated project manager is particularly important. The experience he has in the specific area and with state agencies is extremely important. However, it is also important to select a consultant with a reputation for doing good work. The project manager will be the key man, but it is also important to know that he has a capable support staff.

Some of the information supplied in the letters of interest may also lead to follow-up questions. It is often very wise to follow up on statements made by the consultant with regard to specific experience directly related to the State or area in which the proposed sanitary landfill will be operated. The State regulatory agency may be found considerably more forthcoming when asked for their views on specific pieces of design work.

Work Selection

On the basis of the foregoing information, a short list of 2 or 3 consultants should be invited to make a brief presentation. The presentation team should include the project manager and a project engineer who will be assigned to the project. Items of particular interest at this point will be:

- the project manager's approach to the project
- specific experience with regulatory personnel
- methods of cost and schedule control
- other work commitments

If time and budget will allow, some consideration should be given to holding these interviews at the offices of each consultant. This arrangement affords an excellent opportunity to meet other members of the consultants' staff and gain some insight into how the consultant operates.

There are certain additional intangible judgments which the client must make at these interviews. It is important that the client should feel comfortable with the idea of working with the individuals involved. Knowing the personalities involved at a State or Federal level, it is important that the manager should "fit-in" with the others in the group.

On the basis of the foregoing evaluation, one consultant should be selected for negotiation. This may seem a curious procedure given the important part which cost may play in the final selection. However, costs and profits are not widely different between one consultant and another. The important thing is that the job should be done well. If price were the initial basis of comparison, it creates an illusion of simplicity. The choice would be simplicity itself, but totally illusory because no two consultants will set about doing the job in precisely the same way. Thus, the scope is different, therefore, the price is different and on the basis of a comparison between apples and bananas, the wrong consultant may be chosen.

If price is left as a negotiable item once the consultant has been selected then the scope itself is the subject of discussion, (not as a competitive item but as part of a genuine group effort to do the best job).

Compensation

Once a decision has been reached, the selected consultant should be notified of his selection subject to negotiation. The other applicants should also be notified of the results. The selected consultant should be invited to submit a detailed scope of work and a budget and attend a negotiation session at some fixed date and place. The method of compensation for sanitary landfill design will generally be one of the three alternatives in Table 62.

TABLE 62

Compensation Methods for Landfill Design

Method	Description
Lump Sum	Negotiated fixed cost for the whole job.
Cost Plus Fixed Fee	Costs based on salary plus overhead according to how much work is carried out plus a profit figure fixed at the beginning of the project.
Salary Cost Times Multiplier	Costs based on salary times a multiplier to cover overhead and profit.

The method of compensation depends upon how precisely the scope can be defined. If the breadth and depth of each item in the scope of work can be defined with a reasonable degree of precision then most consultants would prefer to work on a lump sum basis. However, there are occasionally items which may be ill-defined in terms of how much time the consultant can expect to be involved and in these circumstances it is clearly important to allow some flexibility in the fee basis.

Three major areas where lack of definition is often a problem are:

- public participation
- agency liaison
- geological survey

In the absence of a well-defined scope of these areas, the client may prefer to go with a method of compensation which will allow the consultant to cover all the bases (subject to client approval, of course). Alternatively, the client may insist upon lump sum as the required basis of compensation. Traditionally, the cost under this method may be somewhat higher than the other two, since estimates may be included for contingency items. However, the three methods are basically acceptable to most consultants and in most cases, the quality of work does not vary significantly.

If the client and the consultant both review the project scope and understand it thoroughly, there should be few difficulties.

APPENDIX A

Design Criteria for Outdoor Play Activities

STANDARDS FOR RECREATIONAL ACTIVITIES

Type of Recreational Activity	Space Requirements for Activity Per Population	Ideal Size of Space Required for Activity	Recreational Area Wherein Activity May Be Located
A. Standards for Recreational Activities			
Active Recreation			
1. Children's Play Area (with equipment)	0.5 acre/1,000 pop.	1 acre	Playgrounds-Neighborhood Parks Community Parks, School Playgrounds
2. Field Play Areas for Young Children	1.5 acres/1,000 pop.	3 acres	Playgrounds-Neighborhood Parks Community Parks
3. Older Children-Adult Field Sports Activities	1.5 acres/1,000 pop.	15 acres	Playfield-Community Park District Park
4. Tennis-Outdoor Basketball Other Court Sports	1.0 acres/5,000 pop.	2 acres	Playfield-Community Park
5. Swimming	1 outdoor pool/25,000	Competition size plus wading pool 2 acres	Playfield-Community Park
6. Major Boating Activities	100 acres/50,000	100 acres and over	District Park-Regional Park or Reservation
7. Hiking-Camping-Horseback Riding-Nature Study	10 acres/1,000 pop.	500–1,000 acres	Large District Park-Regional Park
8. Golfing	1–18 hole course per 50,000 pop.	120 acres	Community Park-District Park
Passive Recreation			
1. Picnicking	4 acres/1,000 pop.	varies	All parks
2. Passive Water Sports Fishing-Rowing-Canoeing	1 Lake or Lagoon per 25,000 pop.	20 acre water area	Community Park Special Regional Reservations
3. Zoos, Arboretums, Botanical Gardens	1 acre/1,000 pop.	100 acres	Large District Park or Special Facility
Other			
1. Parking at Recreational Areas	1 acre/1,000 pop.	varies	Playfields, Community, District & Regional Parks
2. Indoor Recreation Centers	1 acre/10,000 pop.	1–2 acres	Community Parks
3. Outdoor Theaters, Band Shells	1 acre/25,000 pop.	5 acres	District Parks

Type of Area	Acres Per 1000 Population	Size of Site Ideal	Size of Site Minimum	Radius of Area Served
B. Standards for Recreation Areas				
Playgrounds	1.5	4 acres	2 acres	0.5 miles
Neighborhood Parks	2.0	10	5	0.5
Playfields	1.5	15	10	1.5
Community Parks	3.5	100	40	2.0
District Parks	2.0	200	100	3.0
Regional Parks and Reservations	15.0	500–1,000	varies	10.0

Source: George Nez, Standards for New Urban Development – The Denver Background. Reprinted by permission of Urban Land, Vol. 20, No. 5, Urban Land Institute, 1200 18th Street, N.W. Washington, D.C.

SOFTBALL PITCH

Shown below is a recommended scale layout for a skinned infield on a softball diamond. Both a 55′ and 60′ radius is used with the front center of the pitcher's rubber as the center point. The 60′ radius is recommended as standard.

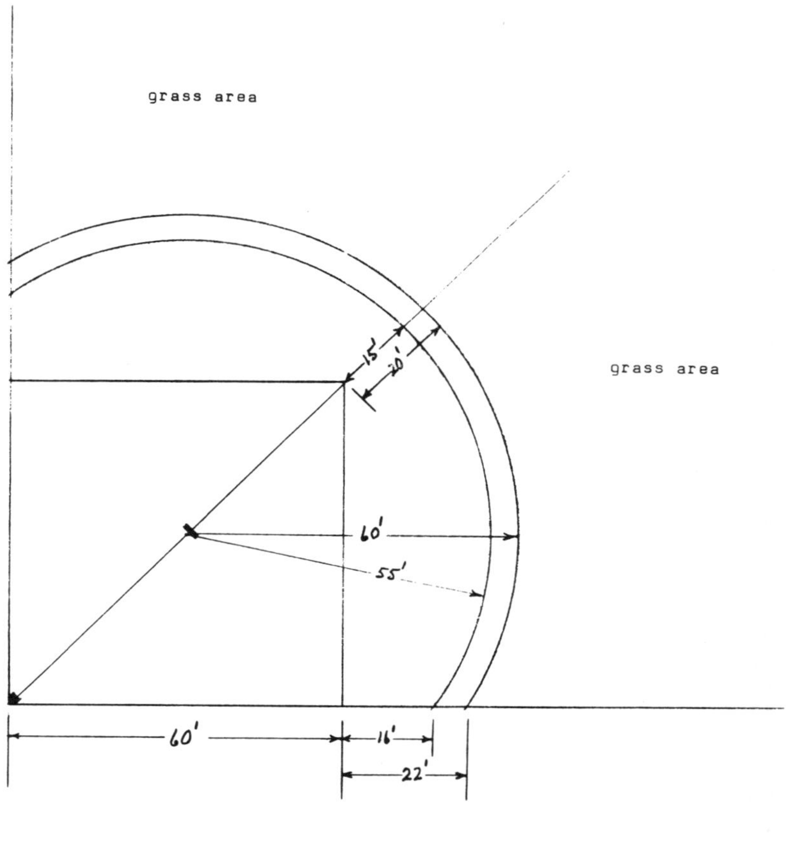

Courtesy: The Amateur Softball Association of America
Oklahoma City, Oklahoma

APPENDIX B

Design Criteria for Golf Courses

GOLF COURSE — 9 HOLE, PAR 3

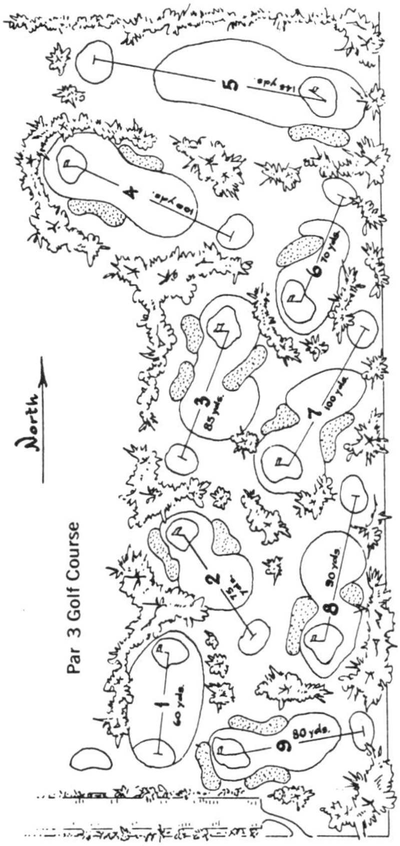

Nine-hole, par 3 course designed for maximum land use at minimum cost. Grassy bunkers and hollows can be substituted for sand traps indicated on plan to further cut cost of construction and maintenance as well as to speed up play for greater traffic capacity. Designed for 15 acre area.

SOURCE: Golf Operators Handbook, edited by Ben Chlevin, National Golf Foundation, Inc.—1956, p 86

	Minimum area required	Maximum area required	No. of parking spaces	Population served	Service radius	Average length
9-hole course	60 acres	80 acres	100 cars	1 hole per 3000 persons or 27,000 persons	½–¾ hour by car or public transportation	Approx. 2250 yards
18-hole course	120 acres	160 acres	200 cars	1 hole per 1500 persons or 25,000 persons	1 hour max. by car or public transportation	6500 yards

There are two general types of golf courses in use today—9-hole and 18-hole, the 18 hole course is the standard layout, the nine hole course is a short course with the fairways and greens smaller than those of a regulation course but similar in every other way.

The average length of an 18-hole course is 6500 yards while a 9-hole course is less than half of this length.

APPENDIX C

25-Year 24 Hour Rainfall

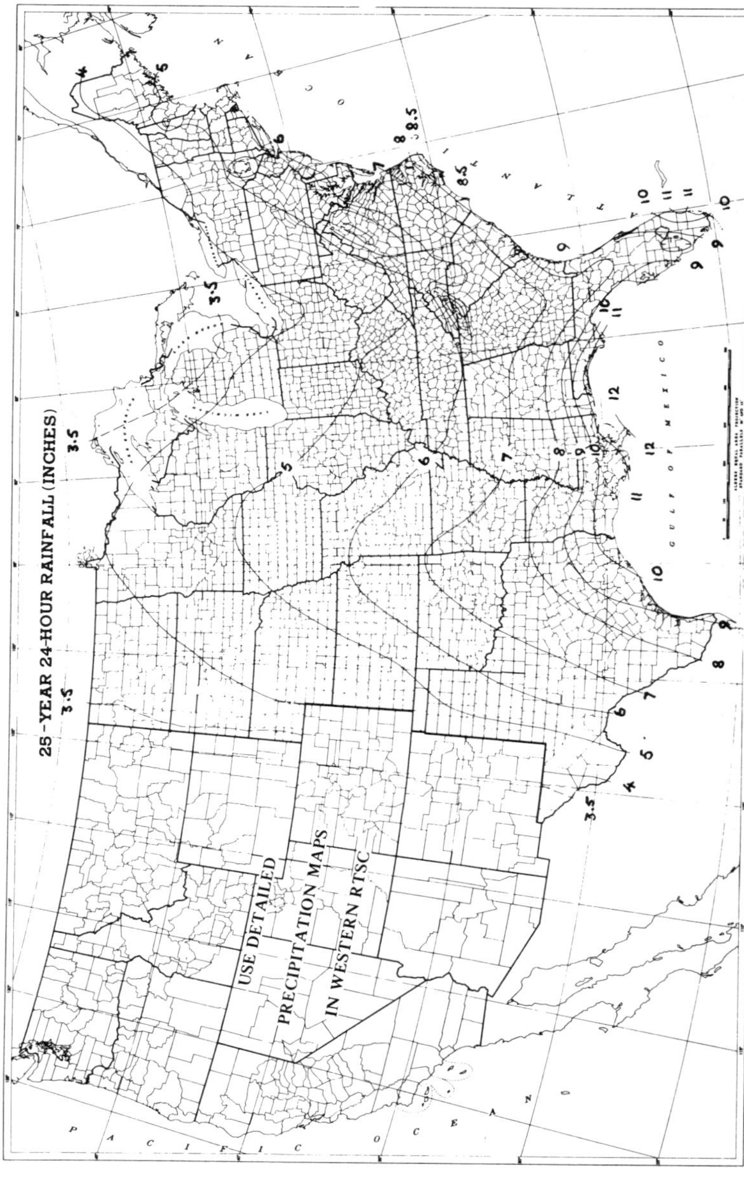

Prepared by U. S. Weather Bureau

Soil Types for Runoff Calculations[1]

EXHIBIT 2-1 (January, 1971) EFM Notice-4, 5/71

		AKAKA	A	AMADOR	D	ARBOR	B	ATLEE	C
AASTAD	B	AKASKA	B	AMAGON	D	ARBUCKLE	B	ATMORE	B/D
ABAJO	C	AKELA	C	AMALU	D	ARCATA	B	ATOKA	C
ABBOTT	D	ALADDIN	B	AMANA	B	ARCH	B	ATSION	C
ABBOTTSTOWN	C	ALAE	A	AMARGOSA	D	ARCHABAL	B	ATTERBERRY	B
ABEGG	B	ALAELOA	B	AMARILLO	B	ARCHER	C	ATTEWAN	A
ABELA	B	ALAGA	A	AMASA	B	ARCHIN	C	ATTICA	B
ABELL	B	ALAKAI	D	AMBERSON		ARCO	B	ATTLEBORO	
ABERDEEN	D	ALAMA	C	AMBOY	C	ARCOLA	C	ATWATER	B
ABES	D	ALAMANCE	B	AMBRAW	C	ARD	C	ATWELL	C
ABILENE	C	ALAMO	D	AMECEE	A	ARDEN	B	ATWOOD	B
ABINGTON	B	ALAMOSA	C	AMELIA	B	ARDENVOIR	B	AUBBEENAUBBEE	B
ABIQUA	C	ALAPAHA	D	AMENIA	B	ARDILLA	C	AUBERRY	B
ABO	B	ALAPAI	A	AMERICUS	A	AREDALE	B	AUBURN	C
ABRA	C	ALBAN	B	AMES	C	ARENA	C	AUBURNDALE	D
ABRAHAM	B	ALBANO	D	AMHERST	C	ARENALES	A	AUDIAN	B
ABSAROKEE	C	ALBANY	C	AMITY	C	ARENDTSVILLE	B	AU GRES	C
ABSCOTA	B	ALBATON	D	AMMON	B	ARENOSA	A	AUGSBURG	B
ABSHER	D	ALBEE	C	AMOLE	C	ARENZVILLE	B	AUGUSTA	C
ACACIO	C	ALBEMARLE	B	AMOR	B	ARGONAUT	D	AULD	D
ACADEMY	C	ALBERTVILLE	C	AMOS	C	ARGUELLO	B	AURA	B
ACADIA	D	ALBIA	C	AMSTERDAM	B	ARGYLE	B	AURORA	C
ACANA	D	ALBION	B	AMTOFT	D	ARIZO	A	AUSTIN	C
ACEITUNAS	B	ALBRIGHTS	C	AMY	D	ARKABUTLA	C	AUXVASSE	D
ACEL	D	ALCALDE	C	ANACAPA	B	ARKPORT	B	AUZQUI	B
ACKER	B	ALCESTER	B	ANAHUAC	D	ARLAND	B	AVA	C
ACKMEN	B	ALCOA	B	ANAHITE	D	ARLING	D	AVALANCHE	B
ACME	C	ALCONA	B	ANAPRA	B	ARLINGTON	A	AVALON	B
ACO	B	ALCOVA	B	ANATONE	D	ARLOVAL	C	AVERY	B
ACOLITA	B	ALDA	C	ANAVERDE	B	ARMAGH	D	AVON	C
ACOVE	C	ALDAX	D	ANCHO	C	ARMINGTON	D	AVONBURG	D
ACTON	B	ALDEN	D	ANCHORAGE	A	ARMO	B	AVONDALE	B
ACUFF	B	ALDER	B	ANCHOR BAY	D	ARMOUR	B	AWBREY	D
ACWORTH	B	ALDERDALE	C	ANCHOR POINT	B	ARMSTER	C	AXTELL	D
ADA	B	ALDERWOOD	C	ANCLOTE	D	ARMSTRONG	D	AYAR	C
ADAIR	D	ALDINO	C	ANCO	C	ARMUCHEE	D	AYCOCK	D
ADAMS	A	ALEKNAGIK	B	ANDERS	C	ARNEGARD	B	AYR	B
ADAMSON	B	ALEX	B	ANDERSON	B	ARNHART	C	AYRES	D
ADAMSTOWN		ALEXANDRIA	C	ANDES	C	ARNHEIM	C	AYRSHIRE	C
ADAMSVILLE	C	ALEXIS	B	ANDORINIA	C	ARNO	D	AYSEES	B
ADATON	D	ALFORD	B	ANDOVER	D	ARNOLD	B	AZTALAN	B
ADAVEN	D	ALGANSEE	B	ANDRES	B	ARNOT	C/D	AZTEC	B
ADDISON	D	ALGIERS	C/D	ANDREWS	C	AROOSTOOK		AZULE	C
ADDY	C	ALGOMA	B/D	ANED	D	AROSA	D	AZWELL	B
ADE	A	ALICE	A	ANETH	A	ARP	D		
ADEL	A	ALICEL	B	ANGELICA	D	ARRINGTON	B	BABB	A

(continued)

[1] This list comprises over 8,000 soil types classified according to their infiltration and transmission rates.

EXHIBIT 2-1 (continued)

ADELAIDE	D	ALICIA	C	ANGELINA	B/D	ARROLIME	C	BABBINGTON	B
ADELANTO	B	ALIDA	B	ANGIE	C	ARRON	C	BABCOCK	B
ADELPHIA	C	ALIKCHI	B	ANGLE	A	ARROW	B	BABYLON	A
ADENA	C	ALKO	D	ANGLEN	B	ARROWSMITH	B	BACA	C
ADILLIS	A	ALLAGASH	B	ANGOLA	C	ARTA	C	BACH	D
ADIRONDACK		ALLARD	B	ANGOSTURA	B	ARTOIS	C	BACHUS	B
ADKINS	B	ALLEGHENY	B	ANIAK	D	ARVADA	D	BACKBONE	A
ADLER	C	ALLEMANDS	D	ANITA	D	ARVANA	C	BADENAUGH	B
ADOLPH	D	ALLEN	B	ANKENY	A	ARVESON	C	BADGER	C
ADRIAN	A/D	ALLENDALE	C	ANLAUF	C	ARVILLA	B	BADGERTON	B
AENEAS	B	ALLENSVILLE	C	ANNABELLA	A	ARZELL	C	BADO	D
AETNA	B	ALLENTINE	D	ANNANDALE	C	ASA	B	BADUS	C
AFTON	D	ALLENWOOD	B	ANNISTON	B	ASBURY		BAGDAD	B
AGAR	B	ALLEY	C	ANOKA	A	ASCALON	B	BAGGOTT	D
AGASSIZ	D	ALLIANCE	B	ANONES	C	ASCHOFF	B	BAGLEY	B
AGATE	D	ALLIGATOR	D	ANSELMO	A	ASHCROFT	B	BAHEM	B
AGAWAM	B	ALLIS	D	ANSON	B	ASHBY	C	BAILE	D
AGENCY	C	ALLISON	C	ANTELOPE SPRINGS	C	ASHDALE	B	BAINVILLE	C
AGER	D	ALLOUEZ	C	ANTERO	C	ASHE	B	BAIRD HOLLOW	C
AGNER	B	ALLOWAY		ANT FLAT	C	ASHKUM	C	BAJURA	D
AGNEW	B/C	ALMAC	B	ANTHONY	B	ASHLEY	A	BAKEOVEN	D
AGNOS	B	ALMENA	C	ANTIGO	B	ASH SPRINGS	C	BAKER	C
AGUA	B	ALMONT	D	ANTILON	B	ASHTON	B	BAKER PASS	B
AGUADILLA	A	ALMY	B	ANTIOCH	D	ASHUE	B	BALAAM	A
AGUA DULCE	C	ALOHA	C	ANTLER	C	ASHUELOT	C	BALCH	D
AGUA FRIA	C	ALONSO	B	ANTOINE	C	ASHWOOD	C	BALCOM	B
AGUEDA	B	ALOVAR	C	ANTY	B	ASKEW	C	BALD	C
AGUILITA	B	ALPENA	B	ANWAY	B	ASO	D	BALDER	C
AGUIRRE	D	ALPON	B	ANZA	B	ASOTIN	C	BALDOCK	B/C
AGUSTIN	B	ALPOWA	B	APACHE	D	ASPEN	B	BALDWIN	D
AHATONE	D	ALPS		APAKUIE	A	ASPERMONT	B	BALDY	B
AHL	C	ALSEA	B	APISHAPA	C	ASSINNIBOINE	B	BALE	C
AHLSTROM	C	ALSTAD	B	APISON	B	ASSUMPTION	B	BALLARD	B
AHMEEK	B	ALSTOWN	B	APPIAN	D	ASTATULA	A	BALLINGER	C
AHOLT	D	ALTAMONT	D	APPLEGATE	C	ASTOR	A/D	BALM	B/C
AHTANUM	C	ALTAVISTA	C	APPLETON	C	ASTORIA	B	BALMAN	B/C
AHWAHNEE	C	ALTDORF	D	APPLING	B	ATASCADERO	C	BALON	B
AIBONITO	C	ALTMAR	B	APRON	B	ATCO	B	BALTIC	D
AIKEN	B	ALTO	C	APT	C	ATEPIC	C	BALTIMORE	B
AIKMAN	D	ALTOGA	C	APTAKISIC	B	ATHELWOLD	B	BAMBER	B
AILEY	B	ALTON	B	ARABY		ATHENA	B	BAMFORTH	B
AINAKEA	B	ALTUS	B	ARADA	B	ATHENS	B	BANCAS	B
AIRMONT	C	ALTVAN	B	ARAPIEN	C	ATHERTON	B/D	BANCROFT	B
AIROTSA	B	ALVIN	B	ARAVE	D	ATHOL	B	BANDERA	B
AIRPORT	D	ALVIRA	C	ARAVETON	B	ATKINSON	B	BANGO	C
AITS	B	ALVISO	D	ARBELA	C	ATLAS	D	BANGOR	B

BANGSTON	A	BEATTY		BERTELSON	A	BLAKENEY	C	BORDA	D
BANKARD	A	BEAUCOUP	B	BERTHOUD	B	BLAKEPORT	B	BORDEAUX	B
BANKS	A	BEAUFORD	D	BERTIE	C	BLAMER	C	BORDEN	B
BANNER	C	BEAUMONT	D	BERTOLOTTI	B	BLANCA	B	BORDER	B
BANNERVILLE	C/D	BEAUREGARD	C	BERTRAND	B	BLANCHARD	A	BORNSTEDT	B
BANNOCK	B	BEAUSITE	B	BERVILLE	D	BLANCHESTER	B/D	BORREGO	C
BANQUETE	D	BEAVERTON	B	BERYL	B	BLAND	C	BORUP	B
BARABOO	B	BECK	C	BESSEMER	B	BLANDFORD	C	BORVANT	D
BARRAGA	C	BECKER	B	BETHANY	C	BLANDING	B	BORZA	C
BARBARY	D	BECKET	C	BETHEL	D	BLANEY	B	BOSANKO	D
BARBOUR	B	BECKLEY	B	BETTERAVIA	C	BLANKET	C	BOSCO	B
BARBOURVILLE	B	BECKTON	D	BETTS	B	BLANTON	A	BOSKET	B
BARCLAY	C	BECKWITH	C	BEULAH	B	BLANYON	C	BOSLER	B
BARCO	B	BECKWOURTH	B	BEVENT	B	BLASINGAME	C	BOSQUE	B
BARCUS	B	BECREEK	B	BEVERLY	B	BLENCOE	C	BOSS	D
BARD	D	BEDFORD	C	BEW	D	BLENC	D	BOSTON	C
BARDEN	C	BEDINGTON	B	BEWLEYVILLE	B	BLENCON	B	BOSTWICK	B
BARDLEY	C	BEDNER	C	BEWLIN	D	BLETHEN	B	BOSWELL	D
BARELA	C	BEEBE	A	BEXAR	C	BLEVINS	B	BOSWORTH	D
BARFIELD	D	BEECHER	C	BEZZANT	B	BLICHTON	D	BOTELLA	B
BARFUSS	B	BEECHY		BIBB	B/D	BLISS	D	BOTHWELL	C
BARKER	C	BEEHIVE	B	BIBON	A	BLOCKTON	C	BOTTINEAU	C
BARKERVILLE	C	BEEZAR	B	BICKELTON	B	BLODGETT	A	BOTTLE	A
BARKLEY	B	BEHANIN	B	BICKMORE	C	BLOMFORD	B	BOULDER	B
BARLANE	D	BEHEMOTOSH	B	BICONDOA	C	BLOOM	C	BOULDER LAKE	D
BARLOW	B	BEJUCOS	B	BIDDEFORD	D	BLOOMFIELD	A	BOULDER POINT	B
BARNARD	D	BELDEN	D	BIDDLEMAN	C	BLOOMING	B	BOULFLAT	D
BARNES	B	BELDING	B	BIDWELL	B	BLOOR	D	BOURNE	C
BARNESTON	B	BELFAST	B	BIEBER	D	BLOSSOM	C	BOW	C
BARNEY	A	BELFIELD	B	BIENVILLE	A	BLOUNT	C	BOWBELLS	B
BARNHARDT	B	BELFORE	B	BIG BLUE	D	BLUCHER	C	BOWDOIN	D
BARNSTEAD		BELGRADE	B	BIGEL	A	BLUEBELL	C	BOWDRE	C
BARNUM	B	BELINDA	D	BIGETTY	C	BLUE EARTH	D	BOWERS	C
BARRADA	D	BELKNAP	C	BIGGS	A	BLUEJOINT	B	BOWIE	B
BARRINGTON	B	BELLAMY	B	BIGGSVILLE	B	BLUE LAKE	A	BOWMAN	B
BARRON	B	BELLAVISTA	D	BIG HORN	C	BLUEPOINT	B	BOWMANSVILLE	C
BARRONETT	C	BELLE	B	BIG TIMBER	D	BLUE STAR	D	BOX ELDER	C
BARROWS	D	BELLEFONTAINE		BIGWIN	A	BLUEWING	B	BOXWELL	C
BARRY	D	BELLICUM	B	BIJOU	A	BLUFFDALE	C	BOY	A
BARSTOW	B	BELLINGHAM	C	BILLETT	A	BLUFFTON	D	BOYCE	B/D
BARTH	C	BELLPINE	C	BILLINGS	C	BLUFORD	D	BOYD	D
BARTLE	D	BELMONT	B	BINFORD	B	BLY	D	BOYER	B
BARTON	B	BELMORE	B	BINGHAM	B	BLYTHE	D	BOYNTON	
BARTONFLAT	B	BELT	D	BINNSVILLE	D	BOARDTREE	C	BOYSAG	D
BARVON	C	BELTED	D	BINS	B	BOBS	D	BOYSEN	B
BASCOM	B	BELTRAMI	B	BIPPUS	B	BOBTAIL	B	BOZARTH	C
BASEHOR	D	BELTSVILLE	C	BIRCH	A	BOCK	B	BOZE	B
BASHAW	D	BELUGA	C	BIRCHWOOD	C	BODENBURG	B	BOZEMAN	A
BASHER	B	BELVOIR	C	BIRDS	C	BODINE	B	BRACEVILLE	C

(continued)

EXHIBIT 2-1 (continued)

BASILE	D	BENCLARE	C	BIRDSALL	D	BOEL	A	BRACKEN	D
BASIN	C	BENEVOLA	C	BIRDSBORO	B	BOELUS	A	BRACKETT	C
BASINGER	C	BENEWAH	C	BIRDSLEY	D	BOETTCHER	C	BRAD	D
BASKET	C	BENFIELD	C	BIRKBECK	B	BOGAN	C	BRADDOCK	C
BASS	A	BENGE	B	BISBEE	A	BOGART	B	BRADENTON	B/D
BASSEL	B	BEN HUR	B	BISCAY	C	BOGUE	D	BRADER	C
BASSETT	B	BENIN	D	BISHOP	B/C	BOHANNON	C	BRADFORD	B
BASSLER	D	BENITO	D	BISPING	B	BOHEMIAN	B	BRADSHAW	A
BASTIAN	D	BENJAMIN	D	BISSELL	B	BOISTFORT	C	BRADWAY	C
BASTROP	B	BEN LOMOND	B	BIT	D	BOLAR	C	BRACY	B
BATAVIA	B	BENMAN	A	BITTERON	A	BOLD	B	BRACYVILLE	C
BATES	B	BENNDALE	B	BITTERROOT	C	BOLES	C	BRAHAM	B
BATH	C	BENNETT	C	BITTER SPRING	C	BOLIVAR	B	BRAINERD	B
BATTLE CREEK	C	BENNINGTON	D	BITTERSPRING	C	BOLIVIA	B	BRALLIER	D
BATZA	C	BENOIT	C	BIXBY	B	BOLTON	B	BRAM	B
BAUDETTE	B	BENSON	C/D	BJORK	C	BOMBAY	B	BRAMARD	B
BAUER	C	BENTONVILLE		BLACHLY	C	BON	B	BRAMBLE	C
BAUGH	B/C	BENZ	D	BLACK BUTTE	C	BONACCORD	D	BRAMWELL	D
BAXTER	B	BELTIA	B	BLACK CANYON	D	BONAPARTE	A	BRAND	D
BAXTERVILLE	B	BEOWAWE	C	BLACKCAP	A	BOND	C	BRANDENBURG	A
BAYAMON	B	BERCAIL	C	BLACKETT	B	BONDRANCH	D	BRANDON	B
BAYARD	A	BERDA	B	BLACKFOOT	B	BONDURANT	B	BRANDYWINE	C
BAYBORO	D	BEREA	C	BLACKHALL	D	BONE	D	BRANFORD	B
BAYSHORE	B/C	BERENICETON	B	BLACKHAWK	D	BONG	B	BRANTFORD	B
BAYSIDE	C	BERENT	A	BLACKLEAF	B	BONHAM	C	BRASHEAR	C
BAYWOOD	A	BERGLAND	D	BLACKLOCK	D	BONILLA	B	BRASSFIELD	B
BAZETTE	C	BERGSTROM	B	BLACKMAN	C	BONITA	D	BRATTON	B
BEAD	C	BERINO	B	BLACK MOUNTAIN	B	BONN	D	BRAXTON	C
BEADLE	C	BERKELEY		BLACKOAR	C	BONNER	B	BRAYMILL	B/D
BEALES	A	BERKS	C	BLACKPIPE	C	BONNET	B	BRAYS	D
BEAR BASIN	B	BERKSHIRE	B	BLACK RIDGE	C	BONNEVILLE	B	BRAYTON	C
BEAR CREEK	C	BERLIN	C	BLACKROCK	B	BONNICK	A	BRAZITO	A
BEARDALL	C	BERMUDIAN	B	BLACKSTON	B	BONNIE	D	BRAZOS	A
BEARDEN	C	BERNAL	C	BLACKTAIL	B	BONO	D	BRECKENRIDGE	D
BEARDSTOWN	C	BERNALDO	B	BLACKWATER	D	BONSALL	D	BRECKNOCK	B
BEAR LAKE	D	BERNARD	D	BLACKWELL	B/D	BONTA	C	BREECE	B
BEARMOUTH	A	BERNARDINO	C	BLADEN	D	BONTI	C	BREGAR	D
BEARPAW	B	BERNARDSTON	C	BLAGO	D	BOOKER	D	BREMEN	B
BEAR PRAIRIE	B	BERNHILL	B	BLAINE	B	BOOMER	B	BREMER	B
BEARSKIN	D	BERNICE	A	BLAIR	C	BOONE	A	BREMO	C
BEASLEY	C	BERNING	C	BLAIRTON	C	BOONESBORO	B	BREMS	A
BEASON	C	BERRENDOS	D	BLAKE	C	BOOTH	C	BRENDA	C
BEATON	C	BERRYLAND	D	BLAKELAND	A	BORAH	A	BRENNAN	B

BRENNER	C/D	BUCKLEY	B/C	CAID	B	CAPUTA	C	CATLIN	B
BRENT	C	BUCKLON	D	CAIRO	D	CARACO	C	CATNIP	D
BRENTON	B	BUCKNER	A	CAJALCO	C	CARALAMPI	B	CATOCTIN	C
BRENTWOOD	B	BUCKNEY	A	CAJON	A	CARBO	C	CATLOSA	B
BRESSER	B	BUCKS	B	CALABAR	D	CARBOL	D	CATSKILL	A
BREVARD	B	BUCKSKIN	C	CALABASAS	C	CARBONDALE	D	CATTARAUGUS	C
BREVORT	B	BUCODA	C	CALAIS	C	CARBURY	B	CAUDLE	B
BREWER	C	BUDD	B	CALAMINE	D	CARDIFF	B	CAVE	D
BREWSTER	D	BUDE	C	CALAPOOYA	C	CARDINGTON	C	CAVE ROCK	A
BREWTON	C	BUDE	C	CALAWAH	B	CARDON	D	CAVO	D
BRICKEL	C	BUELL	A	CALCO	C	CAREY	B	CAVODE	C
BRICKTON	C	BUENA VISTA	B	CALDER	D	CAREY LAKE	B	CAVOUR	D
BRIDGE	C	BUFFINGTON	B	CALDWELL	B	CAREYTOWN	D	CAWKER	B
BRIDGEHAMPTON	B	BUFF PEAK	C	CALEAST	C	CARGILL	C	CAYAGUA	C
BRIDGEPORT	B	BUICK	C	CALEB	B	CARIBE	B	CAYLOR	B
BRIDGER	A	BUKREEK	B	CALERA	C	CARIBEL	B	CAYUGA	C
BRIDGESON	B/C	BULLION	D	CALHI	A	CARIBOU	B	CAZADERO	C
BRIDGEVILLE	B	BULLREY	B	CALHOUN	D	CARLIN	D	CAZADOR	B
BRIDGPORT	B	BULL RUN	B	CALICO	D	CARLINTON	B	CAZENOVIA	B
BRIEDWELL	B	BULL TRAIL	B	CALIFON	C	CARLISLE	A/D	CEBOLIA	C
BRIEF	B	BULLY	B	CALIMUS	B	CARLOTTA	B	CECIL	B
BRIENSBURG		BUMGARD	B	CALITA	B	CARLOW	D	CEDARAN	D
BRIGGS	A	BUNCOMBE	A	CALIZA	A	CARLSBAD	C	CEDAR BUTTE	C
BRIGGSDALE	C	BUNDO	B	CALKINS	C	CARLSBORG	A	CEDAREDGE	B
BRIGGSVILLE	C	BUNEJUG	C	CALLAHAN	C	CARLSON	C	CEDAR MT.	D
BRIGHTON	A/D	BUNKER	D	CALLEGUAS	D	CARLTON	B	CEDARVILLE	B
BRIGHTWOOD	C	BUNSELMEIER	C	CALLINGS	C	CARMI	B	CEDONIA	B
BRILL	B	BUNTINGVILLE	B/C	CALLOWAY	C	CARNEGIE	C	CEDRON	C/D
BRIM	C	BUNYAN	B	CALMAR	B	CARNERO	C	CELAYA	B
BRIMFIELD	C/D	BURBANK	A	CALNEVA	C	CARNEY	D	CELETON	D
BRIMLEY	B	BURCH	B	CALOUSE	B	CAROLINE	C	CELINA	C
BRINEGAR	B	BURCHARD	B	CALPINE	B	CARR	B	CELIO	A
BRINKERTON	D	BURCHELL	B/C	CALVERT	D	CARRISALITOS	D	CELLAR	D
BRISCOT	B	BURDETT	C	CALVERTON	C	CARRIZO	A	CENCOVE	B
BRITE	C	BUREN	C	CALVIN	C	CARSO	D	CENTER	C
BRITTON	C	BURGESS	B	CALVISTA	D	CARSON	D	CENTER CREEK	B
BRIZAM	A	BURGI	B	CAM	B	CARSTAIRS	B	CENTERFIELD	B
BROAD	C	BURGIN	D	CAMAGUEY	D	CARSTUMP	C	CENTERVILLE	D
BROADALBIN	C	BURKE	C	CAMARGO	B	CARTAGENA	D	CENTRALIA	B
BROADAX	B	BURKHARDT	B	CAMARILLO	B/C	CARTECAY	C	CENTRAL POINT	B
BROADBROOK	C	BURLEIGH	D	CAMAS	A	CARUSO	C	CERESCO	A
BROAD CANYON	B	BURLESON	D	CAMASCREEK	B/D	CARUTHERSVILLE	B	CERRILLOS	C
BROADHEAD	C	BURLINGTON	A	CAMBERN	C	CARVER	A	CERRO	C
BROADHURST	D	BURMA		CAMBRIDGE	C	CARWILE	D	CHACRA	C
BROCK	D	BURMESTER	D	CAMDEN	B	CARYVILLE	B	CHAFFEE	C
BROCKLISS	C	BURNAC	C	CAMERON	D	CASA GRANDE	C	CHAGRIN	B
BROCKMAN	C	BURNETTE	B	CAMILLUS	B	CASCADE	C	CHAIX	B
BROCKPORT	D	BURNHAM	D	CAMP	B	CASCAJO	B	CHALFONT	C
BROCKTON	D	BURNSIDE	B	CAMPBELL	B/C	CASCILLA	B	CHALMERS	C
BROCKWAY	B	BURNSVILLE	B	CAMPHORA	B	CASCO	B	CHAMA	B

(continued)

EXHIBIT 2-1 (continued)

BRODY	C	BURNT LAKE	B	CAMPIA	B	CASE	B	CHAMBER	C
BROGAN	B	BURRIS	D	CAMPO	C	CASEBIER	D	CHAMBERINO	C
BROGDON	B	BURT	D	CAMPONE	B/C	CASEY	C	CHAMISE	B
BROLLIAR	D	BURTON	B	CAMPSPASS	C	CASHEL	C	CHAMOKANE	B
BROMO	B	BUSE	B	CAMPUS	B	CASHION	D	CHAMPION	B
BRONAUGH	B	BUSHNELL	C	CAMRODEN	C	CASHMERE	B	CHANCE	B/D
BRONCHO	B	BUSHVALLEY	D	CANA	C	CASHMONT	B	CHANDLER	B
BRONSON	B	BUSTER	C	CANAAN	C/D	CASINO	A	CHANEY	C
BRONTE	C	BUTANO	C	CANADIAN	B	CASITO	C	CHANNAHON	B
BROOKE	C	BUTLER	D	CANADICE	D	CASPAR	B	CHANNING	B
BROOKFIELD	B	BUTLERTOWN	C	CANANDAIGUA	D	CASPIANA	B	CHANTA	B
BROOKINGS	B	BUTTE	C	CANASERAGA	C	CASS	A	CHANTIER	D
BROOKLYN	D	BUTTERFIELD	C	CANAVERAL	C	CASSADAGA		CHAPIN	C
BROOKSIDE	C	BUXIN	D	CANDELERO	C	CASSIA	C	CHAPMAN	
BROOKSTON	B/D	BUXTON	C	CANE	C	CASSCLARY	B	CHAPPELL	B
BROOKSVILLE	D	BYARS	D	CANEADEA	D	CASSVILLE		CHARD	B
BROSELEY	B	BYRON	A	CANEEK	B	CASTAIC	C	CHARITON	D
BROSS	B			CANEL	B	CASTALIA	C	CHARITY	D
BROUGHTON	D	CABALLO	C	CANELOX	C	CASTANA	B	CHARLESTON	C
BROWARD	C	CABARTON	C	CANEY	C	CASTELL	C	CHARLEVOIX	B
BROWNELL	B	CABBA	C	CANEYVILLE	C	CASTILE	B	CHARLOS	A
BROWNFIELD	A	CABBART	D	CANFIELD	C	CASTINO	C	CHARLOTTE	A/D
BROWNLEE	B	CABEZON	D	CANISTEO	C	CASTLE	D	CHARLTON	B
BROYLES	C	CABIN	C	CANNINGER	B	CASTLE VALLEY	D	CHASE	C
BRUCE	D	CABINET	C	CANOE	B	CASTNER	C	CHASEBURG	B
BRUIN	C	CABLE	D	CANONCITO	C	CASTO	C	CHASEVILLE	A
BRUNEEL	B	CABO ROJO	C	CANOVA	B/D	CASTRO	C	CHASKA	C
BRUNO	A	CABOT	D	CANTON	B	CASTROVILLE	B	CHASTAIN	D
BRUNT	C	CACAPON	B	CANTRIL	B	CASUSE	D	CHATBURN	B
BRUSETT	B	CACHE	D	CANTUA	B	CASWELL	B	CHATFIELD	
BRUSH		CACIQUE	B	CANUTIO	B	CATALINA	B	CHATHAM	B
BRUSSETT	B	CADDO	D	CANYON	D	CATALPA	C	CHATSWORTH	D
BRYAN	A	CADEVILLE	D	CAPAC	B	CATANO	A	CHAUNCEY	C
BRYCAN	B	CADMUS	B	CAPAY	D	CATARINA	D	CHAVIES	B
BRYCE	D	CADOMA	C	CAPE	D	CATAULA	C	CHAWANAKEE	C
BUCAN	D	CADOR	C	CAPE FEAR	D	CATAWBA	B	CHEADLE	C
BUCHANAN	C	CAGEY	C	CAPERS	D	CATH	D	CHECKETT	D
BUCHENAU	C	CAGUABO	D	CAPILLO	D	CATHCART	B	CHEDAHAP	B
BUCHER	C	CAHABA	B	CAPLES	C	CATHEDRAL	C	CHEEKTOWAGA	D
BUCKINGHAM		CAHILL	B	CAPPS	B	CATHERINE	B/D	CHEESMAN	B
BUCKLAND	C	CAHONE	C	CAPSHAW	C	CATHRO	D	CHEHALEM	C
BUCKLEBAR	B	CAHTO	C	CAPULIN	C	CATLETT	C/D	CHEHALIS	B

CHEHULPUM	C	CHUTE	A	COACHELLA	B	CONALB	B	COTITO	C
CHELAN	B	CIALES	D	COAD	B	CONANT	C	COTO	C
CHELSEA	A	CIALITOS	B	COAL CREEK	C	CONASAUGA	C	COTOPAXI	A
CHEMAWA	B	CIBEQUE	B	COALMONT	C	CONATA	D	COTT	B
CHEMUNG		CIBO	D	COAMC	C	CONBOY	D	COTTER	B
CHEN	D	CIBOLA	B	COARSEGOLD	B	CONCHAS	C	COTTERAL	B
CHENA	A	CICERO	D	COATICOOK	C	CONCHO	C	COTTIER	B
CHENANGO	A	CIDERCONE	B	COATSBURG	D	CONCONULLY	B	COTTONWOOD	C
CHENEY	B	CIDRAL	C	COBB	B	CONCORD	D	COTTRELL	C
CHENNEBY	C	CIENEBA	B	COBEN	D	CONCREEK	B	COUCH	C
CHENOWETH	B	CIMA	C	COBEY	B	CONDA	C	COUGAR	C
CHEQUEST	C	CIMARRON	C	COBURG	C	CONDIT	D	COULSTONE	B
CHEREETE	A	CINCINNATI	C	COCHETOPA	C	CONDON	C	COUNTS	C
CHEROKEE	D	CINCO	A	COCCA	A	CONE	A	COUPEVILLE	B
CHERRY	C	CINEBAR	B	COCOLALLA	C	CONEJO	C	COURT	B
CHERRYHILL	C	CIRCLE	C	CODORUS	C	CONESTOGA	B	COURTHOUSE	D
CHERRY SPRINGS	D	CIRCLEVILLE	C	CODY	A	CONESUS	B	COURTLAND	B
CHESAW	B	CISNE	D	COE	A	CONGAREE	B	COURTNEY	D
CHESHIRE	B	CISPUS	A	COEBURN	C	CONI	D	COURTROCK	B
CHESHNINA	D	CITICO	B	COFF	D	CONLEN	B	COUSE	C
CHESNIMNUS	B	CLACKAMAS	C	COGGON	B	CONLEY	C	COUSHATTA	B
CHESTER	B	CLAIBORNE	B	COGSWELL	C	CONNEAUT	C	COVE	D
CHESTERTON	C	CLAIRE	A	COHASSET	B	CONNECTICUT		COVEILO	B
CHETCO	D	CLAIREMONT	B	COHOCTAH	D	CONNER	B	COVELAND	C
CHETEK	B	CLALLAM	B	COHOE	B	CONOTTON	B	COVENTRY	B
CHEVELON	C	CLAM GULCH	C	COIT	C	CONOVER	B	COVEYTOWN	C
CHEWACLA	C	CLAMO	C	COKEDALE	C	CONOWINGO	C	COVINGTON	D
CHEWELAH	B	CLANTON	C	COKEL	C	CONRAD	B	COWAN	A
CHEYENNE	B	CLAPPER	B	COKER	D	CONROE	B	COWARTS	C
CHIARA	D	CLAREMORE	D	COKESBURY	D	CONSER	C/D	COWDEN	D
CHICKASHA	B	CLARENCE	D	COKEVILLE	B	CONSTABLE	A	COWDRY	C
CHICOPEE	B	CLARESON	C	COLBATH	C/D	CONSUMO	B	COWEEMAN	C
CHICOTE	D	CLAREVILLE	C	COLBERT	D	CONTINENTAL	C	COWERS	B
CHIGLEY	C	CLARINDA	D	COLBURN	B	CONTRA COSTA	C	COWICHE	B
CHILCOTT	D	CLARION	B	COLBY	B	CONVENT	C	COWOOD	C
CHILDS	B	CLARITA	D	COLCHESTER	B	COOK	D	COX	D
CHILGREN	C	CLARK	B	COLDEN		COOKPORT	C	COXVILLE	D
CHILHOWIE	C	CLARK FORK	A	COLD SPRINGS	C	COOLBRITH	B	COYATA	C
CHILI	B	CLARKSBURG	C	COLE	B/C	COOLVILLE	C	COZAD	B
CHILLICOTHE	C	CLARKSDALE	C	COLEBROCK	B	COOMBS	B	CRABTON	B
CHILLISQUAQUE		CLARKSON	B	COLEMAN	C	COONEY	B	CRADDOCK	B
CHILLUM	B	CLARKSVILLE	B	COLEMANTOWN	D	COOPER	C	CRADLEBAUGH	D
CHILMARK	B	CLARNO	B	COLETO	A	COOTER	C	CRAFTON	C
CHILO	B/D	CLARY	B	COLFAX	C	COPAKE	B	CRAGO	B
CHILOQUIN	B	CLATO	B	COLINAS	B	COPALIS	B	CRAIG	C
CHILSON	D	CLATSOP	D	COLLAMER	C	COPELAND	B/D	CRAIGMONT	C
CHILTON	B	CLAVERACK	C	COLLARD	B	COPITA	B	CRAPER	D
CHIMAYO	C	CLAWSON	C	COLLBRAN	C	COPLAY		CRANE	B
CHIMNEY	B	CLAYBURN	B	COLLEEN	C	COPPER RIVER	D	CRANSTON	B
CHINA CREEK	B	CLAYSPRINGS	D	COLLEGIATE	C	COPPERTON	B	CRARY	C

(continued)

EXHIBIT 2-1 (continued)

CHINCHALLO	B/D	CLAYTON	B	COLLETT	C	COPPOCK	B	CRATER LAKE	B
CHINIAK	A	CLEARFIELD	C	COLLIER	A	COPSEY	D	CRAVEN	C
CHINO	B/C	CLEAR LAKE	D	COLLINGTON	B	COQUILLE	C/D	CRAWFORD	D
CHINOOK	B	CLEEK	C	COLLINS	C	CORA	D	CREAL	D
CHIPETA	D	CLE ELUM	B	COLLINSTON	C	CORAL	C	CREBBIN	C
CHIPLEY	C	CLEGG	B	COLLINSVILLE	C	CORBETT	C	CREEDMAN	D
CHIPMAN	C	CLEMAN	B	COLMA	B	CORBIN	B	CREEDMOOR	C
CHIPPENY	D	CLEMVILLE	B	COLMOR	C	CORCEGA	C	CREIGHTON	B
CHIPPEWA	B/D	CLEORA	B	COLO	B	CORD	C	CRELDON	B
CHIQUITO	C	CLERF	B	COLOCKUM	B	CORDES	B	CRESBARD	C
CHIRICAHUA	D	CLERMONT	D	COLOMA	A	CORDOVA	C	CRESCENT	B
CHITINA	B	CLEVERLY	A	COLOMBO	B	CORINTH	C	CRESCO	C
CHITTENDEN	C	CLIFFDOWN	C	COLONA	C	CORKINDALE	B	CREST	C
CHITWOOD	C	CLIFFHOUSE	C	COLONIE	A	CORLENA	A	CRESTLINE	B
CHIVATO	D	CLIFFORD	B	COLORADO	B	CORLETT	B	CRESTMORE	
CHIWAWA	B	CLIFFWOOD	C	COLOROCK	D	CORLEY	C	CRESTON	A
CHO	C	CLIFTERSON	B	COLOSO	D	CORMANT	C	CRESWELL	C
CHOBEE	D	CLIFTON	C	COLOSSE	A	CORNHILL	B	CRETE	D
CHOCK	B/D	CLIFTY	B	COLP	D	CORNING	D	CREVA	D
CHOCOLOCCO	B	CLIMARA	D	COLRAIN	B	CORNUTT	C	CREVASSE	A
CHUPAKA	C	CLIMAX	C	COLTON	A	CORNVILLE	B	CREWS	D
CHOPTANK	A	CLIME	C	COLTS NECK	B	COROZAL	C	CRIDER	B
CHOPTIE	D	CLINTON	B	COLUMBIA	B	CORPENING	D	CRIP	B
CHORALMONT	B	CLODINE	D	COLUMBINE	A	CORRALITOS	A	CRISFIELD	B
CHOTEAU	C	CLONTARF	B	COLUSA	C	CORRECO	C	CRITCHELL	B
CHRISTIAN	C	CLOQUALLUM	C	COLVILLE	B	CORRERA	D	CRIVITZ	A
CHRISTIANA	B	CLOQUATO	B	COLVIN	C	CORSON	C	CROCKER	A
CHRISTIANBURG	D	CLOQUET	B	COLWOOD	B/D	CORTEZ	D	CROCKETT	D
CHRISTY	B	CLOUD	D	COLYER	C/D	CORTINA	A	CROFTON	B
CHROME	C	CLOUDCROFT	D	COMERIO	B	CORUNNA	C	CROGHAN	B
CHUALAR	B	CLOUD PEAK	B	COMETA	C	CORVALLIS	B	CROOKED	C
CHUBBS	C	CLOUD RIM	B	COMFREY	C	CORWIN	B	CROOKED CREEK	D
CHUCKAWALLA	B	CLOUGH	D	COMITAS	A	CORY	C	CROCKSTON	B
CHULITNA	B	CLOVERDALE	D	COMLY	C	CORYDON	C	CROOM	B
CHUMMY	C	CLOVER SPRINGS	B	COMMERCE	C	COSAD	C	CROPLEY	C
CHUMSTICK	C	CLOVIS	B	COMO	A	COSH	C	CROSBY	C
CHUPADERA	B	CLUFF	C	COMODORE	B	COSHOCTON		CROSS	D
CHURCH	D	CLUNIE	D	COMORO	B	COSKI	B	CROSSVILLE	B
CHURCHILL	D	CLURD	C	COMPTCHE	B	COSSAYUNA	C	CROSWELL	A
CHURCHVILLE	D	CLURO	C	COMPTON	C	COSTILLA	A	CROT	D
CHURN	B	CLYDE	D	COMSTOCK	C	COTACO	C	CROTON	D
CHURNDASHER	B	CLYMER	B	COMUS	B	COTATI	C	CROUCH	B

CROW	C
CROW CREEK	B
CROWFOOT	C
CROWHEART	B
CROW HILL	C
CROWLEY	D
CROWN	B
CROWSHAW	B
CROZIER	C
CRUCKTON	B
CRUICKSHANK	C
CRUME	B
CRUMP	D
CRUTCH	B
CRUTCHER	D
CRUZE	C
CRYSTAL LAKE	B
CRYSTAL SPRINGS	D
CRYSTOLA	B
CUBA	B
CUBERANT	B
CUCHILLAS	D
CUDAHY	D
CUDIYO	B
CUERO	B
CUEVA	D
CUEVITAS	D
CULLEN	C
CULLEOKA	B
CULLO	C
CULPEPER	C
CULVERS	C
CUMBERLAND	B
CUMLEY	C
CUMMINGS	B/D
CUNICO	C
CUPPER	B
CUROLI	C
CURECANTI	B
CURLEW	C
CURRAN	C
CURRANT	B
CURTIS CREEK	D
CURTIS SIDING	A
CUSHING	B
CUSHMAN	C
CUSTER	C
CUTTER	D
CUTZ	D
CUYAMA	B
DANZ	B
DARGOL	D
DARIEN	C
DARLING	B
DARNELL	C
DARNEN	B
DARR	A
DARRET	C
DARROCH	C
DART	A
DARVADA	D
DARWIN	D
DASSEL	D
DATEMAN	C
DATINO	C
DATWYLER	C
DAULTON	D
DAUPHIN	
DAVEY	A
DAVIDSON	B
DAVIS	B
DAVISON	B
DAWES	C
DAWHOO	B/D
DAWSON	D
DAY	D
DAYBELL	A
DAYTON	D
DAYVILLE	B/C
DAZE	D
DEACON	B
DEADFALL	B
DEAMA	C
DEAN	B
DEAN LAKE	C
DEARDURFF	B
DEARY	C
DEARYTON	B
DEATMAN	C
DEAVER	C
DEBENGER	C
DECAN	D
DECATHON	D
DECATUR	B
DECCA	B
DECKER	C
DECKERVILLE	C
DECLO	B
DECORRA	B
DECROSS	B
DEL REY	C
DEL RIO	C
DELTA	C
DELTON	B
DELWIN	A
DELYNDIA	B
DEMAST	B
DEMASTERS	B
DE MAYA	C
DEMERS	D
DEMKY	D
DEMONA	C
DEMOPOLIS	C
DEMPSTER	B
DENAY	B
DENISON	C
DENMARK	D
DENNIS	C
DENNY	D
DENROCK	D
DENTON	D
DENVER	C
DEPEW	C
DEPOE	D
DERINDA	C
DESAN	A
DESART	C
DESCALABRADO	D
DESCHUTES	C
DESERET	C
DESHA	D
DESHLER	C
DESOLATION	C
DESPAIN	B
DETER	C
DETLOR	C
DETOUR	C
DETROIT	C
DEV	B
DEVILS DIVE	D
DEVOL	B
DEVON	B
DEVORE	B
DEWART	
DEWEY	B
DEWVILLE	B
DEXTER	B
DIA	C
DIABLO	D
DIAMOND	C
DIXMONT	C
DIXMORE	B
DIXONVILLE	C
DIXVILLE	A
DOAK	C
DOBBS	B
DOBY	D
DOCAS	B
DOCKERY	C
DOCT	B
DODGE	B
DODGEVILLE	B
DODSON	C
DOGER	A
DOGUE	C
DOLAND	B
DOLE	C
DOLLAR	B
DOLLARD	C
DOLORES	B
DOLPH	C
DOMINGO	C
DOMINGUEZ	C
DOMINIC	A
DOMINO	C
DONA ANA	B
DONALD	B
DONEGAL	
DONERAIL	C
DONICA	A
DONLONTON	C
DONNA	D
DONNAN	C
DONNYBROOK	D
DONOVAN	B
DOOLEY	A
DOONE	B
DOOR	B
DORA	D
DORAN	C
DORCHESTER	B
DOROSHIN	D
DOROTHEA	C
DOROVAN	D
DORS	B
DORSET	B
DOS CABEZAS	C
DOSSMAN	B
DOTHAN	B
DOTTA	B
DRY CREEK	C
DRYDEN	B
DRY LAKE	C
DUANE	B
DUBAKELLA	C
DUBAY	D
DUBBS	B
DUBOIS	C
DUBUQUE	B
DUCEY	B
DUCHESNE	A
DUCKETT	C
DUCOR	D
DUDA	A
DUDLEY	D
DUEL	B
DUELM	C
DUFFAU	B
DUFFER	D
DUFFIELD	B
DUFFSON	B
DUFFY	B
DUFUR	B
DUGGINS	D
DUGOUT	D
DUGWAY	D
DUKES	A
DULAC	C
DUMAS	B
DUMECQ	C
DUMONT	C
DUNBAR	D
DUNBARTON	C
DUNBRIDGE	B
DUNCAN	D
DUNCANNON	B
DUNCOM	D
DUNDAS	C
DUNDAY	A
DUNDEE	C
DUNELLEN	B
DUNE SAND	A
DUNGENESS	B
DUN GLEN	C
DUNKINSVILLE	B
DUNKIRK	B
DUNLAP	B
DUNMORE	B
DUNNING	D
DUNPHY	D

(continued)

EXHIBIT 2-1 (continued)

CYLINDER	B	DEE	C	DIAMOND SPRINGS	C	DOTY	B	DUNVILLE	B
CYNTHIANA	C/D	DEEPWATER	C	DIAZ	C	DOUBLETOP	B	DU PAGE	B
CYPREMORT	C	DEER CREEK	C	DIBBLE	C	DOUDS	B	DUPEE	C
CYRIL	B	DEERFIELD	B	DICK	A	DOUGHERTY	A	DUPLIN	C
		DEERFORD	D	DICKEY	A	DOUGHTY	A	DUPO	C
DABOB	B	DEERING	C	DICKINSON	A	DOUGLAS	B	DUPONT	C
DACONA	C	DEERLODGE	D	DICKSON	C	DOURO	B	DUPREE	D
DADE	A	DEER PARK	A	DIGBY	C	DOVER	B	DURALDE	C
DAFTER	B	DEERTON	B	DIGGER	C	DOVRAY	D	DURAND	B
DAGGETT	A	DEERTRAIL	C	DIGHTON	B	DOW	B	DURANT	D
DAGLUM	D	DEFIANCE	D	DILL	B	DOWAGIAC	B	DURELLE	B
DAGOR	B	DEFORD	D	DILLARD	C	DOWDEN	C	DURHAM	B
DAGUAO	C	DEGARMO	B/C	DILLDOWN		DOWELLTON	D	DURKEE	C
DAGUEY	C	DEGNER	C	DILLINGER	B	DOWNER	B	DURCO	B
DAHLQUIST	B	DE GREY	D	DILLON	D	DOWNEY	B	DURRSTEIN	D
DAIGLE	C	DEJARNET	B	DILLWYN	A	DOWNS	B	DUTCHESS	B
DAILEY	A	DEKALB	C	DILMAN	C	DOXIE	C	DUTSON	D
DAKOTA	B	DEKOVEN	D	DILTS	D	DOYCE	C	DUTTON	D
DALBO	B	DELAKE	B	DILWORTH	C	DOYLE	A	DUVAL	B
DALBY	D	DELANCO	C	DIMAL	C	DOYLE	A	DUZEL	B
DALE	B	DELANEY	A	DIMYAW	C	DOYLESTOWN	D	DWIGHT	D
DALHART	B	DELANO	B	DINGLE	B	DOYN	C	DWYER	A
DALIAN	B	DELECO	C	DINGLISHNA	C	DRA	C	DYE	D
DALLAM	B	DELENA	D	DINKELMAN	B	DRACUT	C	DYER	
DALTON	C	DELFINA	B	DINKEY	A	DRAGE	B	DYKE	B
DALUPE	B	DELHI	A	DINNEN	B	DRAGOON	B	DYRENG	D
DAMASCUS	D	DELICIAS	B	DINSDALE	B	DRAGSTON	C		
DAMON	D	DELKS	B/D	DINUBA	B/C	DRAIN	D	EAD	C
DANA	B	DELL	C	DINZER	B	DRAKE	B	EAGAR	B
DANBURY	C	DELLEKER	C	DIOXICE	C	DRANYON	B	EAGLECONE	B
DANEY		DELLO	A/C	DIQUE	B	DRAPER	C	EAKIN	B
DANDREA	C	DELLROSE	B	DISAPEL	D	DRESDEN	B	EAMES	B
DANDRIDGE	D	DELM	D	DISAUTEL	B	DRESSLER	C	EARLE	D
DANGBERG	D	DELMAR	D	DISCO	B	DREWS	B	EARLMONT	B/C
DANIELS	B	DELMITA	C	DISHNER	D	DRIFTON	C	EARP	B
DANKO	D	DELMONT	A	DISTERHEFF	D	DRIGGS	B	EASLEY	D
DANLEY	C	DELNORTE	C	DITCHCAMP	C	DRUM	C	EAST FORK	C
DANNEMORA	D	DELPHI	B	DIVERS	B	DRUMMER	B	EAST LAKE	A
DANSKIN	B	DELPHILL	C	DIVICE	B	DRUMMOND	D	EASTLAND	C
DANT	D	DELPIEDRA	C	DIX	A	DRURY	B	EASTON	C
DANVERS	C	DELPINE	D	DIX	A	DRYAD	C	EASTONVILLE	A
DANVILLE	C	DELRAY	A/D	DIXIE	C	DRYBURG	B	EAST PARK	D

EASTPURT	A	ELLISCN	B.	ESMCND	B	FARNUM	B	FLEISCHMANN	D
EATUNTOWN		ELLOAM	D	ESPARTO	B	FARRAGUT	C	FLEMING	C
EAUGALLIE	B/D	ELLSBERRY		ESPIL	D	FARRAR	B	FLETCHER	B
EBA	C	ELLSWORTH	C	ESPINAL	A	FARRELL	B	FLOKE	D
EBBERT	D	ELMA	B	ESCUATZEL	B	FARRENBURG	B	FLOM	C
EBBS	B	ELMDALE	B	ESS	B	FARROT	C	FLOMATION	A
EBENEZER	C	ELMIRA	A	ESSEN	C	FARSCN	B	FLORENCE	C
ECCLES	B	ELMO	C	ESSEX	C	FARWELL	C	FLORIDANA	B/D
ECHARD	C	ELMONT	B	ESSEXVILLE	D	FATIMA	B	FLORISSANT	C
ECHLER	B	ELMORE	B	ESTACADO	B	FATTIG	C	FLOWELL	C
ECKLEY	B	ELMWOOD	C	ESTELLINE	B	FAUNCE	A	FLOWEREE	B
ECKMAN	B	ELNORA	B	ESTER	D	FAUQUIER	C	FLOYD	B
ECKRANT	D	ELOIKA	B	ESTERBROOK	B	FAWCETT	C	FLUSHING	
ECTOR	C	ELPAN	D	ESTHERVILLE	B	FAWN	B	FLUVANNA	C
EDALGO	C	EL PECO	C	ESTO	C	FAXON	D	FLYGARE	B
EDDS	B	EL RANCHO	B	ESTRELLA	B	FAYAL	C	FLYNN	D
EDDY	C	ELRED	B/D	ETHAN	B	FAYETTE	B	FOARD	D
EDEN	C	ELRED	B/D	ETHETE	B	FAYETTEVILLE	B	FOGELSVILLE	B
EDENTON	C	ELS	A	ETHRIDGE	C	FAYWCOD	C	FOLA	B
EDENVALE	D	ELSAH	B	ETIL	A	FE	D	FOLEY	D
EDGAR	B	ELSINBORO	B	ETNA		FEDORA	B	FONDA	D
EDGECUMBE	B	ELSMERE	A	ETCWAH	B	FELDA	B/D	FONDIS	C
EDGELEY	C	ELSO	D	ETCWN	B	FELIDA	B	FONTAL	C
EDGEMONT	B	EL SCLYO	C	ETTA	C	FELLOWSHIP	D	FONTREEN	B
EDGEWATER	C	ELSTON	B	ETTER	B	FELT	B	FOPIANO	D
EDGEWICK	B	ELTOPIA	B	ETTERSBURG	B	FELTA	C	FORBES	B
EDGEWOOD	A	ELTREE	B	ETTRICK	D	FELTHAM	A	FORD	D
EDGINGTON	C	ELTSAC	D	EUBANKS	B	FELTON	B	FORDNEY	A
EDINA	D	ELWHA	B	EUDORA	B	FELTONIA	B	FORDVILLE	B
EDINBURG	C	ELWOOD	C	EUFAULA	A	FENCE	B	FORE	D
EDISON	B	ELY	B	EUREKA	D	FENDALL	C	FORELAND	C
EDISTO	C	ELYSIAN	B	EUSTIS	A	FENWOOD	B	FORELLE	B
EDITH	A	ELZINGA	B	EUTAW	D	FERDELFORD	C	FGRESMAN	B
EDLOE	B	EMBDEN	B	EVANGELINE	C	FERDIG	C	FORESTDALE	D
EDMONDS	C	EMDENT	C	EVANS	B	FERGUS	B	FORESTER	C
EDMORE	D	EMER	C	EVANSTON	B	FERGUSON	B	FORGAY	A
EDMUND	C	EMERALD	B	EVARO	A	FERNANDO	C	FORMAN	B
EDNA	D	EMERSON	B	EVART	D	FERNDALE	B	FORNEY	D
EDNEYVILLE	B	EMIDA	D	EVENDALE		FERNLEY	C	FORREST	C
EDOM	C	EMIGRANT	B	EVERETT	B	FERNOW	B	FORSEY	C
EDSON	C	EMIGRATION	D	EVERGLADES	A/D	FERNPOINT	C	FURSGREN	C
EDWARDS	B/D	EMILY	B	EVERLY	B	FERRELO	B	FORT COLLINS	B
EEL	C	EMLIN	B	EVERMAN	C	FERRIS	D	FORT DRUM	C
EFFINGTON	D	EMMA	C	EVERSON	C	FERRON	B	FORT LYON	B
EFWUN	A	EMMERT	A	EVESBORC	A	FERTALINE	D	FORT MEADE	A
EGAM	C	EMMET	B	EWA	B	FESTINA	B	FORT MOTT	A
EGAN	B	EMMONS	C	EWAIL	A	FETTIC	D	FORT PIERCE	C
EGBERT	B/C	EMORY	B	EWINGSVILLE	B	FIANDER	C	FORT ROCK	C
EGELAND	B	EMPEY	B	EXCHEQUER	D	FIBEA	D	FORTUNA	D

(continued)

EXHIBIT 2-1 (continued)

EGGLESTON	B	EMPEYVILLE	C	EXETER	C	FIDALGO	C	FORTWINGATE	C
EGNAR	C	EMPIRE	C	EXLINE	D	FIDDLETOWN	C	FORWARD	C
EICKS	C	EMRICK	B	EXRAY	D	FIDDYMENT	C	FOSHOME	B
EIFORT		ENCE	B	EXUM	C	FIELDING	B	FOSSUM	B
EKAH	C	ENCIERRO	D	EYERBCW	D	FIELDON	B	FOSTER	B/C
EKALAKA	B	ENCINA	B	EYRE	B	FIELDSON	A	FOSTORIA	B
ELAM	A	ENDERS	C			FIFE	B	FOUNTAIN	D
ELBERT	D	ENDICOTT	B	FABIUS	B	FIFER	D	FOURLOG	D
ELBURN	B	ENET	B	FACEVILLE	B	FILLMORE	D	FOURMILE	B
ELCO	B	ENFIELD	B	FAHEY	B	FINCASTLE	C	FOUR STAR	B/C
ELD	B	ENGLE	B	FAIM	C	FINGAL	C	FOUTS	B
ELDER	B	ENGLESIDE	B	FAINES	A	FINLEY	B	FOX	B
ELDER HOLLOW	D	ENGLEWOOD	C	FAIRBANKS	B	FIRESTEEL	B	FOXCREEK	C
ELDERON	B	ENGLUND	D	FAIRDALE	B	FIRGRELL	B	FOXMOUNT	C
ELDON	B	ENNIS	B	FAIRFAX	B	FIRMAGE	B	FOXOL	D
ELDORADO	C	ENOCHVILLE	B/D	FAIRFIELD	B	FIRO	C	FOXPARK	B
ELDRIDGE	C	ENOLA	B	FAIRHAVEN	B	FIRTH	B	FOXTON	C
ELEPHANT	D	ENON	C	FAIRMOUNT	D	FISH CREEK	B	FRAILEY	C
ELFROY	B	ENOS	B	FAIRPORT	C	FISHERS	B	FRAM	B
ELFRIDA	B	ENOSBURG	D	FAJARDO	C	FISHHOOK	D	FRANCIS	A
ELIJAH	C	ENSIGN	D	FALAYA	C	FISHKILL		FRANK	D
ELIOAK	C	ENSLEY	D	FALCON	C	FITCH	A	FRANKFORT	D
ELK	B	ENSTROM	B	FALFURRIAS	A	FITCHVILLE	C	FRANKIRK	C
ELKADER	B	ENTERPRISE	B	FALK	B	FITZGERALD	B	FRANKLIN	B
ELKCREEK	C	ENTIAT	D	FALKNER	C	FITZHUGH	B	FRANKSTOWN	B
ELK HOLLOW	B	ENUMCLAW	C	FALL	B	FIVE DOT	B	FRANKTOWN	D
ELKHORN	B	EPHRAIM	C	FALLBROCK	B	FIVEMILE	B	FRANKVILLE	B
ELKINS	D	EPHRATA	B	FALLON	C	FIVES	B	FRATERNIDAD	D
ELKINSVILLE	B	EPOUFETTE	D	FALLSBURG		FLAGG	B	FRAZER	C
ELKMOUND	C	EPPING	D	FALLSINGTON	D	FLAGSTAFF	C	FRED	C
ELK MOUNTAIN	B	EPSIE	C	FANCHER	C	FLAK	B	FREDENSBORG	C
ELKTON	D	ERA	B	FANG	C	FLAMING	B	FREDERICK	B
ELLABELLE	B/D	ERAM	C	FANNIN	B	FLAMINGO	D	FREDON	C
ELLEDGE	D	ERBER	C	FANNO	C	FLANAGAN	B	FREDONIA	C
ELLERY	D	ERIC	B	FANU	C	FLANDREAU	B	FREDRICKSON	C
ELLETT	D	ERIE	C	FARADAY	C	FLASHEP	A	FREEBURG	C
ELLIBER	A	ERIN	B	FARALLONE	B	FLATHEAD	A	FREECE	D
ELLICOTT	A	ERNEST	C	FARAWAY	D	FLAT HORN	B	FREEHOLD	B
ELLINGTON	B	ERRAMOUSPE	C	FARGO	D	FLATTOP	D	FREEL	B
ELLINOR	B	ESCAL	B	FARISTA	B	FLAXTON	A	FREEMAN	C
ELLIOTT	C	ESCALANTE	B	FARLAND	B	FLEAK	A	FREEMANVILLE	B
ELLIS	D	ESCAMBIA	C	FARMINGTON	C/D	FLECHADO	C	FREEON	B
ELLISFORDE	C	ESCONDIDO	C	FARNUF	B	FLEETWOOD		FREER	C

FREESTONE	C
FREEZENER	B
FREMONT	C
FRENCH	C
FRENCHTOWN	D
FRENEAU	
FRESNO	C
FRIANA	D
FRIANT	D
FRIDLO	C
FRIEDMAN	B
FRIES	D
FRIO	B
FRIZZELL	C
FROBERG	D
FROHMAN	C
FRONHOFER	C
FRONTON	D
FROST	D
FRUITA	B
FRUITLAND	B
FRYE	D
FUEGO	C
FUERA	C
FULDA	C
FULLERTON	B
FULMER	B/D
FULSHEAR	C
FULTON	D
FUQUAY	B
FURNIS	B/D
FURY	B/D
GAASTRA	C
GABALDON	C
GABICA	D
GACEY	D
GADDES	C
GADES	C
GADSDEN	D
GAGE	
GAGEBY	B
GAGETOWN	C
GAHEE	B
GAINES	C
GAINESVILLE	A
GALATA	D
GALE	B
GALEN	B
GALENA	C
GASCONADE	D
GAS CREEK	C
GASKELL	C
GASS	D
GASSET	D
GATESBURG	A
GATEVIEW	B
GATEWAY	C
GATEWOOD	D
GAULDY	B
GAVINS	C
GAVIOTA	D
GAY	D
GAYLORD	B
GAYNOR	C
GAYVILLE	B
GAZELLE	D
GAZOS	B
GEARHART	A
GEARY	B
GEE	B
GEEBURG	C
GEER	C
GEFO	A
GELKIE	B
GEM	C
GEMID	C
GEMSON	C
GENESEE	B
GENEVA	C
GENOA	D
GENOLA	B
GEORGEVILLE	B
GEORGIA	B
GERALD	D
GERBER	D
GERIG	B
GERING	B
GEPLAND	C
GERMANIA	
GERMANY	B
GESTRIN	B
GETTA	C
GETTYS	C
GEYSEN	D
GHENT	C
GIBBLER	C
GIBBON	B
GIBBS	D
GIBBSTOWN	A
GLENFIELD	D
GLENFORD	C
GLENHALL	B
GLENHAM	B
GLENMORA	C
GLENNALLEN	C
GLENOMA	B
GLENROSE	B
GLENSTED	D
GLENTON	B
GLENVIEW	B
GLENVILLE	C
GLIDE	B
GLIKON	B
GLORIA	C
GLOUCESTER	A
GLOVER	C/D
GLYNDON	B
GLYNN	C
GOBLE	C
GODDARD	B
GODDE	D
GODECKE	D
GODFREY	C
GODWIN	D
GOEGLEIN	C
GOESSEL	D
GOFF	C
GOGEBIC	B
GOLBIN	C
GOLCONDA	D
GOLDENDALE	B
GOLDFIELD	B
GOLDHILL	B
GOLDMAN	C
GOLDRIDGE	B
GOLDRUN	A
GOLDSBORO	C
GOLDSTON	C
GOLDSTREAM	D
GOLDVALL	C
GOLDVEIN	C
GOLIAD	C
GOLLAHER	A
GOMEZ	B
GONVICK	B
GOOCH	D
GOODALE	C
GOODING	C
GOODINGTON	C
GRANGER	C
GRANGEVILLE	B/C
GRANILE	B
GRANO	D
GRANT	B
GRANTSBURG	C
GRANTSDALE	A
GRANVILLE	B
GRAPEVINE	C
GRASMERE	B
GRASSNA	B
GRASSY BUTTE	A
GRATZ	C
GRAVDEN	C
GRAVE	B
GRAVITY	C
GRAYCALM	A
GRAYFORD	B
GRAYLING	A
GRAYLOCK	B
GRAYPOINT	B
GRAYS	B
GREAT BEND	B
GREELEY	B
GREEN BLUFF	B
GREEN CANYON	B
GREENCREEK	B
GREENDALE	B
GREENFIELD	B
GREENHORN	D
GREENLEAF	B
GREENOUGH	C
GREENPORT	
GREEN RIVER	B
GREENSBORO	
GREENSON	C
GREENTON	C
GREENVILLE	B
GREENWATER	A
GREENWICH	B
GREENWOOD	D
GREER	C
GREGORY	A
GRELL	C
GRENADA	C
GRENVILLE	B
GRESHAM	C
GREWINGK	C
GREYBACK	B
GREYBULL	C
GUNN	B
GUNTER	A
GURABO	D
GURNEY	C
GUSTAVUS	B
GUSTIN	C
GUTHRIE	D
GUYTON	D
GWIN	D
GWINNETT	B
GYMER	C
HACOKE	C
HACIENDA	C
HACK	B
HACKERS	B
HACKETTSTOWN	B
HADLEY	B
HADO	B
HAGEN	B
HAGENBARTH	B
HAGENER	A
HAGER	C
HAGERMAN	B
HAGERSTOWN	C
HAGGA	B
HAIG	C
HAIKU	B
HAILMAN	B
HAINES	B/C
HAIRE	C
HALAWA	B
HALDER	C
HALE	B
HALEIWA	B
HALEY	B
HALF MOON	B
HALFORD	A
HALFWAY	D
HALII	B
HALIIMAILE	B
HALIS	B
HALL	B
HALLECK	B
HALL RANCH	C
HALLVILLE	B
HALSEY	D
HAMAKUAPOKO	B
HAMAN	B
HAMAR	B

(continued)

EXHIBIT 2-1 (continued)

GALEPPI	C	GIFFIN	C	GOODLOW	B	GREYCLIFF	C	HAMBLEN	C
GALESTOWN	A	GIFFORD	C	GOODMAN	B	GRIFFY	B	HAMBRIGHT	D
GALEY	B	GILA	C	GOODRICH	B	GRIGSTON	B	HAMBURG	B
GALISTEO	D	GILBY	B	GOODSPRINGS	D	GRIMSTAD	B	HAMEL	C
GALLAGHER	B	GILCHRIST	B	GOOSE CREEK	B	GRISWOLD	B	HAMERLY	C
GALLATIN	A	GILCREST	B	GOOSE LAKE	C	GRIVER	C	HAMILTON	A
GALLEGOS	B	GILEAD	C	GOOSMUS	B	GRIZZLY	C	HAMLET	B
GALLINA	C	GILES	B	GORDO	C	GROGAN	B	HAMLIN	B
GALLION	B	GILFORD	B/D	GORE	D	GROSECLOSE	C	HAMPDEN	
GALVA	B	GILHOULY	B	GORGONIO	A	GROSS	C	HAMPSHIRE	C
GALVESTON	A	GILISPIE	C	GORHAM	B	GROTON	A	HAMPTON	C
GALVIN	C	GILLIAM	C	GORIN	C	GROVE	A	HAMTAH	C
GAMBLER	A	GILLIGAN	B	GORING	C	GROVELAND	B	HANA	A
GANNETT	D	GILLS	C	GORMAN	B	GROVER	B	HANALEI	C
GANSNER	D	GILMORE	D	GORUS	A	GROVETON	B	HANAMAULU	A
GAPO	D	GILPIN	C	GORZELL	B	GRUBBS	D	HANCEVILLE	B
GAPPMAYER	B	GILROY	C	GOSHEN	B	GRULLA	D	HAND	B
GARA	B	GILSON	B	GOSHUTE	D	GRUMMIT	D	HANDFORD	B
GARBER	B	GILT EDGE	C	GOSPORT	C	GRUNDY	C	HANEY	B
GARBUTT	B	GINAT	D	GOTHAM	A	GRUVER	C	HANGAARD	C
GARCENO	C	GINGER	C	GOTHARD	D	GRYGLA	C	HANGER	B
GARDENA	B	GINI	B	GOTHIC	C	GUADALUPE	B	HANIPOE	B
GARDINER	A	GINSER	C	GOTHO	C	GUAJE	A	HANKINS	C
GARDNERS FORK	B	GIRD	A	GOULDING	D	GUALALA	D	HANKS	B
GARDNERVILLE	D	GIVEN	C	GOVAN	C	GUAMANI	B	HANLY	A
GARCONE	A	GLADDEN	A	GOVE	B	GUANAJIBO	C	HANNA	B
GAREY	C	GLADSTONE	B	GOWEN	B	GUANICA	D	HANOVER	C
GARFIELD	C	GLADWIN	A	GRABE	B	GUAYABO	B	HANS	C
GARITA	C	GLAMIS	C	GRABLE	B	GUAYABOTA	D	HANSEL	C
GARLAND	B	GLANN	B/C	GRACEMONT	B	GUAYAMA	D	HANSKA	C
GARLET	A	GLASGOW	C	GRACEVILLE	B	GUBEN	B	HANSON	A
GARLOCK	C	GLEAN	B	GRACY	D	GUCKEEN	C	HANTHO	B
GARMON	C	GLEASON	C	GRAFTON	B	GUELPH	B	HANTZ	D
GARMORE	B	GLEN	B	GRAHAM	D	GUENOC	C	HAP	B
GARNER	D	GLENBERG	B	GRAIL	C	GUERNSEY	C	HAPGOOD	B
GARO	D	GLENBROOK	D	GRAMM	B	GUERRERO	C	HAPNEY	C
GARR	D	GLENCOE	C	GRANATH	B	GUEST	D	HARBORD	B
GARRARD	B	GLENDALE	B	GRANBY	A/D	GUIN	A	HARBOURTON	
GARRETSON	B	GLENDALE	B	GRANDE RONDE	D	GULER	B	HARCO	B
GARRETT	B	GLENDIVE	B	GRANDFIELD	B	GULKANA	B	HARDEMAN	B
GARRISON	B	GLENDORA	C	GRANDVIEW	C	GUMBOOT	C	HARDESTY	B
GARWIN	C	GLENFLG	B	GRANER	C	GUNBARREL	A	HARDING	D

HARDSCRABBLE	B	HEBER	B	HILDRETH	D	HONEYGROVE	C	HUMBARGER	B
HARDY	D	HEBERT	C	HILEA	D	HONEYVILLE	C	HUMBIRD	C
HARGREAVE	B	HEBGEN	A	HILES	B	HONN	B	HUMBOLDT	C
HARKERS	C	HEBO	D	HILGER	B	HONOKAA	A	HUMDUN	B
HARKEY	B	HEBRON	C	HILGRAVE	B	HONOLUA	B	HUME	C
HARLAN	B	HECHT	B	HILLEMANN	C	HONOMANU	B	HUMESTON	C
HARLEM	C	HECKI	C	HILLET	D	HONOULIULI	D	HUMPINGTON	C
HARLESTON	C	HECLA	B	HILLFIELD	B	HONUAULU	A	HUMPHREYS	B
HARLINGEN	D	HECTOR	D	HILLGATE	C	HOOD	B	HUMPTULIPS	B
HARMEHL	C	HEDDON	C	HILLIARD	B	HOODLE	B	HUNSAKER	B
HARMONY	C	HEDRICK	B	HILLON	B	HOODSPORT	B	HUNTERS	B
HARNEY	C	HEDVILLE	D	HILLSBORO	B	HOODVIEW	B	HUNTING	C
HARPETH	B	HEGNE	D	HILLSDALE	B	HOOKTON	C	HUNTINGTON	B
HARPS	B	HEIDEN	D	HILMAR	C/D	HOOLEHUA	B	HUNTSVILLE	B
HARPSTER	C	HEIDTMAN	C	HILO	A	HOOPAL	D	HUPP	B
HARPT	B	HEIL	D	HILT	B	HOOPER	D	HURLEY	D
HARQUA	B	HEIMDAL	B	HILTON	B	HOOPESTON	B	HURON	C
HARQUA	B	HEISETON	B	HINCKLEY	A	HOOSIC	A	HURST	D
HARRIET	D	HEISLER	B	HINDES	C	HOOT	D	HURWAL	B
HARRIMAN	B	HEIST	B	HINESBURG	C	HOOTEN	D	HUSE	C
HARRIS	D	HEITT	C	HINKLE	D	HOOVER	B	HUSSA	B/D
HARRISBURG	D	HEITZ	D	HINMAN	C	HOPETON	C	HUSSMAN	D
HARRISON	C	HEIZER	D	HINSDALE		HOPEWELL		HUTCHINSON	C
HARRISVILLE	C	HELDT	C	HINTZE	D	HOPGOOD	C	HUTSON	B
HARSTENE	B	HELEMANO	C	HISLE	D	HOPKINS	B	HUXLEY	D
HART	D	HELENA	C	HITT	B	HOPLEY	B	HYAM	D
HART CAMP	C	HELMER	C	HI VISTA	C	HOPPER	B	HYATT	A
HARTFORD	A	HELVETIA	C	HIWASSEE	B	HOQUIAM	B	HYATTVILLE	B
HARTIG	B	HELY	B	HIWOOD	A	HORATIO	D	HYDABURG	D
HARTLAND	B	HEMBRE	B	HIXTON	B	HORD	B	HYDE	D
HARTLETON	B	HEMMI	C	HOBACKER	B	HOREB	B	HYDRO	C
HARTLINE	B	HEMPFIELD		HOBAN	C	HORNELL	D	HYMAS	D
HARTSBURG	B	HEMPSTEAD	C	HOBBS	B	HORNING	A	HYRUM	B
HARTSELLS	B	HENCRATT	B	HOBSON	C	HORNITOS	D	HYSHAM	D
HARTSHORN	B	HENDERSON	B	HOCHHEIM	B	HORROCKS	B		
HARVARD	B	HENDRICKS	B	HOCKING	B	HORSESHOE	B	IAO	C
HARVEL	B	HENEFER	C	HOCKINSON	C	HORTON	B	IBERIA	D
HARVEY	C	HENKIN	B	HOCKLEY	C	HORTONVILLE	B	ICENE	C
HASKILL	A	HENLEY	C	HODGE	B	HOSKIN	C	IDA	B
HASKINS	C	HENLINE	C	HODGINS	C	HOSLEY	D	IDABEL	B
HASSELL	D	HENNEKE	C	HODGSON	C	HOSMER	C	IDANA	C
HASTINGS	B	HENNEPIN	B	HOEBE	B	HOTAW	C	IDEON	D
HAT	B	HENNINGSEN	C	HOELZLE	C	HOT LAKE	C	IDMON	B
HATBORO	D	HENRY	D	HOFFMAN	C	HOUDEK	B	IGNACIO	B
HATCH	C	HENSEL	B	HOFFMANVILLE	C	HOUGHTON	A/D	IGO	D
HATCHERY	C	HENSHAW	C	HOGANSBURG	B	HOUK	C	IGUALDAD	D
HATFIELD	C	HENSLEY	D	HOGELAND	B	HOULKA	D	IHLEN	D
HATHAWAY	B	HEPLER	D	HOGG	D	HOULTON	C/D	IJAM	C
HATTIE	C	HERBERT	B	HOGRIS	B	HOUNDBY	D	ILDEFONSO	B

(continued)

EXHIBIT 2-1 (continued)

HATTON	C	HEREFORD	B	HOCH	B	HOURGLASS	B	ILKA	B
HAUBSTADT	C	HERKIMER	B	HOHMANN	C	HOUSATONIC	D	ILLION	B/D
HAVANA	B	HERLONG	D	HOKO	C	HOUSE MOUNTAIN	D	IMA	B
HAVEN	B	HERMISTON	B	HOLBROOK	B	HOUSEVILLE	C	IMBLER	B
HAVERLY	B	HERMON	A	HOLCOMB	D	HOUSTON	D	IMLAY	C
HAVERSON	B	HERNDON	B	HOLDAWAY	D	HOUSTON BLACK	D	IMMOKALEE	B/D
HAVILLAH	B	HERO	B	HOLDEN	A	HOVDE	A/C	IMPERIAL	C
HAVINGDON	C	HERRERA	A	HOLDERNESS	C	HOVEN	D	INAVALE	A
HAVRE	B	HERRICK	C	HOLDREGE	B	HOVENWEEP	C	INDIAHOMA	C
HAVRELON	B	HERRON	B	HOLLAND	B	HOVERT	D	INDIAN	
HAW	B	HERSH	A	HOLLINGER	B	HOVEY	C	INDIAN CREEK	D
HAWES	A	HERSHAL	B/D	HOLLIS	C/D	HOWARD	B	INDIANO	C
HAWI	B	HESCH	B	HOLLISTER	D	HOWELL	C	INDIANOLA	A
HAWKEYE	A	HESPER	C	HOLLOMAN	C	HOWLAND	C	INDIO	B
HAWKSELL	A	HESPERIA	B	HOLLOWAY	A	HOYE	B	INGA	B
HAWKSPRINGS	B	HESPERUS	B	HOLLY	D	HOYLETON	C	INGALLS	B
HAXTUN	A	HESSE	C	HOLLY SPRINGS	D	HOYPUS	A	INGARD	B
HAYBOURNE	B	HESSEL	D	HOLLYWOOD	D	HOYTVILLE	D	INGENIO	C
HAYBRO	C	HESSELBERG	C	HOLMDEL	C	HUBBARD	A	INGRAM	D
HAYDEN	B	HESSELTINE	B	HOLMES	B	HUBER	D	INKLER	B
HAYESTON	B	HESSON	C	HOLOMUA	B	HUBERT	B	INKS	D
HAYESVILLE	B	HETTINGER	D	HOLOPAW	B/D	HUBLERSBURG	C	INMAN	C
HAYFIELD	B	HEXT	B	HOLROYD	B	HUCKLEBERRY	C	INMO	A
HAYFORD	C	HEZEL	B	HOLSINE	B	HUDSON	C	INSKIP	C
HAYMOND	B	HIALEAH	D	HOLST	B	HUECO	C	INVERNESS	D
HAYNESS	D	HIAWATHA	A	HOLSTON	B	HUEL	A	INWOOD	C
HAYNIE	B	HIBBARD	C	HOLT	B	HUENEME	B/C	IO	B
HAYPRESS	A	HIBBING	C	HOLTLE	B	HUERFUERO	D	IOLA	A
HAYSPUR	B/D	HICKORY	C	HOLTVILLE	C	HUEY	D	IOLEAU	C
HAYTER	B	HICKS	B	HOLYOKE	C/D	HUFFINE	A	IONA	B
HAYTI	D	HIDALGO	B	HOMA	C	HUGGINS	C	IONIA	B
HAYWOOD	B	HIDEAWAY	D	HOME CAMP	C	HUGHES	C	IOSCO	B
HAZEL	C	HIDEWOOD	C	HOMELAKE	B	HUGHESVILLE	B	IPAVA	B
HAZELAIR	D	HIGHAMS	D	HOMER	C	HUGO	B	IRA	C
HAZEN	B	HIGHFIELD	B	HOMESTAKE	D	HUICHICA	C	IREDELL	D
HAZLEHURST	C	HIGH GAP	C	HOMESTEAD	B	HUIKAU	A	IRETEBA	C
HAZLETON	B	HIGHLAND	B	HONAUNAU	C	HULETT	B	IRIM	C
HEADLEY	B	HIGHMORE	B	HONCUT	B	HULLS	C	IROCK	B
HEADQUARTERS	B	HIGH PARK	B	HONDALE	D	HULLT	B	IRON BLOSSOM	D
HEAKE	D	HIHIMANU	A	HONDO	C	HULUA	D	IRON MOUNTAIN	D
HEATH	C	HIIBNER	C	HONCHO	B	HUM	B	IRON RIVER	B
HEATLY	A	HIKO PEAK	B	HONEOYE	B	HUMACAO	B	IRONTON	C
HEBBRONVILLE	B	HIKO SPRINGS	C	HONEY	D	HUMATAS	C	IRVINGTON	C

IRWIN	D
ISAAC	C
ISAAQUAH	B/C
ISABELL	C
ISAN	D
ISANTI	D
ISHAM	C
ISHI PISHI	C
ISLAND	B
ISOM	B
ISTOKPOGA	D
ITSWOOT	B
IUKA	C
IVA	C
IVAN	B
IVES	B
IVIE	A
IVINS	C
IZAGORA	C
IZEE	C
JABU	C
JACAGUAS	B
JACANA	D
JACINTO	B
JACK CREEK	B
JACKLIN	B
JACKNIFE	C
JACKS	D
JACKSON	B
JACKSONVILLE	C
JACOB	D
JACOBSEN	D
JACORY	C
JACQUES	C
JACQUITH	C
JACWIN	B
JAFFREY	A
JAGUEYES	B
JAL	B
JAMES CANYON	B/C
JAMESTOWN	C
JANE	C
JANISE	C
JANSEN	A
JARBOE	C
JARITA	C
JARRE	B
JARVIS	B

JONUS	B
JOPLIN	B
JOPPA	B
JORDAN	D
JORNADO	B
JORY	C
JOSE	C
JOSEPHINE	B
JOSIE	B
JOY	B
JUBILEE	C
JUDD	D
JUDITH	B
JUDKINS	C
JUDSON	B
JUDY	C
JUGET	D
JUGHANDLE	B
JULES	B
JULESBURG	A
JULIAETTA	B
JUMPE	B
JUNCAL	C
JUNCOS	D
JUNCTION	B
JUNEAU	B
JUNIATA	
JUNIUS	C
JUNO	B
JUNQUITOS	C
JURA	C
JUVA	B
JUVAN	D
KAALUALU	A
KACHEMAK	B
KADASHAN	B
KADE	C
KADOKA	B
KAUCKA	B
KAENA	D
KAHALUU	D
KAHANA	B
KAHANUI	B
KAHLER	B
KAHOLA	B
KAH SHEETS	D
KAHUA	D
KAIKLI	D

KARLO	D
KARLUK	C
KARNAK	D
KARNES	B
KARRO	B
KARS	A
KARTA	C
KARTAR	B
KASHWITNA	B
KASILOF	A
KASKI	B
KASOTA	C
KASSLER	A
KASSON	C
KATAMA	B
KATEMCY	C
KATO	C
KATRINE	B
KATULA	B
KATY	C
KAUFMAN	D
KAUPO	A
KAVETT	D
KAWAIHAE	C
KAWAIHAPAI	B
KAWBAWGAM	C
KAWICH	A
KAWKAWLIN	C
KEAAU	D
KEAHUA	B
KEALAKEKUA	C
KEALIA	D
KEANSBURG	D
KEARNS	B
KEATING	C
KEAUKAHA	D
KEAWAKAPU	B
KEBLER	B
KECH	C
KECKO	B
KEDRON	C
KEEFERS	C
KEEGAN	
KEEI	D
KEEKEE	B
KEENE	C
KEENO	C
KEG	B
KEHENA	C

KERMIT	A
KERR	B
KERRICK	B
KERRTOWN	
KERSHAW	A
KERSICK	D
KERSTON	A/D
KERT	C
KERWIN	C
KESSLER	C
KESWICK	D
KETCHLY	B
KETTLE	B
KETTLEMAN	B
KETTNER	C
KEVIN	C
KEWAUNEE	C
KEWEENAW	A
KEYA	B
KEYES	D
KEYPORT	C
KEYTESVILLE	D
KEZAR	B
KIAWAH	C
KIBBIE	B
KICKERVILLE	B
KIDD	C
KIDMAN	B
KIEHL	A
KIEV	B
KIKONI	B
KILARC	D
KILAUEA	B
KILBOURNE	A
KILBURN	A
KILCHIS	C
KILDOR	C
KILGORE	B/D
KILKENNY	B
KILLBUCK	C/D
KILLEY	B
KILLINGWORTH	
KILLPACK	C
KILMERQUE	C
KILOA	A
KILOHANA	A
KILWINNING	C
KIM	B
KIMAMA	B

KITSAP	C
KITTANNING	
KITTITAS	B
KITTREDGE	C
KITTSON	C
KIUP	B
KIVA	B
KIWANIS	A
KIZHUYAK	B
KJAR	D
KLABER	C
KLAMATH	B/D
KLAUS	A
KLAWASI	C
KLEJ	B
KLICKER	C
KLICKITAT	C
KLINE	B
KLINESVILLE	C/D
KLINGER	B
KLONDIKE	D
KLONE	B
KLOOCHMAN	B
KLOTEN	B
KLUTINA	B
KNAPPA	B
KNEELAND	C
KNIFFIN	C
KNIGHT	C
KNIK	B
KNIPPA	D
KNOB HILL	B
KNOWLES	B
KNOX	B
KNULL	B
KNUTSEN	A
KUBAR	C
KOCH	C
KODAK	C
KODIAK	B
KOEHLER	C
KOELE	B
KOEPKE	B
KOERLING	B
KOGISH	D
KOHALA	A
KOKEE	B
KOKO	B
KOKOKAHI	D

(continued)

EXHIBIT 2-1 (continued)

JASPER	B	KAILUA	A	KEIGLEY	C	KIMBALL	C	KOKOMO	B/D
JAUCAS	A	KAIMU	A	KEISER	B	KIMBERLY	B	KOLBERG	B
JAVA	B	KAINALIU	A	KEITH	B	KIMBROUGH	C	KOLEKOLE	C
JAY	C	KAIPOIOI	B	KEKAHA	B	KIMMERLING	C	KOLLS	D
JAYEM	B	KAIWIKI	A	KEKAKE	D	KIMO	C	KOLOA	C
JAYSON	D	KALAE	B	KELLER	C	KINA	D	KOLOKOLO	B
JEAN	A	KALAMA	C	KELLY	D	KINCO	A	KONA	D
JEANERETTE	C	KALAMAZOO	B	KELN	C	KINGFISHER	B	KONAWA	B
JEAN LAKE	B	KALAPA	B	KELSEY	C	KINGHURST	B	KONNER	C
JEDDO	D	KALAUPAPA	D	KELSO	C	KINGMAN	C	KONOCTI	C
JEFFERSON	B	KALIFONSKY	C	KELTNER	B	KINGS	C/D	KOOLAU	C
JEKLEY	C	KALIHI	D	KELVIN	C	KINGSBURY	D	KOOSKIA	C
JELM	D	KALISPELL	A	KEMOO	B	KINGSLEY	B	KOOTENAI	A
JENKINS	B	KALKASKA	A	KEMPSVILLE	B	KINGS RIVER	C	KOPIAH	C
JENKINSON	D	KALMIA	B	KEMPTON	B	KINGSTON	B	KOPP	B
JENNESS	B	KALOKO	D	KENAI	C	KINGSVILLE	C	KOPPES	B
JENNINGS	C	KALOLOCH	E	KENANSVILLE	A	KINKEAD	D	KORCHEA	B
JENNY	D	KALSIN	C	KENDAIA	C	KINKEL	B	KORNMAN	B
JERAULD	D	KAMACK	B	KENDALL	B	KINKORA	D	KOSMOS	C
JERICHO	C	KAMAKOA	A	KENDALLVILLE	B	KINMAN	C	KOSSE	D
JEROME	B	KAMAOA	B	KENESAW	B	KINNEY	B	KOSTER	C
JERRY	C	KAMAOLE	B	KENMOOR	B	KINNICK	C	KOSZTA	B
JESBEL	D	KAMRAR	B	KENNALLY	B	KINREAD	D	KOUTS	B
JESSE CAMP	C	KANABEC	B	KENNAN	B	KINROSS	D	KOVICH	C
JESSUP	C	KANAKA	B	KENNEBEC	B	KINSTON	C	KOYEN	B
JETT	B	KANAPAHA	A/D	KENNEDY	B	KINTON	C	KOYUKUK	B
JIGGS	C	KANDIK	B	KENNEWICK	B	KINZEL	B	KRADE	B
JIM	C	KANE	B	KENNEY	A	KIONA	B	KRANZBURG	B
JIMENEZ	C	KANEOHE	B	KENNEY LAKE	C	KIPLING	D	KRATKA	C
JIMTOWN	C	KANEPUU	B	KENO	D	KIPP	C	KRAUSE	A
JOB	C	KANLEE	B	KENOMA	D	KIPPEN	B	KREAMER	
JOBES	C	KANOSH	C	KENSAL	B	KIPSON	C	KREMLIN	B
JOCITY	B	KANZA	D	KENSPUR	A	KIRK	B/D	KRENTZ	B
JOEL	B	KAPAA	A	KENT	D	KIRKHAM	C	KRESSON	C
JOES	B	KAPAPALA	B	KENYON	C	KIRKLAND	D	KRUM	D
JOHNS	C	KAPOO	B	KEO	B	KIRKTON	B	KRUSE	B
JOHNSBURG	D	KAPOWSIN	B	KEOLDAR	B	KIRTLEY	B	KRUZOF	B
JOHNSON	B	KAPUHIKANI	C	KEOMAH	C	KIRVIN	C	KUBE	B
JOHNSTON	B/D	KARAMIN	B	KEOTA	C	KISRING	D	KUBLER	C
JOHNSWOOD	B	KARDE	C	KEOWNS	D	KISSICK	D	KUBLI	C
JOICE	D	KARHEEN	D	KEPLER	C	KISTLER	C/D	KUCERA	B
JOLIET	C	KARLAN	C	KERBY	B	KITCHELL	B	KUCK	C
JONESVILLE	A	KARLIN	A	KERMEL	B	KITCHEN CREEK	B	KUHL	D

LUN	C	MALABAR	A/D	MARKSBORO		MAYFLOWER	C	MCVICKERS	C
LUNA	C	MALABON	C	MARLA	A	MAYHEW	D	MEAD	D
LUNCH	C	MALACHY	B	MARLBORO	B	MAYLAND	B	MEADIN	A
LUNDIMO	C	MALAGA	B	MARLEAN	B	MAYMEN	C	MEADOWVILLE	B
LUNDY	B	MALAMA	A	MARLETTE	B	MAYNARD LAKE	B	MEADVILLE	
LUNT	C	MALAYA	D	MARLEY	B	MAYO	B	MEANDER	D
LUPTON	D	MALCOLM	B	MARLIN	D	MAYODAN	B	MECAN	B
LURA	D	MALEZA	B	MARLOW	C	MAYOWORTH	D	MECCA	B
LURAY	C/D	MALIBU	D	MARLTON	C	MAYSDORF	B	MECKESVILLE	C
LUTE	D	MALIN	C/D	MARMARTH	B	MAYSVILLE		MECKLENBURG	C
LUTH	C	MALJAMAR	A	MARNA	D	MAYTOWN	C	MEDA	B
LUTHER	B	MALLOT	A	MARPA	B	MAYVILLE	B	MEDANO	C
LUTIE	B	MALM	C	MARQUETTE	A	MAYWOOD	B	MEDARY	C
LUTON	D	MALO	B	MARR	B	MAZEPPA	B	MEDFORD	D
LUVERNE	C	MALONE	B	MARRIOTT	B	MAZON	C	MEDFRA	D
LUXOR	C	MALOTERRE	D	MARSDEN	C	MAZUMA	C	MEDICINE LODGE	B
LUZENA	D	MALPAIS	C	MARSELL	B	MCAFEE	C	MEDINA	B
LYCAN	B	MALPOSA	C	MARSHALL	B	MCALLEN	B	MEDWAY	B
LYCOMING	C	MALVERN	C	MARSHAN	D	MCALLISTER	C	MEEKS	A
LYDICK	B	MAMALA	D	MARSHDALE	C	MCALPIN	C	MEETEETSE	C
LYFORD	C	MAMOU	C	MARSHFIELD	C	MCBEE	B	MEGGETT	D
LYLES	B	MANAHAA	C	MARSING	B	MCBETH	C	MEGON	C
LYMAN	C/D	MANALAPAN		MART	C	MCBRIDE	B	MEHL	C
LYNCH	D	MANANA	C	MARTELLA	B	MCCABE	B	MEHLHORN	C
LYNCHBURG	B/D	MANASSA	C	MARTIN	C	MCCAFFERY	A	MEIGS	
LYNDEN	A	MANASSAS	B	MARTINA	A	MCCAIN	C	MEIKLE	D
LYNNDYL	A	MANASTASH	C	MARTINECK	D	MCCALEB	B	MEISS	D
LYNN HAVEN	B/D	MANATEE	B/D	MARTINEZ	D	MCCALLY	D	MELBOURNE	B
LYNNVILLE	C	MANAWA	C	MARTINI	B	MCCAMMON	C	MELBY	C
LYNX	B	MANCELONA	A	MARTINSBURG	B	MCCARRAN	D	MELITA	B
LYONMAN	C	MANCHESTER	A	MARTINSDALE	B	MCCARTHY	B	MELLENTHIN	D
LYONS	D	MANDAN	B	MARTINSON	D	MCCLAVE	C	MELLOR	D
LYONSVILLE	B	MANDERFIELD	B	MARTINSVILLE	B	MCCLEARY	C	MELLOTT	B
LYSINE	D	MANDEVILLE	B	MARTINTON	C	MCCLELLAN	B	MELOLAND	C
LYSTAIR	B	MANFRED	C	MARTY	B	MCCLOUD	C	MELROSE	C
LYTELL	B	MANGUM	D	MARVAN	D	MCCOIN	D	MELSTONE	A
		MANHATTAN	A	MARVIN	C	MCCOLL	D	MELTON	B
MABEN	C	MANHEIM	C	MARY	C	MCCONNEL	B	MELVILLE	B
MABI	D	MANI	C	MARYDEL	B	MCCOOK	B	MELVIN	D
MABRAY	D	MANILA	C	MARYSLAND	D	MCCORMICK	B	MEMALOOSE	D
MACAY	B	MANISTEE	B	MASADA	C	MCCOY	C	MEMPHIS	B
MACEDONIA	C	MANITOU	C	MASCAMP	D	MCCREE	B	MENAHGA	A
MACHETE	C	MANLEY	B	MASCOTTE	D	MCCRORY	D	MENAN	C
MACHIAS	B	MANLIUS	C	MASHEL	B	MCCROSKIE	D	MENARD	B
MACK	C	MANLOVE	B	MASHULAVILLE	B/D	MCCULLOUGH	C	MENCH	C
MACKEN	D	MANNING	B	MASON	B	MCCULLY	C	MENDEBOURE	C
MACKINAC	B	MANOR	B	MASONVILLE	C	MCCUNE	D	MENDOCINO	B
MACKSBURG	B	MANSFIELD		MASSACK	B	MCCUTCHEN	C	MENDON	B
MACOMB	B	MANSIC	B	MASSENA	C	MCDOLE	B	MENDOTA	B

(continued)

EXHIBIT 2-1 (continued)

MACOMBER	B	MANSKER	B	MASSILLON	B	MCDONALD	B	MENEFEE	D
MACON	B	MANTACHIE	C	MASTERSON	B	MCDONALDSVILLE	C	MENFRO	B
MACY	B	MANTEO	C/D	MATAMOROS	C	MCEWEN	B	MENLO	D
MADALIN	D	MANTER	B	MATANUSKA	C	MCFADDEN	B	MENO	C
MADAWASKA	B	MANTON	B	MATANZAS	B	MCFAIN	C	MENOKEN	C
MADDOCK	A	MANTZ	B	MATAPEAKE	B	MCFAUL	C	MENOMINEE	B
MADDOX		MANU	C	MATAWAN	C	MCGAFFEY	C	MENTO	C
MADELIA	C	MANVEL	C	MATCHER	A	MCGARY	C	MENTOR	B
MADELINE	D	MANWOOD	D	MATFIELD	C	MCGEHEE	C	MEQUON	C
MADERA	D	MANZANITA	C	MATHERS	B	MCGILVERY	D	MERCED	C/D
MADISON	B	MANZANO	C	MATHERTON	B	MCGINTY	B	MERCEDES	D
MADONNA	C	MANZANOLA	C	MATHESON	B	MCGIRK	C	MERCER	C
MADRAS	C	MAPES	C	MATHEWS		MCGOWAN	B	MERCEY	C
MADRID	B	MAPLE MT.	B	MATHISTON	C	MCGRATH	B	MEREDITH	B
MADUREZ	B	MAPLETON	C/D	MATLOCK	D	MCGREW	A	MERETA	C
MAGALLON	B	MARATHON	B	MATMON	D	MCHENRY	B	MERGLE	B
MAGENS	B	MARBLE	A	MATTAPEX	C	MCILWAINE	A	MERIDIAN	B
MAGINNIS	C	MARBLEMOUNT	B	MATTOLE	C	MCINTOSH	B	MERINO	D
MAGNA	D	MARCETTA	A	MAUDE	B	MCINTYRE	B	MERKEL	B
MAGNOLIA	B	MARCUM	B	MAUGHAN	C	MCKAMIE	D	MERLIN	D
MAGNUS	C	MARCUS	C	MAUKEY	C	MCKAY	D	MERMILL	B/D
MAGUAYO	D	MARCY	D	MAUMEE	A/D	MCKENNA	C/D	MERNA	C
MAHAFFY	C/D	MARDEN	C	MAUNABO	D	MCKENZIE	D	MEROS	A
MAHALA	C	MARDIN	C	MAUPIN	C	MCKINLEY	C	MERRIFIELD	B
MAHALASVILLE	B/D	MARENGO	C/D	MAUREPAS	D	MCKINNEY	D	MERRILL	C
MAHANA	B	MARESUA	B	MAURINE	D	MCLAIN	C	MERRILLAN	C
MAHASKA	B	MARGERUM	B	MAURY	B	MCLAURIN	B	MERRIMAC	A
MAHER	C	MARGUERITE	B	MAVERICK	C	MCLEAN	C	MERRITT	B/C
MAHONING	D	MARIA	B/C	MAVIE	D	MCLEOD	B	MER ROUGE	B
MAHUKONA	B	MARIANA	C	MAWAE	A	MCMAHON	C	MERTON	B
MAIDEN	B	MARIAS	D	MAX	B	MCMEEN	C	MERTZ	
MAILE	A	MARICAO	B	MAXEY	C	MCMULLIN	C	MESA	B
MAJADA	B	MARICOPA	B	MAXFIELD	C	MCMURDIE	C	MESCAL	C
MAKAALAE	B	MARIETTA	C	MAXSON	A	MCMURPHY	B	MESCALERO	C
MAKALAPA	D	MARILLA	C	MAXTON	B	MCMURRAY	D	MESITA	C
MAKAPILI	A	MARINA	A	MAXVILLE	A	MCNARY	D	MESKILL	C
MAKAWAO	B	MARION	D	MAXWELL	D	MCPAUL	B	MESMAN	C
MAKAWELI	B	MARIPOSA	C	MAY	B	MCPHERSON	C	MESSER	C
MAKENA	B	MARISSA	C	MAYBERRY	C	MCPHIE	B	MET	D
MAKIKI	B	MARKES	D	MAYBESO	C	MCQUARRIE	D	METALINE	B
MAKOTI	C	MARKEY	D	MAY DAY	D	MCQUEEN	C	METAMORA	B
MAL	B	MARKHAM	C	MAYER	D	MCRAE	B	METEA	B
MALA	B	MARKLAND	C	MAYFIELD	B	MCTAGGART	B	METIGOSHE	A

METOLIUS	B	MISSION	B	MORGANFIELD	B	NABESNA	C	NESS	D
METRE	D	MITCH	B	MORGNEC		NACEVILLE	C	NESSEL	B
METZ	A	MITCHELL	B	MORIARTY	D	NACHES	B	NESSOPAH	B
MEXICO	D	MITIWANGA	C	MORICAL	C	NACIMIENTO	C	NESTER	C
MHOON	D	MIZPAH	C	MORLEY	C	NACOGDOCHES	B	NESTUCCA	C
MIAMI	B	MOANO	D	MORMON MESA	C	NADEAN	B	NETARTS	A
MIAMIAN	C	MOAPA	D	MOROCCO	A/C	NADINA	B	NETC	B
MICCO	A/D	MOAULA	A	MORONI	D	NAFF	B	NETTLETON	C
MICHELSON	B	MOBEETIE	B	MOROP	C	NAGEESI	B	NEUBERT	B
MICHIGAMME	C	MOCA	D	MORRILL	B	NAGITSY	C	NEUNS	B
MICK	B	MOCHO	B	MORRIS	C	NAGLE	B	NEUSKE	B
MIDAS	D	MODA	D	MORRISON	B	NAHMA	C	NEVADOR	C
MIDDLE	C	MODALE	C	MORROW	C	NAHUNTA	C	NEVILLE	B
MIDDLEBURY	B	MODEL	C	MORSE	D	NAIWA	B	NEVIN	C
MIDESSA	B	MODENA	B	MORTENSON	C	NAKAI	B	NEVINE	B
MIDLAND	D	MODESTO	C	MORTON	B	NAKNEK	D	NEVOYER	D
MIDNIGHT	C	MODOC	C	MORVAL	C	NAMBE	B	NEVTAH	C
MIDVALE	C	MOENKOPIE	D	MOSBY	C	NANAMKIN	A	NEVU	D
MIDWAY	D	MOFFAT	B	MOSCA	A	NANCY	B	NEWARK	C
MIFFLIN	B	MOGOLLON	B	MOSEL	C	NANNY	B	NEWART	B
MIFFLINBURG	B	MOGUL	B	MOSHANNON	B	NANNYTON	B	NEWAYGO	B
MIGUEL	C	MOHAVE	B	MOSHER	D	NANSENE	B	NEWBERG	B
MIKE	D	MOHAVE	B	MOSHERVILLE	C	NANTUCKET	C	NEWBERRY	C
MIKESELL	C	MOHAWK	B	MOSIDA	B	NANUM	C	NEWBY	B
MILACA	B	MOIRA	C	MOSQUET	D	NAPA	C	NEW CAMBRIA	C
MILAN	B	MOKELUMNE	D	MOSSYROCK	B	NAPIER	B	NEW CASTLE	B
MILES	B	MOKENA	C	MOTA	B	NAPLES	B	NEWCOMB	A
MILFORD	C	MOKULEIA	B	MOTTSVILLE	A	NAPPANEE	D	NEWDALE	B
MILHAM	C	MOLAND	B	MOULTON	B	NAPTOWNE	B	NEWELL	B
MILHEIM	C	MOLCAL	B	MOUND	C	NARANJITO	C	NEWELLTON	D
MILL	B	MOLENA	A	MOUNTAINBURG	D	NARANJO	C	NEWFANE	
MILLARD	B	MOLINOS	B	MOUNTAINVIEW	B/D	NARCISSE	B	NEWFORK	D
MILLBORO	D	MOLLY	B	MOUNTAINVILLE	B	NARD	B	NEWKIRK	B
MILLBROOK	B	MOLOKAI	B	MOUNT AIRY	A	NARLON	C	NEWLANDS	B
MILLBURNE	B	MOLSON	B	MOUNT CARROLL	B	NARON	B	NEWLIN	B
MILLCREEK	B	MOLYNEUX	B	MOUNT HOPE	B	NARRAGANSETT	B	NEWMARKET	B
MILLER	D	MONAD	A	MOUNT HOOD	B	NARROWS	D	NEWPORT	C
MILLERLUX	D	MONAHAN	D	MOUNT LUCAS	C	NASER	B	NEWRY	B
MILLERTON	D	MONAHANS	B	MOUNT OLIVE	D	NASH	B	NEWSKAH	B
MILLETT	B	MONARDA	D	MOUNTVIEW	B	NASHUA	A	NEWSTEAD	D
MILLGROVE	B/D	MONCLOVA	B	MOVILLE	C	NASHVILLE	B	NEWTON	A/D
MILL HOLLOW	B	MONDAMIN	C	MOWATA	D	NASON	C	NEWTONIA	B
MILLICH	D	MONDOVI	B	MOWER	C	NASSAU	C/D	NEWTOWN	C
MILLINGTON	B	MONEE	D	MOYINA	D	NASSET	B	NEWVILLE	C
MILLIS	C	MONICO	B	MUCARA	D	NATALIE	C	NEZ PERCE	C
MILLRACE	B	MONIDA	B	MUCET	C	NATCHEZ	B	NIAGARA	C
MILLSAP	C	MONITEAU	D	MUDRAY	D	NATHROP	B	NIART	B
MILLSDALE	B/D	MONMOUTH	C	MUD SPRINGS	C	NATIONAL	B	NIBLEY	C
MILLSHOLM	C	MONO	D	MUGHOUSE	C	NATRONA	B	NICHOLSON	C

(continued)

EXHIBIT 2-1 (continued)

MILLVILLE	B	MONOLITH	C	MUIR	B	NATURITA	B	NICHOLVILLE	C
MILLWOOD	D	MONONA	B	MUIRKIRK	B	NAUKATI	D	NICKEL	B
MILNER	C	MONONGAHELA	C	MUKILTEO	D	NAUMBURG	C	NICKLE	B
MILPITAS	C	MONROE	B	MULDROW	D	NAVAJO	D	NICODEMUS	B
MILROY	D	MONROEVILLE	C/D	MULKEY	C	NAVAN	D	NICOLAUS	C
MILTON	C	MONSE	B	MULLINS	D	NAVARCO	B	NICOLLET	B
MIMBRES	C	MONSERATE	C	MULT	C	NAVESINK		NIELSEN	D
MIMOSA	C	MONTAGUE	D	MULTORPOR	A	NAYLOR		NIGHTHAWK	B
MINAM	B	MONTALTO	C	MUMFORD	B	NAZ	B	NIHILL	B
MINATARE	D	MONTARA	D	MUNDELEIN	B	NEAPOLIS	B/D	NIKISHKA	B
MINCHEY	B	MONTAUK	C	MUNISING	B	NEBEKER	C	NIKLASON	B
MINCO	B	MONTCALM	A	MUNK	C	NEBISH	B	NIKOLAI	D
MINDALE	B	MONTE CRISTO	D	MUNSON	D	NEBO		NILAND	C
MINDEGO	B	MONTEGRANDE	D	MUNUSCONG	D	NECHE	C	NILES	C
MINDEMAN	B	MONTELL	D	MURDO	B	NEDERLAND	B	NIMROD	C
MINDEN	C	MONTELLO	C	MURDOCK	C	NEEDHAM	D	NINCH	C
MINE	B	MONTEVALLO	D	MUREN	B	NEEDLE PEAK	C	NINEMILE	D
MINEOLA		MONTGOMERY	D	MURRILL	B	NEEDMORE	C	NINEVEH	B
MINER	D	MONTICELLO	B	MUSCATINE	B	NEELEY	B	NINIGRET	B
MINERAL	A	MONTIETH	A	MUSE	C	NEGITA	B	NININGER	B
MINERAL MT.	C	MONTMORENCI	B	MUSELLA	B	NEGLEY	B	NINNESCAH	B
MINERVA	B	MONTOSA	C	MUSICK	B	NEHALEM	B	NIOBELL	C
MING	B	MONTOUR	D	MUSINIA	B	NEILTON	A	NIOTA	D
MINGO	B	MONTOYA	D	MUSKINGUM	C	NEISSON	B	NIPE	B
MINIDOKA	C	MONTPELLIER	C	MUSKOGEE	C	NEKIA	C	NIPPERSINK	B
MINNEISKA	C	MONTROSE	B	MUSSELSHELL	B	NELLIS	B	NIPSUM	C
MINNEOSA	A	MONTVALE	D	MUSSEY	D	NELSCOTT	B	NIRA	B
MINNEQUA	B	MONTVERDE	A/D	MUSTANG	A/D	NELSON	B	NISHNA	C
MINNETONKA	D	MONTWELL	C	MUTNALA	B	NEMAH	C	NISHON	D
MINNEWAUKAN	B	MOODY	B	MUTUAL	B	NENANA	B	NISQUALLY	A
MINNIECE	D	MOOHOO	B	MYAKKA	A/D	NENNO	B	NISSWA	B
MINOA	C	MOOSE RIVER	C	MYATT	B/D	NEOLA	D	NIU	B
MINORA	C	MOPA	B	MYERS	D	NEOTOMA	B	NIULII	C
MINTO	B	MORADO	C	MYERSVILLE	B	NEPESTA	C	NIVLOC	C
MINU	D	MORALES	D	MYLREA	B	NEPHI	B	NIWOT	C
MINVALE	B	MORD	C	MYRICK	D	NEPPEL	B	NIXA	C
MIRABAL	C	MOREAU	D	MYRTLE	B	NEPTUNE	A	NIXON	B
MIRACLE	B	MOREHEAD	C	MYSTEN	A	NERESON	B	NIXONTON	B
MIRAMAR	B	MOREHOUSE	C	MYSTIC	D	NESDA	A	NIZIAN	A
MIRANDA	D	MORELAND	D	MYTON	B	NESHAMINY	B	NOBLE	B
MIRES	B	MORELANDTON	A			NESIKA	B	NOBSCOTT	A
MIRROR	B	MOREY	D	N-BAR	B	NESKOWIN	B	NODAWAY	B
MIRROR LAKE	A	MORFITT	B	NAALEHU	B	NESPELEM	B	NOEL	D

NOHILI	D	OCILLA	C	ONSLOW	B	OWOSSO	B	PARALOMA	C
NOKASIPPI	D	OCKLEY	B	• ONTARIO	B	OWYHEE	B	PARAMORE	D
NOKAY	C	OCOEE	A/D	ONTKO	B/D	OXALIS	C	PARASOL	B
NOKOMIS	B	OCONEE	C	ONTONAGON	D	OXBOW	C	PARCELAS	D
NOLAM	B	OCONTO	B	ONYX	B	OXERINE	C	PARDEE	D
NOLICHUCKY	B	OCOSTA	C	OOKALA	A	OXFORD	D	PAREHAT	B
NOLIN	B	OCQUEOC	B	OPAL	D	OZAMIS	B/D	PARENT	C
NOLO	B	OCTAGON	B	OPEQUON	C/D	OZAN	D	PARIETTE	C
NOME	C	ODELL	B	OPHIR	C	OZAUKEE	C	PARIS	
NONDALTON	B	ODERMOTT	C	OPIHIKAO	D			PARISHVILLE	C
NONOPAHU	D	ODESSA	D	OQUAGA	C	PAAIKI	B	PARKAY	B
NOOKACHAMPS	C/D	ODIN	C	ORA	C	PAALOA	B	PARKDALE	B
NOOKSACK	B	ODNE	C	ORAN	B	PAAUHAU	A	PARKE	B
NOONAN	D	O'FALLON	D	ORANGE	D	PACHAPPA	B	PARKER	B
NORA	B	OGDEN	D	ORANGEBURG	B	PACHECO	B/C	PARKFIELD	C
NORAD	B	OGEECHEE	C	ORCAS	D	PACK	C	PARKHILL	D
NORBORNE	B	OGEMAW	C	ORCHARD	B	PACKARD	B	PARKHURST	
NORBY	B	OGILVIE	C	ORD	A	PACKER	C	PARKINSON	B
NORD	B	OGLALA	B	ORDNANCE	C	PACKHAM	B	PARKVILLE	C
NORDEN	B	OGLE	B	ORDWAY	D	PACKSADDLE	B	PARKWOOD	A/D
NORDNESS	B	OHAYSI	C	ORELIA	D	PACKWOOD	D	PARLEYS	B
NORFOLK	B	OHIA	A	ORELLA	D	PACOLET	B	PARLIN	C
NORGE	B	OJAI	B	OREM	A	PACTOLUS	C	PARLO	B
NORKA	B	OJATA	D	ORESTIMBA	C	PADEN	C	PARMA	C
NORMA	B	OKANOGAN	B	ORFORD	C	PADRONI	B	PARNELL	D
NORREST	C	OKAW	D	ORIDIA	C	PADUCAH	B	PARR	B
NORRIS	C	OKEECHOBEE	A/D	ORIF	A	PADUS	B	PARRAN	D
NORTHDALE	C	OKEELANTA	A/D	ORIO	C	PAESL	B	PARRISH	C
NORTHFIELD	B	OKEMAH	C	ORION	B	PAGET	B	PARSHALL	B
NORTHPORT		OKLARED	B	ORITA	B	PAGODA	C	PARSIPPANY	D
NORTH POWDER	C	OKLAWAHA	A/D	ORLAND	B	PAHRANAGAT	C	PARSONS	D
NORTHUMBERLAND	C/D	OKMOK	C	ORLANDO	A	PAHREAH	D	PARTRI	C
NORTON	C	OKO	D	ORMAN	C	PAHROC	D	PASAGSHAK	B
NORTONVILLE	C	OKOBOJI	C	ORMSBY	B/C	PAIA	C	PASCO	B
NORTUNE	D	OKOLONA	C	ORODELL	C	PAICE	C	PASO SECO	D
NORWALK	B	OKREEK	D	OROFINO	B	PAINESVILLE		PASQUETTI	C/D
NORWAY FLAT	A	OKTIBBEHA	D	ORO GRANDE	C	PAINTROCK	B	PASQUOTANK	B/D
NORWELL	C	OLA	C	ORONO	D	PAIT	B	PASSAR	C
NORWICH	D	OLAA	A	OROVADA	C	PAJARITO	B	PASS CANYON	D
NORWOOD	B	OLALLA	C	ORR	C	PAJARO	C	PASSCREEK	B
NOTI	D	OLANTA	B	ORRVILLE	C	PAKALA	B	PASTURA	D
NOTUS	A	OLATHE	C	ORSA	A	PAKINI	B	PATAHS	B
NOVARA	B	OLD CAMP	C	ORSINO	A	PALA	B	PATENT	C
NOVARY	B	OLDHAM	C	ORTELLO	A	PALACIO	B	PATILLAS	B
NOWOOD	C	OLDS	C	ORTIGALITA	C	PALAPALAI	B	PATILO	C
NOYO	C	OLDSMAR	B/D	ORTING	C	PALATINE	B	PATIT CREEK	B
NUBY	C	OLDWICK	B	ORTIZ	C	PALESTINE	B	PATNA	C
NUCKOLLS	C	OLELO	B	ORWOOD	B	PALISADE	B	PATOUTVILLE	C
NUCLA	B	OLENA	B	OSAGE	D	PALMA	B	PATRICIA	B

(continued)

EXHIBIT 2-1 (continued)

NUECES	C	OLEQUA	B	OSAKIS	B	PALMAREJO	C	PATRICK	B
NUGGET	C	OLETE	B	OSGOOD	B	PALM BEACH	A	PATROLE	C
NUMA	C	OLEX	B	OSHA	B	PALMER	D	PATTANI	C
NUNDA	C	OLGA	C	OSHAWA	D	PALMER CANYON	B	PATTENBURG	B
NUNICA	C	OLI	B	O'SHEA	C	PALMICH	B	PATTERSON	C
NUNN	C	OLIAGA	B	OSHKOSH	C	PALMS	D	PATTON	B/C
NUSS	D	OLINDA	B	OSHTEMO	B	PALMYRA	B	PATWAY	C
NUTLEY	C	OLIPHANT	B	OSIER	B/D	PALO	B	PAUL	B
NUTRAS	C	OLIVENHAIN	C	OSKA	C	PALOMAS	B	PAULDING	D
NUTRIOSO	B	OLIVER	B	OSMUND	B	PALOMINO	D	PAULINA	D
NUVALDE	C	OLIVIER	C	OSO	B	PALOS VERDES	B	PAULSELL	D
NYALA	D	OLMITO	D	OSOBB	D	PALOUSE	B	PAULVILLE	B
NYMORE	A	OLMITZ	B	OSO RIDGE	C	PALSGROVE	B	PAUMALU	B
NYSSA	C	OLMOS	C	OSOTE	B	PAMLICO	D	PAUNSAUGUNT	D
NYSSATON	B	OLMSTED	B/D	OSSIAN	C	PAMLA	C	PAUSANT	B
NYSTROM	C	OLNEY	B	OST	B	PAMSDEL	D	PAUWELA	B
		OLOKUI	C	OSTRANDER	B	PANA	B	PAVAHROO	B
OAHE	B	OLPE	C	OTERO	B	PANACA	D	PAVANT	D
OAKDALE	B	OLSON	C	OTHELLO	D	PANAEWA	D	PAVILLION	B
OAKDEN	D	OLTON	C	OTIS	C	PANASOFFKEE	D	PAWCATUCK	D
OAKFORD	B	OLUSTEE	B/D	OTISCO	A	PANCHERI	B	PAWLET	B
OAK GLEN	B	OLYIC	B	OTISVILLE	A	PANCHUELA	C	PAWNEE	D
OAK GROVE	C	OLYMPIC	B	OTLEY	B	PANCO	B	PAXTON	C
OAK LAKE	B	OMAHA	B	OTSEGO	C	PANDOAH	C	PAYETTE	B
OAKLAND	C	OMAK	C	OTTER	B/D	PANDORA	D	PAYMASTER	B
OAKS RIDGE	C	OMEGA	A	OTTERBEIN	C	PANDURA	D	PAYNE	C
OAKVILLE	A	OMENA	B	OTTERHOLT	B	PANE	B	PAYSON	D
OAKWOOD	D	OMNI	C	OTTOKEE	A	PANGUITCH	B	PEACHAM	C
OANAPUKA	B	ONA	A/D	OTWAY	D	PANHILL	B	PEARL HARBOR	D
OASIS	B	ONALASKA	B	OTWELL	C	PANIOGUE	B	PEARMAN	
OATMAN	B	ONAMIA	B	OUACHITA	C	PANKY	C	PEARSOLL	D
OBAN	C	ONARGA	B	OURAY	A	PANOCHE	B	PEAVINE	C
OBARO	B	ONAWA	C	OUTLET	C	PANOLA	D	PECATONICA	B
OBRAY	D	ONAWAY	B	OVALL	C	PANSEY	D	PECOS	C
OBURN	D	ONDAWA	B	OVERGAARD	C	PANTHER	C	PEDEE	C
OCALA	D	ONEIDA	B	OVERLY	C	PANTON	C	PEDERNALES	C
OCEANET	D	O'NEILL	B	OVERTON	C	PAOLA	A	PEDIGO	B/C
OCEANO	A	ONEONTA	B	OVID	C	PAOLI	B	PEDLAR	D
OCHEYEDAN	B	ONITA	C	OVINA	B	PAPAA	D	PEDRICK	B
OCHLOCKONEE	B	ONITE	B	OWEGO	D	PAPAI	A	PEEBLES	C
OCHO	D	ONOTA	C	OWEN CREEK	C	PAPAKATING	D	PEEL	C
OCHOCO	C	ONOVA	C	OWENS	D	PAPOOSE	C	PEELER	B
OCHOPEE	B/D	ONRAY	D	OWHI	B	PARADISE	C	PEEVER	C

PEGLER	D	PIE CREEK	D	POE	B/C	PREBLE	C	QUINN	D
PEGRAM	B	PIERIAN	A	POGANEAB	C	PRENTISS	C	QUINNEY	C
PEKIN	C	PIERPONT	C	POGUE	B	PRESQUE ISLE	B	QUINTON	
PELHAM	B/D	PIERRE	D	POHAKUPU	A	PRESTO	A	QUITMAN	C
PELIC	C	PIIHONUA	A	POINSETT	B	PRESTON	A	QUONSET	A
PELLA	D	PIKE	B	POINT	B	PREWITT	C		
PELONA	C	PILCHUCK	A	POINT ISABEL	C	PREY	D	RABER	C
PEMBERTON	A	PILGRIM	B	POJOAQUE	B	PRICE	C	RABEY	A
PEMBINA	C	PILOT	B	POKEGEMA	B	PRIDA	D	RABIDEUX	B
PEMBROKE	B	PILOT ROCK	C	POKER	B	PRIDHAM	D	RABUN	B
PENA	B	PIMA	B	POLAND		PRIETA	C	RACE	D
PENCE	A	PINAL	D	POLAR	B	PRIMEAUX	C	RACHERT	D
PENDEN	B	PINALENO	B	POLATIS	C	PRIMGHAR	B	RACINE	B
PEND OREILLE	B	PINATA	C	POLE	A	PRINCETON	B	RACOON	D
PENDROY	D	PINAVETES	A	POLEBAR	C	PRINEVILLE	C	RAD	C
PENISTAJA	B	PINCHER	C	POLELINE	B	PRING	B	RADFORD	B
PENITENTE	B	PINCKNEY	C	POLEO	C	PRINS	C	RADLEY	C
PENN	C	PINCONNING	D	POLEY	C	PROCTOR	B	RADNOR	D
PENNEL	C	PINCUSHION	B	POLICH	C	PROGRESSO	C	RAFAEL	D
PENNINGTON	B	PINEDA	B/D	POLLARD	C	PROMISE	D	RAGLAN	D
PENNISULA	C	PINEDALE	B	POLLASKY	C	PROMO	D	RAGNAR	B
PENO	C	PINEGUEST	B	POLLY	B	PROMONTORY	B	RAGO	C
PENOYER	C	PINELLOS	A/D	POLO	B	PRONG	C	RAGSDALE	B/D
PENROSE	D	PINETOP	C	POLSON	C	PROSPECT	B	RAGTOWN	D
PENTHOUSE	D	PINEVILLE	B	POLVADERA	C	PROSPER	B	RAHM	C
PENTZ	D	PINEY	C	POMAT	C	PROSSER	C	RAIL	C/D
PENWOOD	A	PINICON	B	POMELLO	C	PROTIVIN	C	RAINBOW	C
PEOGA	C	PINKEL	C	POMPANO	A/D	PROUT	C	RAINEY	B
PEOH	C	PINKSTON	B	POMPONIC	C	PROVIDENCE	C	RAINS	B/D
PEONE	B/C	PINNACLES	C	POMPTON	B	PROVO	C	RAINSBORO	C
PEOTONE	C	PINO	C	POMROY	B	PROVO BAY	D	RAKE	D
PEPOON	B	PINOLA	C	PONCENA	D	PROWERS	B	RALSEN	B/C
PEQUEA	C	PINOLE	B	PONCHA	A	PTARMIGAN	B	RAMADA	C
PERCHAS	D	PINON	C	POND	B/C	PUAULU	A	RAMADERO	B
PERCIVAL	C	PINONES	D	POND CREEK	B	PUCHYAN	A	RAMBLER	B
PERELLA	C	PINTAS	D	PONCILLA	A	PUDDLE	D	RAMELLI	C
PERHAM	C	PINTLAR	A	PONIL	D	PUERCO	D	RAMIRES	D
PERICO	B	PINTO	C	PONTOTOC	B	PUETT	C	RAMPEL	C
PERKINS	C	PINTURA	A	POOZER	D	PUGET	C	RAMO	C
PERKS	A	PINTWATER	C	POCKU	A	PUGSLEY	B	RAMONA	B
PERLA	C	PIOPOLIS	D	POOLE	B/D	PUHI	A	RAMPART	B
PERMA	A	PIPER	B/C	POOLER	D	PUHIMAU	D	RAMPARTAR	A
PERMANENTE	C	PIROUETTE	C	POORMA	B	PULASKI	B	RAMSEY	D
PERRIN	B	PISGAH	C	POPE	B	PULEHU	B	RAMSHORN	B
PERKINE	D	PISHKUN	B	POPPLETON	A	PULLMAN	D	RANCE	C
PERROT	D	PISTAKEE	B	POQUONOCK	C	PULS	D	RANCHERIA	B
PERRY	D	PIT	D	PORRETT	B/D	PULSIPHER	D	RAND	B
PERRYVILLE	B	PITTMAN	C	PORT	B	PULTNEY	C	RANDADO	C
PERSAYO	D	PITTSFIELD	B	PORTAGEVILLE	D	PUMPER	C	RANDALL	D

(continued)

EXHIBIT 2-1 (continued)

PERSHING	C	PITTSTOWN	C	PORTALES	C	PUNA	A	RANDOLPH	D
PERSIS	B	PITTWOOD	B	PORT BYRON	B	PUNALUU	D	RANDS	C
PERT	D	PLACENTIA	C	PORTERS	B	PUNOHU	A	RANGER	D
PERU	C	PLACERITOS	C	PORTERVILLE	D	PURDAM	C	RANIER	C
PESCADERO	C/D	PLACID	A/D	PORTHILL	C	PURDY	D	RANKIN	C
PESET	C	PLACK	D	PORTINO	C	PURGATORY	D	RANTOUL	D
PESHASTIN	B	PLAINFIELD	A	PORTLAND	D	PURNER	D	RANYHAN	B
PESO	C	PLAINVIEW	C	PORTNEUF	B	PURSLEY	B	RAPELJE	C
PETEETNEET	D	PLAISTED	C	PORTOLA	C	PURVES	D	RAPHO	B
PETERBORO	B	PLANO	B	PORTSMOUTH	D	PUSTOI	A	RAPIDAN	B
PETERS	D	PLATA	B	POSANT	C	PUTNAM	C	RARDEN	C
PETOSKEY		PLATEA	C	POSEY	B	PUUKALA	D	RARICK	B
PETRIE	D	PLATEAU	B	POSITAS	D	PUUONE	C	RARITAN	C
PETROLIA	D	PLATNER	C	POSKIN	C	PUU OO	A	RASBAND	B
PETTONS	C	PLATO	C	POSOS	C	PUU OPAE	B	RASSET	B
PEWAMO	B/D	PLATTE	C	POST	D	PUU PA	B	RATHBUN	C
PEYTON	B	PLATTVILLE	B	POTAMO	D	PUYALLUP	B	RATLIFF	B
PHAGE	B	PLAZA	B/C	POTLATCH	C	PYLE	A	RATON	C
PHARO	B	PLEASANT	C	POTRATZ	C	PYLON	C	RATTLER	B
PHARCLIO	D	PLEASANT GROVE	B	POTSDAM	C	PYOTE	A	RAUB	B
PHEBA	C	PLEASANTON	B	POTTER	C	PYRAMID	C	RAUVILLE	D
PHEENEY	B	PLEASANT VALE	B	POTTER	C	PYRMONT	D	RAUZI	B
PHELAN	B	PLEASANT VIEW	B	POTTS	B			RAVALLI	C
PHELPS	B	PLEDGER	D	POUCRE	B	QUAKER	C	RAVENDALE	D
PHIFERSON	B	PLEEK	C	POULTNEY	B	QUAKERTOWN	B	RAVENNA	C
PHILBON	B/D	PLEINE	C	POVERTY	A	QUAMBA	D	RAVOLA	B
PHILLIPS	C	PLEVNA	D	POWDER	B	QUANAH	B	RAWAH	B
PHILLIPSBURG	A	PLOME	C	POWDERHORN	C	QUANDAHL	B	RAWHIDE	D
PHILO	B	PLOVER	B	POWELL	C	QUARLES	C	RAWSON	B
PHILOMATH	D	PLUMAS	B	POWER	B	QUARTZBURG	C	RAY	B
PHIPPS	C	PLUMMER	B/D	POWHITE	C	QUATAMA	C	RAYADO	C
PHOEBE	B	PLUSH	B	POWLEY	D	QUAY	C	RAYENOUF	B
PHOENIX	D	PLUTH	B	POWWATKA	C	QUEBRADA	C	RAYMONDVILLE	D
PIASA	D	PLUTOS	C	POY	D	QUEETS	B	RAYNE	B
PICACHO	C	PLYMOUTH	A	POYGAN	D	QUEMADO	C	RAYNESFORD	B
PICAYUNE	B	POALL	C	POZO	C/D	QUENZER	D	RAYNHAM	C
PICKAWAY	C	POARCH	B	POZO BLANCO	B	QUICKSELL	C	RAYNOR	D
PICKENS	D	POCALLA	A	PRAG	C	QUIGLEY	B	RAZOR	C
PICKETT	B	POCATELLO	B	PRATHER	B	QUILCENE	C	RAZORT	B
PICKFORD	D	POCKER	D	PRATLEY	C	QUILLAYUTE	B	READING	C
PICKWICK	B	POCOMOKE	C	PRATT	A	QUIMBY	C	READINGTON	C
PICO	B	PODO	D	PREACHER	B	QUINCY	A	READLYN	B
PICTOU	B	PODUNK	B	PREBISH	D	QUINLAN	C	REAGAN	B

REAKOR	B	RHOADES	D	ROCK RIVER	B	RUDYARD	D	SALMON	B
REAL	D	RIB	C	ROCKTON	B	RUELLA	B	SALOL	D
REAP	D	RICCO	D	ROCKWELL	B	RUGGLES	B	SALONIE	B
REARDAN	C	RICETON	B	ROCKWOOD	B	RUIDOSO	C	SALTAIR	D
REAVILLE	C	RICEVILLE	C	ROCKY FORD	B	RUKO	D	SALT CHUCK	A
REBA	C	RICHARDSON	B	RODDY		RULE	B	SALTER	B
REBEL	B	RICHEAU	D	RODMAN	A	RULICK	C	SALTERY	D
REBUCK		RICHEY	C	ROE		RUMBO	C	SALT LAKE	D
RECLUSE	D	RICHFIELD	C	ROEBUCK	D	RUMFORD	B	SALUDA	C
REDBANK	B	RICHFORD	A	ROELLEN	D	RUMNEY	C	SALUVIA	
RED BAY	B	RICHLIE	A	ROESIGER	B	RUMPLE	C	SALVISA	C
RED BLUFF	B	RICHMOND	D	ROHNERVILLE	B	RUM RIVER	C	SALZER	D
RED BUTTE	B	RICHTER	B	ROHRERSVILLE	C	RUNE	C	SAMBA	D
REDBY	C	RICHVALE	B	ROKEBY	D	RUNNELLS	C	SAMISH	C/D
REDCHIEF	C	RICHVIEW	C	ROLETTE	C	RUNNYMEDE	B	SAMMAMISH	C
REDCLOUD	B	RICHWOOD	B	ROLFE	C	RU[illegible]T	A	SAMPSEL	D
REDDICK	C	RICKMORE	C	ROLISS	D	R[illegible]	C	SAMPSON	B
REDDING	D	RICKS	A	ROLLA	C	RUSE	D	SAMSIL	D
REDFIELD	B	RICREST	B	ROLLIN	D	RUSH	C	SAN ANDREAS	C
RED HILL	C	RIDD	C	ROLOFF	C	RUSHTOWN	A	SAN ANTON	B
RED HOOK	C	RIDGEBURY	C	ROMBO	C	RUSHVILLE	D	SAN ANTONIO	C
REDLAKE	D	RIDGECREST	C	ROMEO	C	RUSS	D	SAN ARCACIO	B
REDLANDS	B	RIDGEDALE	B	ROMNEY	C	RUSSELL	B	SAN BENITO	B
REDMANSON	B	RIDGELAND	D	ROMULUS	D	RUSSELLVILLE	C	SANCHEZ	D
REDMOND	C	RIDGELAWN	A	ROND	D	RUSSLER	C	SANDALL	C
REDNUN	C	RIDGELY	B	RONNEBY	B	RUSTON	B	SANDERSON	B
REDOLA	B	RIDGEVILLE	B	RONSON	B	RUTLAND	C	SANDLAKE	D
REDONA	B	RIDGEWAY	D	ROSACHI	C	RUTLEGE	D	SANDLEE	A
REDRIDGE	D	RIFTBROCK	C	ROSAMOND	B	RYAN	D	SANELI	D
REDROB	B	RIFFE	B	ROSANE	C	RYAN PARK	B	SAN EMIGDIO	B
RED ROCK	B	RIFLE	A/D	ROSARIO	C	RYDE	B/D	SANGER	B
RED SPUR	B	RIGA	D	ROSCOE	D	RYDER	C	SAN GERMAN	D
REDSTOE	B	RIGGINS	A	ROSCOMMON	D	RYEGATE	B	SANGO	C
REDTHAYNE	B	RILEY	C	ROSEBERRY	B	RYEPATCH	D	SANGREY	A
REDTOM	C	RILLA	B	ROSEBLOOM	D	RYER	C	SANILAC	C
REDVALE	C	RILLITO	B	ROSEBUD	B	RYUS	C	SAN ISABEL	B
REDVIEW	C	RIMER	C	ROSEBURG	B			SAN JOAQUIN	D
REE	B	RIMINI	A	ROSE CREEK	C	SABANA	D	SAN JON	C
REED	C	RIMROCK	D	ROSEGLEN	D	SABANA SECA	D	SAN JOSE	B
REEDER	B	RIN	B	ROSEHILL	D	SABENYO	B	SAN JUAN	A
REEDPOINT	C	RINCON	C	ROSELAND	D	SABINA	C	SAN LUIS	B
REEDY	C	RINCONADA	C	ROSELMS	D	SABINE	A	SAN MATEO	C
REELFOOT	C	RINGLING	C	ROSEMOUNT	B	SABLE	D	SAN MIGUEL	C
REESER	C	RINGO	D	ROSENDALE	B	SAC	B	SANPETE	A
REESVILLE	C	RINGOLD	B	ROSEVILLE	B	SACO	D	SANPITCH	C
REFUGE	C	RINGWOOD	B	ROSEWORTH	C	SACRAMENTO	C/D	SAN POIL	B
REGAN	B	RIO	D	ROSHE SPRINGS	C	SACUL	D	SAN SABA	D
REGENT	C	RIO ARRIBA	D	ROSITAS	A	SADDLE	B	SAN SEBASTIAN	B
REHM	C	RIO GRANDE	B	ROSLYN	B	SADDLEBACK	B	SANTA	C

(continued)

EXHIBIT 2-1 (continued)

REICHEL	B	RIO KING	C	ROSMAN	B	SADIE	B	SANTA CLARA	C
REIFF	B	RIO LAJAS	A	ROSNEY	C	SADLER	C	SANTA FE	D
REILLY	A	RIO PIEDRAS	B	ROSS	B	SAFFELL	B	SANTA ISABEL	D
REINACH	B	RIPLEY	B	ROSS FORK	C	SAGANING	D	SANTA LUCIA	C
RELAN	A	RIPON	B	ROSSI	C	SAGE	D	SANTA MARTA	C
RELAY	B	RIRIE	B	ROSSMOYNE	C	SAGEHILL	B	SANTANA	C
RELIANCE	C	RISTA	C	ROSS VALLEY	C	SAGEMOOR	C	SANTAQUIN	A
RELIZ	D	RISUE	D	ROTAN	C	SAGERTON	C	SANTA YNEZ	C
RELSE	B	RITCHEY	B	ROTHIEMAY	B	SAGINAW		SANTEE	D
REMBERT	D	RITNER	C	ROTHSAY	B	SAGO	D	SANTIAGO	B
REMMIT	A	RITTER	B	ROUBIDEAU	C	SAGOUSPE	C	SANTIAM	C
REMSEN	D	RITTMAN	C	ROUEN	C	SAGUACHE	A	SAN TIMOTEO	C
REMUDAR	B	RITZCAL	B	ROUND BUTTE	D	SAHALI	B	SANTONI	D
REMUNDA	C	RITZVILLE	B	ROUNDTOP	C	SAINT ALBANS	B	SANTOS	C
RENFROW	D	RIVERHEAD	B	ROUNDUP	C	SAINT CHARLES	B	SANTO TOMAS	B
RENO	D	RIVERSIDE	A	ROUNDY	C	SAINT CLAIR	D	SAN YSIDRO	D
RENOHILL	C	RIVERTON	B	ROUSSEAU	A	SAINT ELMO	A	SAPP	D
RENOVA	B	RIVERVIEW	B	ROUTON	D	SAINT GEORGE	C	SAPPHIRE	B
RENOX	B	RIVRA	A	ROUTT	C	SAINT HELENS	A	SAPPINGTON	B
RENSHAW	B	RIXIE	C	ROVAL	D	SAINT IGNACE	C	SARA	C
RENSLOW	B	RIZ	D	ROWE	D	SAINT JOE	B	SARALEGUI	B
RENSSELAER	C	ROANOKE	C	ROWENA	D	SAINT JOHNS	B/D	SARANAC	C
RENTIDE	C	ROBBINS	B	ROWLAND	C	SAINT LUCIE	A	SARAPH	C
RENTON	B/C	ROBBS	D	ROWLEY	B	SAINT MARTIN	C	SARATOGA	B
RENTSAC	C	ROBERTS	D	ROXBURY	B	SAINT MARYS	B	SARCO	B
REPARADA	D	ROBERTSDALE	C	ROY	B	SAINT NICHOLAS	C	SARDINIA	C
REPPART	B	ROBERTSVILLE	D	ROYAL	B	SAINT PAUL	B	SARGEANT	D
REPUBLIC	B	ROBIN	B	ROYALTON	C	SAINT THOMAS	D	SARITA	A
RESCUF	C	ROBINSON	D	ROYSTONE	B	SALADO	B	SARKAR	C
RESERVE	B	ROBINSONVILLE	B	ROZA	D	SALAL	B	SARPY	A
RESNER	B	ROBLEDO	D	ROZELLVILLE	B	SALAMATOF	C	SARTELL	A
RET	B/C	ROB ROY	C	ROZETTA	B	SALAS	C	SASKA	B
RETRIEVER	D	ROBY	C	ROZLEE	C	SALCHAKET	B	SASSAFRAS	B
RETSOF	C	ROCHE	C	RUARK	C	SALEM	B	SASSER	B
RETSOK	B	ROCHELLE	C	RUBICON	A	SALEMSBURG	B	SATANKA	C
REXBURG	B	ROCHEPORT	C	RUBIO	C	SALGA	C	SATANTA	B
REXOR	A	ROCKAWAY	C	RUBY	B	SALIDA	A	SATELLITE	C
REYES	C/D	ROCKCASTLE	D	RUCH	B	SALINAS	C	SATT	D
REYNOLDS		ROCK CREEK	D	RUCKLES	D	SALISBURY	D	SATTLEY	B
REYNOSA	B	ROCKFORD	B	RUCLICK	C	SALIX	B	SATTRE	B
REYWAT	D	ROCKINGHAM	C/D	RUDD	D	SALKUM	C	SATUS	B
RHAME	B	ROCKLIN	C	RUDEEN	B	SALLISAW	B	SAUDE	B
RHINEBECK	D	ROCKPORT		RUDOLPH	C	SALLYANN	C	SAUGATUCK	C

SAUGUS	B	SELFRIDGE	C	SHIOCTON	B	SKIYOU	B	SPARTA	A
SAUK	B	SELKIRK	D	SHIPLEY	C	SKOKOMISH	B/C	SPEARFISH	B
SAULICH	D	SELLE	B	SHIPROCK	B	SKOOKUMCHUCK	B	SPEARVILLE	C
SAUM	C	SELLERS	A/D	SHIRK	B	SKOWHEGAN	B	SPECK	D
SAUNDERS	C	SELMA	B	SHOALS	C	SKULL CREEK	D	SPECTER	D
SAUVIE	C/D	SEMIAHMOO	D	SHOEFFLER	B	SKUMPAH	C	SPEELYAI	C
SAUVOLA	C	SEMIHMOO	D	SHONKIN	D	SKUTUM	C	SPEIGLE	B
SAVAGE	C	SEMINARIO	D	SHOOK	A	SKYBERG	C	SPENARD	C
SAVANNAH	C	SEN	B	SHOREWOOD	C	SKYHAVEN	D	SPENCER	B
SAVO	C	SENECAVILLE	C	SHOREY	B	SKYKOMISH	B	SPERRY	C
SAVOIA	B	SEQUATCHIE	B	SHORN	B	SKYLINE	C	SPICER	C
SAWABE	D	SEQUIM	A	SHORT CREEK	D	SKYWAY	B	SPILLVILLE	B
SAWATCH	C	SEQUOIA	C	SHOSHONE	D	SLAB	C	SPINKS	A
SAWCREEK	B	SERENE	D	SHOTWELL	D	SLATE CREEK	C	SPIRIT	B
SAWMILL	C	SERNA	D	SHOUNS	B	SLAUGHTER	C	SPIRO	B
SAWYER	C	SEROCO	A	SHOWALTER	C	SLAVEN	C	SPLENDORA	C
SAXBY	D	SERPA	C/D	SHOWLOW	C	SLAWSON	B	SPLITROCK	D
SAXON	B	SERVOSS	D	SHREWSBURY	D	SLAYTON	C	SPOFFORD	C
SAYBROOK	B	SESAME	C	SHRINE	B	SLEETH	C	SPOKANE	B
SAYLESVILLE	C	SESPE	C	SHROUTS	D	SLETTEN	D	SPONSELLER	B
SAYLOR	A	SESSIONS	C	SHUBUTA	C	SLICKROCK	B	SPOON BUTTE	D
SCALA	B	SESSUM	D	SHULE	B	SLIGHTS	C	SPOONER	C
SCAMMAN	C	SETTERS	C	SHULLSBURG	C	SLIGO	B	SPOTTSWOOD	B
SCANDIA	B	SETTLEMEYER	D	SHUMWAY	D	SLIKOK	C	SPRAGUE	B/C
SCANTIC	C	SEVERN	B	SHUPERT	C	SLIP	B	SPRECKELS	C
SCAR	A	SEVILLE	D	SHUWAH	B	SLOAN	C	SPRING	C/D
SCARBORO	D	SEVY	C	SI	B	SLOCUM	B	SPRING CREEK	C
SCAVE	C	SEWARD	B	SIBLEYVILLE	B	SLODUC	C	SPRINGDALE	B
SCHAFFENAKER	A	SEWELL	B	SIBYLEE	D	SLOSS	C	SPRINGER	B
SCHAMBER	A	SEXTON	D	SICILY	B	SLUICE	B	SPRINGERVILLE	D
SCHAMP	C	SEYMOUR	C	SICKLESTEETS	C	SMARTS	B	SPRINGFIELD	D
SCHAPVILLE	C	SHAAK	D	SIDELL	B	SMITH CREEK	A	SPRINGMEYER	C
SCHEBLY	D	SHADELAND	C	SIEANCIA	B	SMITHNECK	B	SPRINGTOWN	C
SCHERRARD	D	SHAFFER	A	SIEBER	A	SMITHTON	D	SPUR	B
SCHLEY	B	SHAKOPEE	C	SIELO	C	SMOLAN	C	SPURLOCK	B
SCHNORBUSH	C	SHALCAR	D	SIEROCLIFF	D	SMOOT	D	SQUALICUM	B
SCHODACK		SHAM	D	SIERRA	B	SNAG	B	SQUAW	B
SCHODSON	C	SHAMBO	B	SIERRAVILLE	B	SNAHOPISH	B	SQUILLCHUCK	B
SCHOFIELD	D	SHAMEL	B	SIESTA	D	SNAKE	C	SQUIM	B
SCHOHARIE	C	SHANAHAN	B	SIFTON	B	SNAKE HOLLOW	B	SQUIRES	B
SCHOLLE	C	SHANDON		SIGNAL	D	SNAKELUM	B	STAATSBURG	
SCHOOLEY	C/D	SHANE	D	SIGURD	B	SNEAD	D	STABLER	B
SCHRIER	B	SHANO	B	SIKESTON	D	SNELL	C	STACY	B
SCHROCK	B	SHANTA	B	SILCOX	B	SNELLING	B	STADY	B
SCHUMACHER	B	SHAPLEIGH	C/D	SILENT	D	SNOHOMISH	C	STAFFORD	C
SCHUYLKILL	B	SHARATIN	B	SILER	B	SNOQUALMIE	B	STAGECOACH	B
SCIO	B	SHARKEY	D	SILERTON	B	SNOW	B	STAHL	C
SCIOTOVILLE	C	SHARON	B	SILI	D	SNOWDEN	C	STALEY	C
SCISM	B	SHARPSBURG	B	SILVER	D	SNOWLIN	B	STAMBAUGH	B

(continued)

EXHIBIT 2-1 (continued)

SCITUATE	C	SHARVANA	C	SILVERBOW	D	SNOWVILLE	D	STAMFORD	D
SCOBEY	C	SHASKIT	B/C	SILVER CREEK	D	SNOWY	A	STAMPEDE	D
SCOOTENEY	B	SHASTA	A	SILVERTON	C	SOAP LAKE	B	STAN	B
SCORUP	C	SHAVANO	B	SILVIES	D	SOBOBA	A	STANDISH	C/D
SCOTT	D	SHAVER	B	SIMAS	C	SOBRANTE	C	STANEY	D
SCOTT LAKE	B	SHAWANO	A	SIMCOE	C	SODA LAKE	B	STANFIELD	C
SCOUT	A	SHAWMUT	B	SIMEON	A	SODHOUSE	C	STANLEY	C
SCOWLALE	C	SHAY	D	SIMPLER	D	SODUS	C	STANSBURY	D
SCRANTON	B/D	SHEAR	C	SIMNER	A	SOELBERG	B	STANTON	D
SCRIBA	C	SHECKLER	C	SIMON	C	SOFIA	C	STAPLETON	B
SCRIVER	B	SHEDD	C	SIMONA	B	SOGN	D	STARBUCK	D
SCROGGIN	C	SHEEGE	D	SIMPERS	B	SOGZIE	B	STARICHKOF	D
SCULLIN	C	SHEEP CREEK	C	SIMPSON	C	SOLANO	D	STARKS	C
SEABROOK	A	SHEEPHEAD	C	SIMS	D	SOLDATNA	B	STARR	B
SEAMAN	C	SHEEPROCK	A	SINAI	C	SOLDIER	D	STASER	B
SEAQUEST	C	SHEETIRON	C	SINCLAIR	B	SOL DUC	B	STATE	B
SEARCHLIGHT	C	SHEFFIELD	D	SINE	C	SOLLEKS	C	STATEN	D
SEARING	B	SHELBURNE	C	SINGLETREE	D	SOLLER	D	STAVE	C
SEARLA	B	SHELBY	B	SINGSAAS	B	SOLOMON	D	STAYTON	D
SEARLES	C	SHELBYVILLE	B	SINNIGAM	C	SOLONA	B	STEAMBOAT	D
SEATON	B	SHELDON		SINUK	B	SOMBRERO	D	STEARNS	D
SEATTLE	D	SHELIKOF	C	SION	B	SOMERS	B	STECUM	A
SEBAGO	D	SHELLABARGER	B	SICUX	A	SOMERSET	C	STEED	A
SEBASTIAN	D	SHELLDRAKE	A	SIPPLE	A	SOMERVELL	B	STEEDMAN	D
SEBASTOPOL	C	SHELLROCK	A	SISKIYOU	B	SOMSEN	C	STEEKEE	C
SEBEKA	D	SHELMADINE	D	SISSETON	B	SONOITA	B	STEELE	B
SEBEWA	B/D	SHELOCTA	B	SISSON	B	SONOMA	D	STEESE	C
SEBREE	D	SHELTON	C	SITES	C	SONTAG	C	STEFF	C
SEBRING	D	SHENA	C	SITKA	B	SOPER	B	STEGALL	C
SECATA	C	SHENANDOAH	C	SIXMILE	B	SOQUEL	B	STEIGER	A
SECRET	C	SHEPPARD	A	SIZEMORE	B	SORF	C	STEINAUER	B
SECRET CREEK	B	SHERIDAN	B	SIZER	B	SORRENTO	B	STEINBECK	B
SEDAN		SHERM	D	SKAGGS	B	SORTER	B/D	STEINMETZ	D
SEECSKADEE	D	SHERRYL	B	SKAGIT	B/C	SOSA	C	STEINSBURG	C
SEES	C	SHIBLE	B	SKAHA	A	SOTELLA	C	STEIWER	C
SEEWEE	B	SHIELDS	C	SKALAN	C	SOUTHFORK	D	STELLAR	D
SEGAL	D	SHIFFER	B	SKAMANIA	B	SOUTHGATE	C	STEMILT	C
SEGNO	C	SHILOH	C	SKAMOKAWA	B	SOUTHWICK	C	STENDAL	C
SEHORN	D	SHINAKU	C	SKANEE	C	SPAA	D	STEPHEN	C
SEJITA	D	SHINGLE	D	SKELLOCK	B	SPACE CITY	A	STEPHENSBURG	B
SEKIU	D	SHINGLETOWN	C	SKERRY		SPADE	B	STEPHENVILLE	B
SELAH	C	SHINN	B	SKILLET	C	SPALDING	C	STERLING	A
SELDEN	C	SHINROCK	C	SKINNER	C	SPANAWAY	B	STERLINGTON	B

STETSON	B	SUNSET	B	TALLAPOOSA	C	TENSAS	D	TIMKEN	D
STETTER	D	SUNSHINE	C	TALLEYVILLE	B	TENSED	C	TIMMERMAN	B
STEUBEN	B	SUNSWEET	C	TALLS	B	TENSLEEP	B	TIMMONS	B
STEVENS	B	SUPAN	B	TALLULA	B	TEOCULLI	B	TIMPAHUTE	D
STEVENSON	B	SUPERIOR	C	TALLY	B	TEPEE	D	TIMPANOGOS	B
STEWART	D	SUPERSTITION	A	TALMAGE	A	TEPETE	B/D	TIMPER	D
STICKNEY	C	SUPERVISOR	C	TALMO	B	TERINO	C	TIMPOONEKE	B
STIDHAM	A	SUPPLEE	B	TALOKA	D	TERMINAL	D	TIMULA	B
STIGLER	C	SUR	B	TALPA	D	TERMO	C	TINA	C
STILLMAN	A	SURGH	B	TAMA	B	TEROUGE	D	TINCAHAY	A
STILLWATER	D	SURPRISE	B	TAMALCO	D	TERRA CEIA	A/D	TINE	A
STILSON	B	SURRENCY	B/D	TAMBA	C/D	TERRAD	D	TINSLEY	A
STIMSON	B/C	SURVYA	C	TAMMANY CREEK	B	TERRERA	C	TINTON	A
STINGAL	B	SUSIE CREEK	D	TAMMS	C	TERRIL	B	TINYTOWN	B
STINSON	C	SUSITNA	B	TAMPICO	B	TERRY	B	TIOCANO	D
STIRK	D	SUSQUEHANNA	D	TANAMA	D	TERWILLIGER	C	TIOGA	B
STIRUM	B	SUTHER	C	TANANA	C	TESAJO	A	TIPPAH	C
STISSING		SUTHERLIN	C	TANBERG	D	TESCOTT	C	TIPPECANOE	B
STIVERSVILLE	B	SUTPHEN	D	TANDY	C	TESUQUE	B	TIPPER	A
STOCKBRIDGE	B	SUTTLER	B	TANEUM	C	TETON	A	TIPPERARY	A
STOCKLAND	B	SUTTON	B	TANEY	C	TETONIA	B	TIPPIPAH	D
STOCKPEN	D	SVEA	B	TANGAIR	C	TETONKA	C	TIPPO	C
STOCKTON	D	SVERDRUP	B	TANNA	C	TETOTUM	C	TIPTON	B
STODICK	D	SVOLD	C	TANNER	C	TEW	B/D	TIPTONVILLE	B
STOKES	D	SWAGER	C	TANSEM	B	TEX	B	TIRA	B
STOMAR	C	SWAKANE	B	TANTALUS	A	TEXLINE	B	TISBURY	B
STONER	B	SWAN	C	TANWAX	D	TEZUMA	C	TISCH	C
STONEWALL	A	SWANBOY	D	TAOPI	C	THACKERY	B	TISH TANG	B
STONO	B/D	SWANNER	D	TAOS	C	THADER	C	TITUSVILLE	C
STONYFORD	D	SWANSON	C	TAPIA	C	THANYON	A	TIVERTON	A
STOOKEY	D	SWANTON	B/D	TAPPEN	D	THATCHER	B	TIVOLI	A
STORDEN	B	SWANTOWN	C	TARA	B	THATUNA	C	TIVY	C
STORLA	B	SWAPPS	C	TARKIO	D	THAYNE	B	TOA	C
STORM KING	D	SWARTSWOOD	C	TARKLIN	C	THEBES	B	TOBICO	D
STORY	C	SWARTZ	D	TARPO	C	THEBO	D	TOBIN	B
STOSSEL	C	SWASEY	D	TARRANT	C	THEDULUND	C	TOBLER	B
STOUGH	C	SWASTIKA	C	TARYALL	B	THENAS	C	TOBOSA	D
STOWELL	D	SWATARA	A	TASCOSA	B	THERESA	B	TOBY	B
STOY	D	SWAUK	C	TASSEL	D	THERIOT	D	TOCCOA	B
STRAIGHT	C	SWAWILLA	A	TATE	B	THERMAL	C	TODD	B
STRAIN	B	SWEATMAN	C	TATIYEE	C	THERMOPOLIS	D	TODDLER	B
STRASBURG	C	SWEDE	B	TATU	C	THETFORD	A	TODDVILLE	B
STRATFORD	B	SWEDEN	B	TATUM	C	THIOKOL	C	TOEHEAD	C
STRAUSS	C	SWEEN	C	TAUNTON	C	THOENY	D	TOEJA	C
STRAW	B	SWEENEY	B	TAVARES	A	THOMAS	C	TOEM	C
STRAWN	B	SWEET	C	TAWAS	A/D	THORNDALE	C	TOGO	B
STREATOR	C	SWEETGRASS	B	TAYLOR	C	THORNDIKE	C/D	TOHUNA	C
STROLE	B	SWEETWATER	D	TAYLOR CREEK	D	THORNOCK	D	TOINE	C

(continued)

EXHIBIT 2-1 (continued)

STRONGHURST	B	SWENODA	B	TAYLORSFLAT	D	THORNTON	C	TOIYABE	C
STRONTIA	B	SWIFTON	A	TAYLORSVILLE	C	THORNWOOD	B	TOKEEN	C
STROUPE	D	SWIMS	A	TAYSOM	B	THOROUGHFARE	B	TOKUL	B
STRYKER	B	SWINGLE	D	TAZLINA	A	THORP	C	TOLBY	B
STUKEL	C	SWISBOB	D	TEAL	D	THORR	B	TOLEDO	D
STUKY	B	SWITCHBACK	C	TEALWHIT	C	THORREL	B	TOLICHA	D
STUMBLE	A	SWITZERLAND	B	TEANAWAY	C	THOW	B	TOLL	A
STUMPP	D	SWOPE	C	TEAPO	B	THREE MILE	C	TOLLGATE	B
STUMP SPRINGS	B	SWYGERT	C	TEAS	C	THUNDERBIRD	D	TOLLHOUSE	D
STUTTGART	D	SYCAMORE	B	TEASDALE	B	THURBER	C	TOLNA	B
STUTZVILLE	B/C	SYCAN	A	TEBO	B	THURLONI	D	TOLO	B
SUBLETTE	B	SYLACAUGA	B/D	TECHICK	B	THURLOW	C	TOLSONA	D
SUDBURY	B	SYLVAN	B	TECOLOTE	B	THURMAN	A	TOLSTOI	D
SUFFIELD	C	SYMERTON	B	TECUMSAH	B	THURMONT	B	TOLT	C
SUGARLOAF	B	SYNAREP	B	TEDROW	B	THURSTON	B	TOLTEC	C
SUISUN	D	SYRACUSE	B	TEEL	B	TIAK	C	TOLUCA	B
SULA	B	SYRENE	D	TEHACHAPI	D	TIBBITTS	B	TOLVAR	B
SULLY	B	SYRETT	C	TEHAMA	C	TICA	C	TOMAH	C
SULPHURA	D			TEJA	D	TICE	C	TOMAS	B
SULTAN	B	TABERNASH	B	TEJON	B	TICHIGAN	C	TOMAST	C
SUMAS	B/C	TABIONA	B	TEKOA	C	TICHNOR	D	TOMERA	D
SUMDUM	D	TABLE MOUNTAIN	B	TELA	B	TICKAPOO	C	TOMICHI	A
SUMMA	B	TABLER	D	TELEFONO	D	TICKASON	B	TOMOKA	A/D
SUMMERFIELD	C	TABOR	D	TELEPHONE	D	TIDWELL	D	TONATA	C
SUMMERS	B	TACLMA	D	TELFER	A	TIERRA	C	TONAWANDA	C
SUMMERVILLE	C	TACOOSH	D	TELL	B	TIETON	B	TONEY	D
SUMMIT	C	TAFT	C	TELLER	B	TIFFANY	C	TONGUE RIVER	B
SUMMITVILLE	B	TAGGERT	C	TELLICO	B	TIFTON	B	TONINI	B
SUMTER	C	TAHOMA	B	TELLMAN	B	TIGER CREEK	B	TONKA	C
SUN	D	TAHQUAMENON	D	TELSTAD	B	TIGERON	A	TONKEY	C
SUNBURST	C	TAHQUATS	C	TEMESCAL	D	TIGIWON	B	TONKS	B
SUNBURY	B	TAINTOR	C	TEMPLE	B/C	TIGHETT		TONOPAH	B
SUNCOOK	A	TAJO	C	TEMVIK	B	TIGUA	D	TONOWEK	B
SUND	C	TAKEUCHI	C	TENABO	D	TIJERAS	B	TONSINA	B
SUNDELL	C	TAKILMA	B	TENAHA	B	TILFORD	B	TONUCO	B
SUNDERLAND	C/D	TAKOTNA	B	TENAS	C	TILLEDA	B	TOOLE	D
SUNDOWN	B	TALANTE	C	TENCEE	C	TILLICUM	B	TOOMES	D
SUNFIELD	B	TALAPUS	B	TENERIFFE	C	TILLMAN	C	TOP	C
SUNNILAND	C	TALBOTT	C	TENEX	A	TILMA	C	TOPPENISH	B/C
SUNNYHAY	D	TALCOT	C	TENIBAC	B	TILSIT	C	TOPTON	
SUNNYSIDE	B	TALIHINA	D	TENINO	B	TILTON	B	TOQUERVILLE	D
SUNNYVALE	C	TALKEETNA	C	TENNC	D	TIMBERG	C	TOQUOP	A
SUNRAY	C	TALLAC	B	TENOT	C	TIMBERLY	B	TORBOY	B
SUNRISE	C	TALLADEGA	C	TENREG	B	TIMENTWA	B	TORCHLIGHT	C

TORHUNTA	C	TRUCKEE	C	UKIAH	C	VASHTI	C	VOLINIA	B
TORNING	B	TRUCKTON	B	ULEN	B	VASQUEZ	B	VOLKE	C
TORODA	B	TRUESDALE	C	ULLOA	B	VASSAR	B	VOLKMAR	B
TORONTO	C	TRULL	C	ULM	B	VASTINE	C	VOLNEY	B
TORPEDO LAKE	C	TRULON	B	ULRICHER	B	VAUCLUSE	C	VOLPERIE	C
TORREON	C	TRUMAN	B	ULUPALAKUA	B	VAUGHNSVILLE	C	VOLTAIRE	D
TORRES	B	TRUMBULL	D	ULYSSES	B	VAYAS	D	VOLUSIA	C
TORRINGTON	B	TRUMP	D	UMA	A	VEAL	B	VONA	B
TORRO	C	TRYON	D	UMAPINE	B/C	VEAZIE	B	VORE	B
TORTUGAS	D	TSCHICOMA	C	UMIKOA	B	VEBAR	B	VROOMAN	B
TOTEM	B	TUB	C	UMIL	D	VEBAR	B	VULCAN	C
TOTTEN	B	TUBAC	C	UMNAK	B	VEGA	C	VYLACH	D
TOUCHET	B	TUCANNON	C	UMPA	B	VEGA ALTA	C		
TOUHEY	B	TUCKERMAN	D	UNA	D	VEGA BAJA	C	WABANICA	D
TOULON	B	TUCUMCARI	C	UNADILLA	B	VEKOL	D	WABASH	D
TOURN	C	TUFFIT	D	UNAWEEP	B	VELMA	B	WABASHA	D
TOURNQUIST	B	TUGHILL	D	UNCOMPAGHRE	C	VELVA	B	WABASSA	B/D
TOURS	B	TUJUNGA	A	UNEEDA	B	VENA	B	WABEK	B
TOUTLE	A	TUKEY	C	UNGERS	B	VENANGO	C	WACA	C
TOWER	D	TUKWILA	D	UNION	C	VENATOR	D	WACCTA	B
TOWNER	B	TULA	C	UNIONTOWN	B	VENETA	C	WACCUSTA	C
TOWNLEY	C	TULANA	C/D	UNIONVILLE	C	VENEZIA	D	WADAMS	B
TOWNSBURY	B	TULARE	C/D	UNISON	C	VENICE	D	WADDELL	B
TOWNSEND	C	TULAROSA	C	UPSAL	C	VENLO	D	WADCOUPS	B
TOWSON	B	TULIA	B	UPSHUR	C	VENUS	B	WADENA	B
TOXAWAY	D	TULLER	D	UPTON	C	VERBOORT	D	WADESBORO	B
TOY	D	TULLOCK	B	URACCA	B	VERDE	C	WADLEIGH	C
TOYAH	B	TULLY	C	URBANA	C	VERDEL	D	WADMALAW	D
TOZE	B	TUMBEZ	C	URBO	D	VERDELLA	D	WADSWORTH	C
TRABUCO	C	TUMEY	D	URICH	D	VERDIGRIS	B	WAGES	B
TRACK	B	TUMITAS	B	URNE	B	VERDUN	D	WAGNER	D
TRACY	B	TUMWATER	A	URSINE	C	VERGENNES	C	WAGRAM	A
TRAER	C	TUNEHEAN	D	URTAH	C	VERHALEN	C	WAHA	C
TRAIL	A	TUNICA	D	URWIL	D	VERMEJO	D	WAHEE	D
TRAIL CREEK	B	TUNIS	D	USAL	B	VERNAL	B	WAHIAWA	B
TRANSYLVANIA	B	TUNKHANNOCK	A	USPAR	B	VERNALIS	B	WAHIKULI	B
TRAPPER	A	TUNNEL	B	USINE	B	VERNON	D	WAHKEENA	B
TRAPPIST	C	TUPELO	D	USKA	D	VERONA	C	WAHKIACUS	B
TRASK	C	TUPUKNUK	D	UTALINE	B	VESSER	C	WAHLUKE	B
TRAVER	B/C	TURBEVILLE	C	UTE	C	VESTON	D	WAHPETON	C
TRAVESSILLA	D	TURBOTVILLE	C	UTICA	A	VETAL	A	WAHTIGUP	B
TRAVIS	C	TURBYFILL	B	UTLEY	B	VETERAN	B	WAHTUM	B/C
TRAWICK	B	TURIN	B	UTUADO	B	VEYO	D	WAIAHA	D
TRAY	C	TURK	D	UVADA	D	VIA	B	WAIAKOA	C
TREADWAY	D	TURKEYSPRINGS	C	UVALDE	C	VIAN	B	WAIALEALE	D
TREASURE	B	TURLEY	C	UWALA	B	VIBORAS	D	WAIALUA	B
TREBLOC	D	TURLIN	B			VIBORG	B	WAIAWA	D
TREGO	C	TURNBOW	C	VACHERIE	C	VICKERY	C	WAIHUNA	D
TREFLONA	D	TURNER	B	VACER	B	VICKSBURG	B	WAIKALOA	B

(continued)

EXHIBIT 2-1 (continued)

TREMBLES	B	TURNERVILLE	B	VADO	A	VICTOR	A	WAIKANE	B
TREMPE	A	TURNEY	B	VAIDEN	D	VICTORIA	D	WAIKAPU	B
TREMPEALEAU	B	TURRET	B	VAILTON	B	VICTORY	B	WAIKOMO	D
TRENARY	B	TURRIA	C	VALCO	C	VICU	D	WAILUKU	B
TRENT	B	TUSCAN	D	VALDEZ	B/C	VIDA	B	WAIMEA	B
TRENTON	D	TUSCARAWAS	C	VALE	B	VIDRINE	C	WAINEE	B
TREP	B	TUSCARORA	C	VALENCIA	B	VIENNA	B	WAINOLA	A
TRES HERMANOS	C	TUSCOLA	B	VALENT	A	VIEQUES	B	WAIPAHU	C
TRES HERMANOS	C	TUSCUMBIA	C	VALENTINE	A	VIEW	C	WAISKA	B
TRETTEN	C	TUSEL	C	VALERA	C	VIGAR	C	WAITS	B
TREVINO	D	TUSKEEGO	C	VALKARIA	B/D	VIGO	D	WAKE	D
TREXLER	C	TUSLER	B	VALLAN	D	VIKING	D	WAKEEN	B
TRIAMI	C	TUSQUITEE	B	VALLECITOS	C	VIL	D	WAKEFIELD	B
TRIASSIC		TUSTIN	B	VALLERS	C	VILAS	A	WAKELAND	B/D
TRICON	C	TUSTUMENA	B	VALMONT	C	VILLA GROVE	B	WAKONDA	C
TRIDELL	B	TUTHILL	B	VALMY	B	VILLARS	B	WALCOTT	B
TRIDENT	D	TUTTLE	B	VALOIS	B	VINA	B	WALDECK	C
TRIGO	C	TUTWILER	B	VANAJO	D	VINCENNES	C	WALDO	D
TRIMBLE	B	TUXEDO		VANANDA	D	VINCENT	C	WALDRON	D
TRIMMER	B	TUXEKAN	B	VAN BUREN		VINEYARD	C	WALDROUP	D
TRINCHERA	C	TWILACKS	A	VANCE	C	VINGO	B	WALES	C
TRINITY	D	TWIN CREEK	B	VANDA	D	VINING	C	WALFORD	C
TRIPLEN	C	TWINING	C	VANDALIA	C	VINITA	C	WALKE	C
TRIPOLI	C	TWISP	B	VANDERDASSON	D	VINLAND	C	WALL	B
TRIPP	B	TWO DOT	C	VANDERGRIFT	C	VINTON	B	WALLACE	B
TRITON	C	TYEE	D	VANDERHOFF	D	VIRA	C	WALLA WALLA	B
TRIX	B	TYGART	D	VANDERLIP	A	VIRATON	C	WALLER	B/D
TROJAN	B	TYLER	D	VAN DUSEN	B	VIRDEN	C	WALLINGTON	C
TROMMALD	D	TYNDALL	B/C	VANET	C	VIRGIL	B	WALLIS	B
TROMP	B/C	TYNER	A	VANG	B	VIRGIN PEAK	D	WALLKILL	C/D
TRONSEN	B	TYRONE	C	VANHORN	B	VIRGIN RIVER	D	WALLMAN	C
TROOK	B	TYSON	C	VAN NOSTERN	B	VIRTUE	B	WALLOWA	C
TROPAL	D			VANNOY	B	VISALIA	B	WALLPACK	
TROSI	D	UBAR	C	VANOSS	B	VISTA	C	WALLROCK	B
TROUP	A	UBLY	B	VANTAGE	C	VIVES	B	WALLSBURG	D
TROUT CREEK	C	UCOLA	C	VAN WAGONER	D	VIVI	B	WALLSON	B
TROUTDALE	B	UCOPIA	B	VARCO	C	VLASATY	C	WALPOLE	C
TROUT LAKE	C	UDEL	D	VARELUM	C	VOCA	C	WALSH	B
TROUT RIVER	A	UDOLPHO	C	VARICK	D	VODERMAIER	B	WALSHVILLE	D
TROUTVILLE	B	UFFENS	D	VARINA	C	VOLADORA	B	WALTERS	A
TROXEL	B	UGAK	D	VARNA	C	VOLENTE	C	WALTON	C
TROY	C	UHLIG	B	VARRO	B	VOLGA	D	WALLM	B
TRUCE	C	UINTA	B	VARYSBURG	B	VOLIN	B	WALVAN	B

WAMBA	B/C	WEHADKEE	D	WHITNEY	B	WINU	C	YAMPA	C
WAMIC	B	WEIKERT	C/D	WHITORE	A	WINZ	C	YAMSAY	D
WAMPSVILLE	B	WEIMER	D	WHITSOL	B	WIOTA	B	YAKA	B
WANATAH	B	WEINBACH	C	WHITSON	D	WISHEYLU	C	YAQUINA	B/D
WANBLEE	D	WEIR	D	WHITWELL	C	WISKAH	C	YARDLEY	C
WANDO	A	WEIRMAN	B	WHOLAN	C	WISNER	D	YATES	D
WANETTA	A	WEISER	C	WIBAUX	D	WITBECK	D	YAWDIM	C
WANN	A	WEISHAUPT	C	WICHITA	C	WITCH	D	YAWKEY	C
WANA	A	WEISS	A	WICHUP	D	WITHAM	C	YAXON	B
WAPAL	B	WEITCHPEC	B	WICKERSHAM	B	WITHEE	C	YEATES HOLLOW	C
WAPATO	C/D	WELBY	B	WICKETT	C	WITT	C	YEGEN	B
WAPELLO	B	WELCH	C	WICKHAM	B	WITZEL	D	YELM	B
WAPINITIA	B	WELD	C	WICKIUP	C	WODEN	B	YENRAB	A
WAPPING	B	WELDA	C	WICKLIFFE		WODSKOW	B	YECMAN	B
WAPSIE	B	WELDON	C	WICKSBURG	B	WOLCOTTSBURG		YETULL	A
WARBA	B	WELDONA	B	WIDTSOE	C	WOLDALE	C/D	YODER	B
WARD	D	WELLER	C	WIEHL	C	WOLF	B	YOKOHL	C
WARDBORO	A	WELLERHORN	C	WIEN	D	WOLFESEN	C	YOLLABOLLY	D
WARDELL	C	WELLINGTON	C	WIGGLETON	B	WOLFORD	B	YOLO	B
WARDEN	B	WELLMAN	B	WILBRAHAM	C	WOLF POINT	D	YOLOGO	D
WARDWELL	C	WELLNER	B	WILBUR	C	WOLFTEVER	C	YOMONT	B
WARE	B	WELLSBORO	C	WILCO	C	WOLVERINE	A	YONCALLA	C
WAREHAM	C	WELLSTON	B	WILCOX	D	WOODBINE	B	YONGES	D
WARMAN	D	WELLSVILLE	B	WILCOXSON	C	WOODBRIDGE	C	YONNA	B/D
WARM SPRINGS	C	WEMPLE	B	WILDCAT	D	WOODBURN	C	YORDY	B
WARNERS	A/D	WENAS	B/C	WILDER	B	WOODBURY	D	YORK	C
WARREN		WENATCHEE	C	WILDERNESS	C	WOODCOCK	B	YORKVILLE	D
WARRENTON	B/D	WENDEL	B/C	WILDROSE	D	WOODENVILLE	C	YOST	C
WARRIOR		WENHAM		WILDWOOD	D	WOODGLEN	D	YOUGA	B
WARSAW	B	WENONA	C	WILEY	C	WOODHURST	A	YOUMAN	C
WARSING	B	WENTWORTH	B	WILKES	C	WOODLY	B	YOUNGSTON	B
WARWICK	A	WERNER	B	WILKESON	C	WOODLYN	C	YOURAME	A
WASATCH	A	WESO	C	WILKINS	D	WOODMANSIE	B	YOVIMPA	D
WASEPI	B	WESSEL	B	WILL	D	WOODMERE	B	YSIDORA	C
WASHBURN		WESTBROOK	D	WILLACY	B	WOOD RIVER	C	YTURBIDE	A
WASHINGTON	B	WESTBURY	C	WILLAKENZIE	C	WOODROCK	B	YUBA	D
WASHOE	C	WESTCREEK	B	WILLAMAR	D	WOODROW	C	YUKON	D
WASHOUGAL	B	WESTERVILLE	C	WILLAMETTE	B	WOODS CROSS	D	YUNES	D
WASHTENAW	C/D	WESTFALL	C	WILLAPA	C	WOODSFIELD	C	YUNQUE	C
WASILLA	C	WESTFIELD		WILLARD	C	WOODSIDE	A		
WASIOJA	B	WESTFORD		WILLETTE	A/D	WOODSON	D	ZAAR	D
WASSAIC	B	WESTLAND	B/D	WILLHAND	B	WOODSTOCK	C/D	ZACA	C
WATAB	C	WESTMINSTER	C/D	WILLIAMS	B	WOODSTOWN	C	ZACHARIAS	B
WATAUGA	B	WESTMORE	B	WILLIAMSBURG	B	WOODWARD	B	ZACHARY	C
WATCHAUG	B	WESTMORELAND	B	WILLIAMSON	C	WOOLMAN	B	ZAFRA	B
WATCHUNG	D	WESTON	D	WILLIS	C	WOOLPER	C	ZAHILL	B
WATERBORO		WESTPHALIA	B	WILLITS	B	WOOLSEY	C	ZAHL	B
WATERBURY	D	WESTPLAIN	C	WILLOUGHBY	B	WOOSLEY	B	ZALESKI	C

(continued)

EXHIBIT 2-1 (continued)

Soil	Group	Soil	Group	Soil	Group	Soil	Group	Soil	Group
WATERINO	C	WESTPORT	A	WILLOW CREEK	B	WOOSTER	C	ZALLA	A
WATERS	C	WESTVILLE	B	WILLOWDALE	B	WOOSTERN	B	ZAMORA	B
WATKINS	B	WETHERSFIELD	C	WILLOWS	D	WOOTEN	A	ZANE	C
WATKINS RIDGE	B	WETHEY	B/C	WILLWOOD	A	WORCESTER	B	ZANEIS	B
WATOPA	B	WETZEL	D	WILMER	C	WORF	D	ZANESVILLE	C
WATROUS	B	WEYMOUTH	B	WILPAR	D	WORK	C	ZANONE	C
WATSEKA	C	WHALAN	B	WILSON	D	WORLAND	B	ZAPATA	C
WATSON	C	WHARTON	C	WILTSHIRE	C	WORLEY	C	ZAVALA	B
WATSONIA	D	WHATCOM	C	WINANS	B/C	WORMSER	C	ZAVCO	C
WATSONVILLE	D	WHATELY	D	WINCHESTER	A	WOROCK	B	ZEB	B
WATT	D	WHEATLEY	D	WINCHUCK	C	WORSHAM	C	ZEESIX	C
WATTON	C	WHEATRIDGE	C	WINDER	B/D	WORTH	C	ZELL	B
WAUBAY	B	WHEATVILLE	B	WINDMILL	B	WORTHEN	B	ZEN	C
WAUBEEK	B	WHEELER	B	WINDOM	B	WORTHING	D	ZENDA	C
WAUBONSIE	B	WHEELING	B	WIND RIVER	B	WORTHINGTON	C	ZENIA	B
WAUCHULA	B/D	WHEELOCK	C	WINDSOR	A	WORTMAN	C	ZENIFF	B
WAUCOMA	B	WHEELON	D	WINDTHORST	C	WRENTHAM	C	ZEONA	A
WAUCONDA	B	WHELCHEL	B	WINDY	C	WRIGHT	C	ZIEGLER	C
WAUKEE	B	WHETSTONE	B	WINEG	B	WRIGHTSVILLE	D	ZIGWEID	B
WAUKEGAN	B	WHIDBEY	B	WINEMA	C	WUNJEY	B	ZILLAH	B/C
WAUKENA	D	WHIPPANY	C	WINETTI	B	WURTSBORO	C	ZIM	D
WAUKON	B	WHIPSTOCK	C	WINFIELD	C	WYALUSING	D	ZIMMERMAN	A
WAUMBEK	B	WHIRLO	C	WING	D	WYARD	B	ZING	C
WAUPIKA	D	WHIT	B	WINGATE	B	WYARNO	C	ZINZER	B
WAUSEON	B/D	WHITAKER	C	WINGER	C	WYATT	C	ZION	C
WAVERLY	B/D	WHITCOMB	C	WINGVILLE	B/D	WYEAST	C	ZIPP	C/D
WAWAKA	C	WHITE BIRD	C	WINIFRED	C	WYEVILLE	C	ZITA	B
WAYCUP	B	WHITECAP	D	WINK	B	WYGANT	B	ZOAR	C
WAYDEN	D	WHITEFISH	B	WINKLEMAN	C	WYKOFF	B	ZOHNER	B/D
WAYLAND	C/D	WHITEFORD	B	WINLO	D	WYMAN	B	ZOOK	C
WAYNE	B	WHITEHORSE	B	WINLOCK	C	WYMORE	C	ZORRAVISTA	A
WAYNESBORO	B	WHITE HOUSE	C	WINN	C	WYNN	B	ZUFELT	B
WAYSIDE		WHITELAKE	B	WINNEBAGO	B	WYNOOSE	D	ZUMBRO	B
WEA	B	WHITELAW	B	WINNEMUCCA	B	WYO	B	ZUMWALT	C
WEAVER	C	WHITEMAN	D	WINNESHIEK	B	WYOCENA	B	ZUNDELL	B
WEBB	C	WHITEROCK	D	WINNETT	D			ZUNHALL	B
WEBER	B	WHITESBURG	C	WINONA	D	YACOLT	B	ZUNI	C
WEBSTER	C	WHITE STORE	D	WINOOSKI	B	YAHARA	B	ZURICH	B
WEDGE	A	WHITE SWAN	C	WINSTON	A	YAHOLA	B		
WEDOWEE	B	WHITEWATER	B	WINTERS	C	YAKIMA	B		
WEED	C	WHITEWOOD	C	WINTERSBURG	B	YAKUS	D		
WEEDING	A	WHITLEY	B	WINTERSET	C	YALLANI	B		
WEEDMARK	B	WHITLOCK	B	WINTHROP	A	YALMER	B		
WEEKSVILLE	B/D	WHITMAN	D	WINTONER	C	YAMHILL	C		

NOTES: A blank hydrologic soil group indicates the soil group has not been determined.
Two soil groups such as B/C indicates the drained/undrained situation.

SOURCE: SCS Engineering Handbook of Conservation.

APPENDIX D

Graphs for Runoff Calculations

PEAK RATES OF DISCHARGE FOR SMALL WATERSHEDS

TYPE STORM DISTRIBUTION

1

Slopes FLAT
Curve Number 50
24 Hour Rainfall

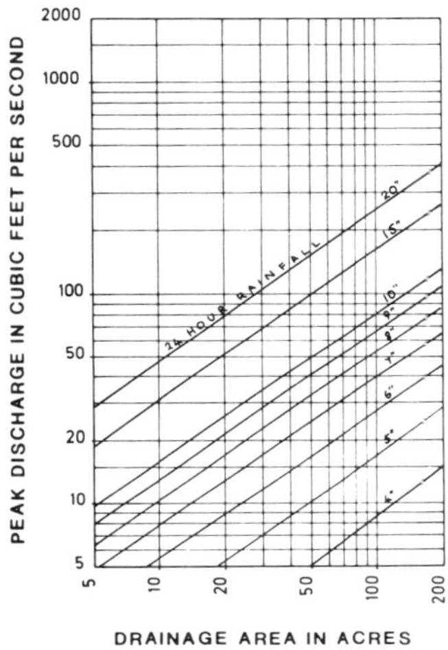

Source: "A Method for Estimating Volume and Rate of Run-off in Small Watersheds", U.S. DA SCS, April, 1973.

PEAK RATES OF DISCHARGE FOR SMALL WATERSHEDS

TYPE STORM DISTRIBUTION

1

Slopes FLAT
Curve Number 65
24 Hour Rainfall

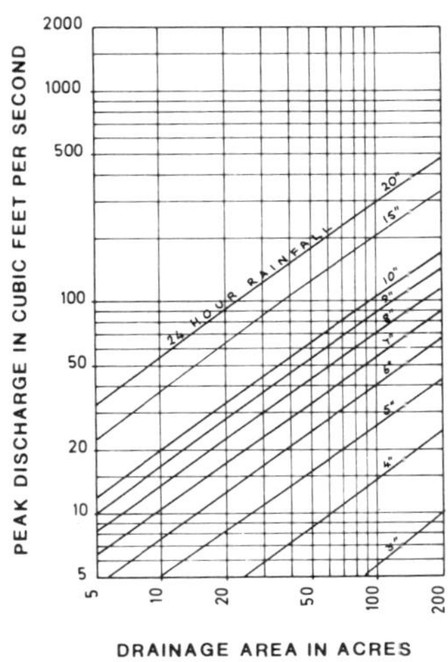

Source: "A Method for Estimating Volume and Rate of Run-off in Small Watersheds", U.S. DA SCS, April, 1973.

PEAK RATES OF DISCHARGE FOR SMALL WATERSHEDS

TYPE STORM DISTRIBUTION

1

Slopes FLAT
Curve Number 70
24 Hour Rainfall

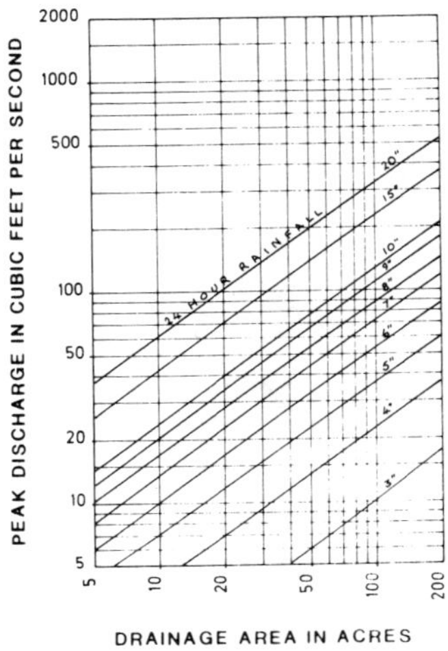

Source: "A Method for Estimating Volume and Rate of Run-off in Small Watersheds", U.S. DA SCS, April, 1973.

PEAK RATES OF DISCHARGE FOR SMALL WATERSHEDS

TYPE STORM DISTRIBUTION

1

Slopes FLAT
Curve Number 75
24 Hour Rainfall

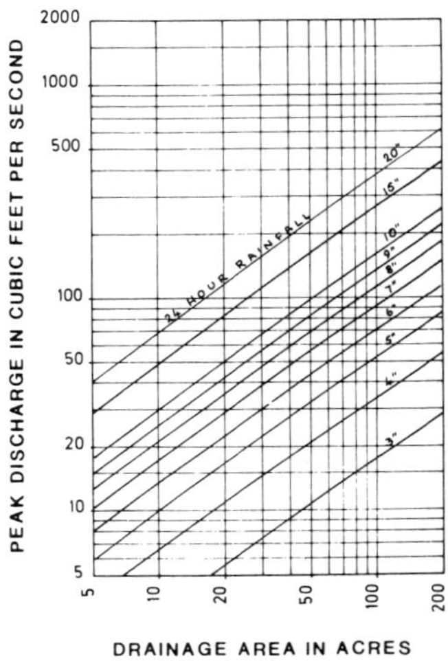

Source: "A Method for Estimating Volume and Rate of Run-off in Small Watersheds", U.S. DA SCS, April, 1973.

PEAK RATES OF DISCHARGE FOR SMALL WATERSHEDS

TYPE STORM DISTRIBUTION

1

Slopes FLAT
Curve Number 80
24 Hour Rainfall

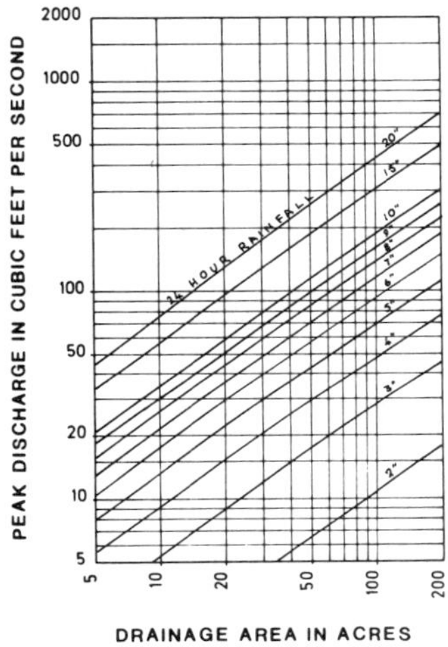

Source: "A Method for Estimating Volume and Rate of Run-off in Small Watersheds", U.S. DA SCS, Apr I, 1973.

PEAK RATES OF DISCHARGE FOR SMALL WATERSHEDS

TYPE STORM DISTRIBUTION

1

Slopes FLAT
Curve Number 85
24 Hour Rainfall

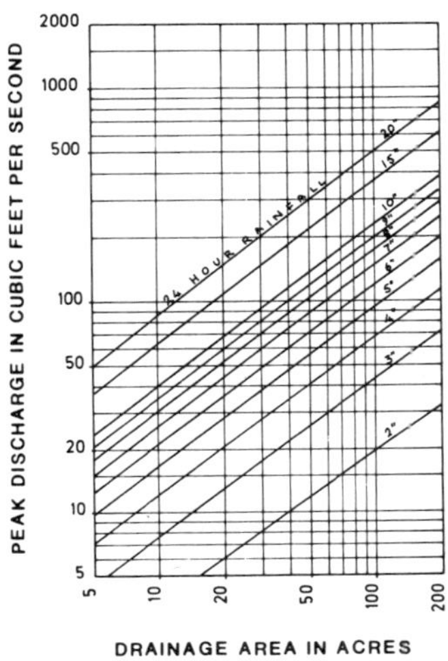

Source: "A Method for Estimating Volume and Rate of Run-off in Small Watersheds", U.S. DA SCS, April, 1973.

PEAK RATES OF DISCHARGE FOR SMALL WATERSHEDS

TYPE STORM DISTRIBUTION

1

Slopes FLAT
Curve Number 90
24 Hour Rainfall

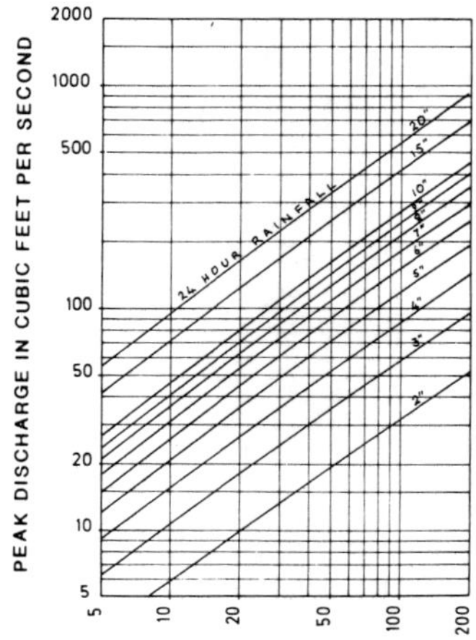

Source: "A Method for Estimating Volume and Rate of Run-off in Small Watersheds", U.S. DA SCS, April, 1973.

PEAK RATES OF DISCHARGE FOR SMALL WATERSHEDS

TYPE STORM DISTRIBUTION

1

Slopes MODERATE
Curve Number 60
24 Hour Rainfall

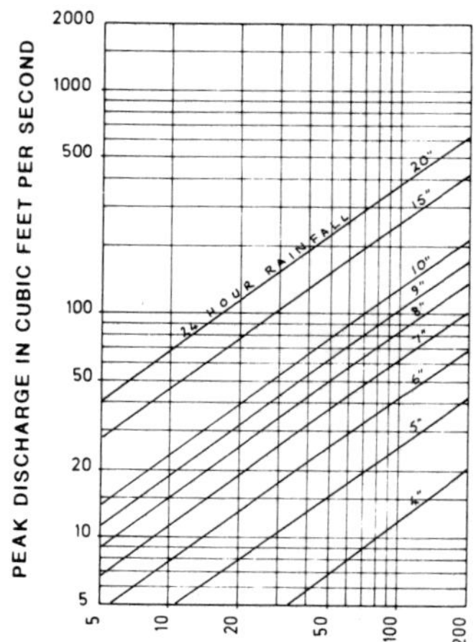

Source: "A Method for Estimating Volume and Rate of Run-off in Small Watersheds", U.S. DA SCS, April, 1973.

PEAK RATES OF DISCHARGE FOR SMALL WATERSHEDS

TYPE STORM DISTRIBUTION

1

Slopes MODERATE
Curve Number 65
24 Hour Rainfall

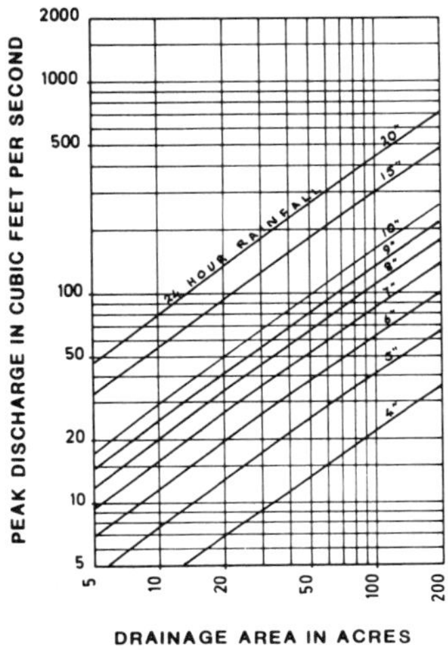

Source: "A Method for Estimating Volume and Rate of Run-off in Small Watersheds", U.S. DA SCS, April, 1973.

PEAK RATES OF DISCHARGE FOR SMALL WATERSHEDS

TYPE STORM DISTRIBUTION

1

Slopes MODERATE
Curve Number 70
24 Hour Rainfall

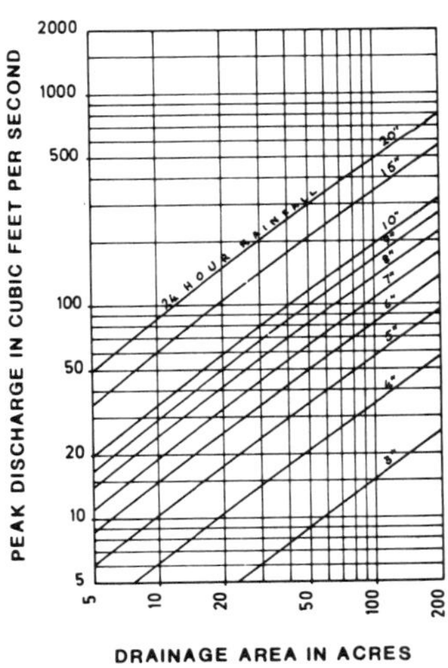

Source: "A Method for Estimating Volume and Rate of Run-off in Small Watersheds", U.S. DA SCS, April, 1973.

PEAK RATES OF DISCHARGE FOR SMALL WATERSHEDS

TYPE STORM DISTRIBUTION

1

Slopes MODERATE
Curve Number 75
24 Hour Rainfall

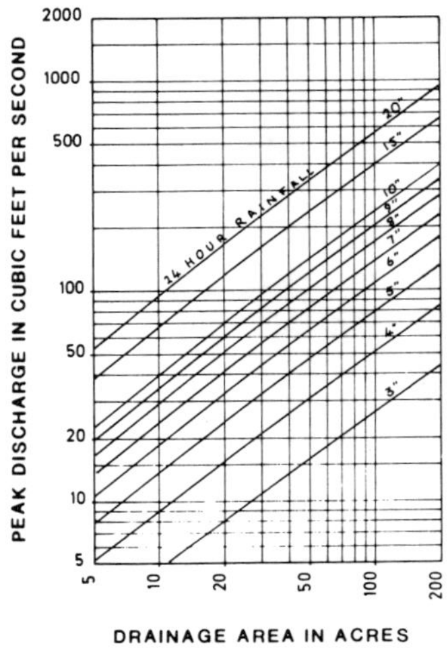

Source: "A Method for Estimating Volume and Rate of Run-off in Small Watersheds", U.S. DA SCS, April, 1973.

PEAK RATES OF DISCHARGE FOR SMALL WATERSHEDS

TYPE STORM DISTRIBUTION

1

Slopes MODERATE
Curve Number 80
24 Hour Rainfall

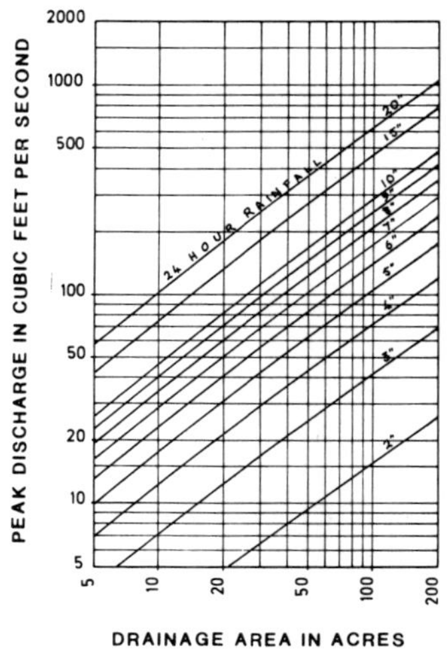

Source: "A Method for Estimating Volume and Rate of Run-off in Small Watersheds", U.S. DA SCS, April, 1973.

PEAK RATES OF DISCHARGE FOR SMALL WATERSHEDS

TYPE STORM DISTRIBUTION

1

Slopes MODERATE
Curve Number 85
24 Hour Rainfall

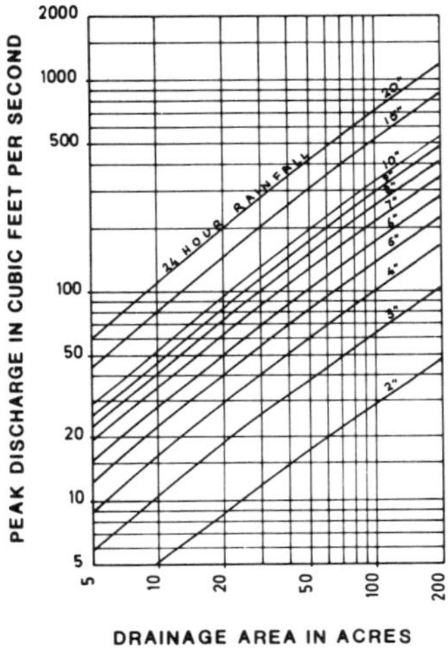

Source: "A Method for Estimating Volume and Rate of Runoff in Small Watersheds", U.S. DA SCS, April, 1973.

PEAK RATES OF DISCHARGE FOR SMALL WATERSHEDS

TYPE STORM DISTRIBUTION

1

Slopes MODERATE
Curve Number 90
24 Hour Rainfall

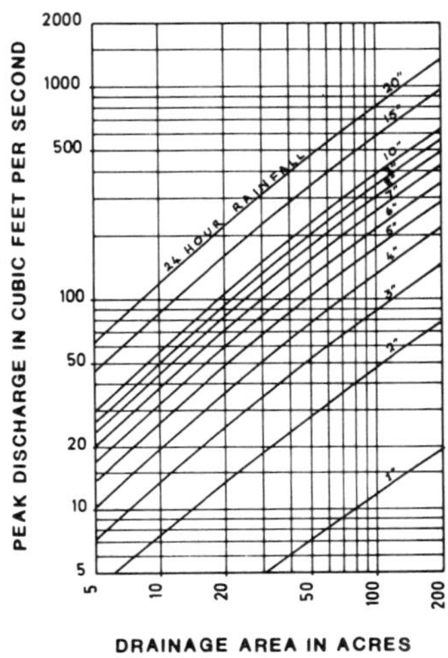

Source: "A Method for Estimating Volume and Rate of Runoff in Small Watersheds", U.S. DA SCS, April, 1973.

PEAK RATES OF DISCHARGE FOR SMALL WATERSHEDS

TYPE STORM DISTRIBUTION

1

Slopes STEEP
Curve Number 60
24 Hour Rainfall

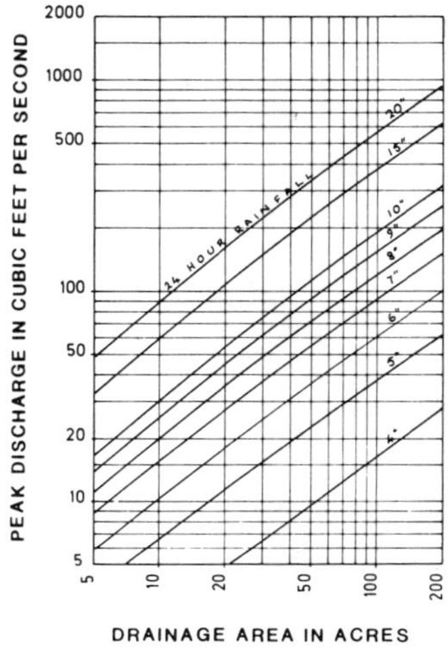

Source: "A Method for Estimating Volume and Rate of Runoff in Small Watersheds", U.S. DA SCS, April, 1973.

PEAK RATES OF DISCHARGE FOR SMALL WATERSHEDS

TYPE STORM DISTRIBUTION

1

Slopes STEEP
Curve Number 65
24 Hour Rainfall

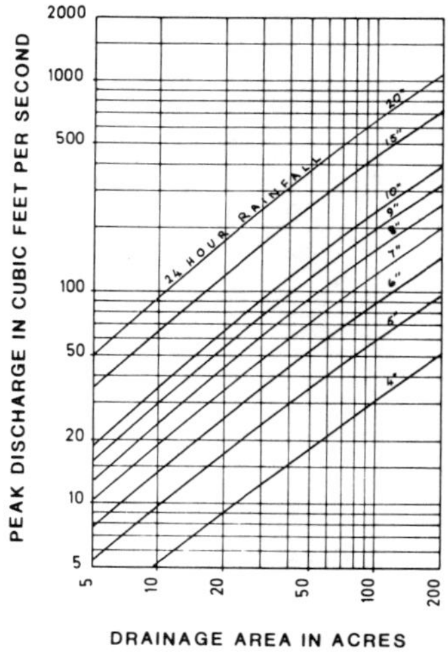

Source: "A Method for Estimating Volume and Rate of Runoff in Small Watersheds", U.S. DA SCS, April, 1973.

PEAK RATES OF DISCHARGE FOR SMALL WATERSHEDS

TYPE STORM DISTRIBUTION

1

Slopes STEEP
Curve Number 70
24 Hour Rainfall

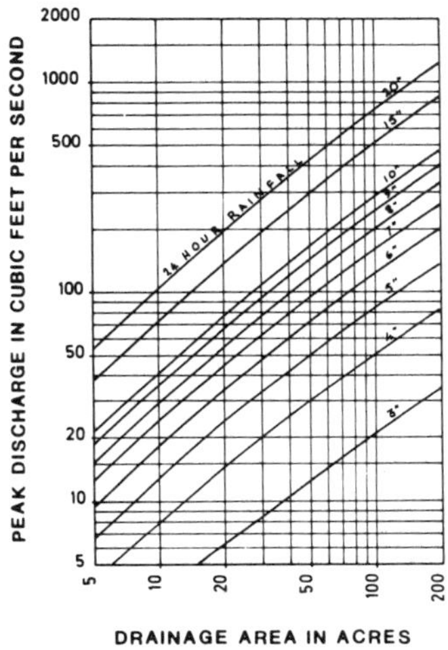

Source: "A Method for Estimating Volume and Rate of Run-off in Small Watersheds", U.S. DA SCS, April, 1973.

PEAK RATES OF DISCHARGE FOR SMALL WATERSHEDS

TYPE STORM DISTRIBUTION

1

Slopes STEEP
Curve Number 75
24 Hour Rainfall

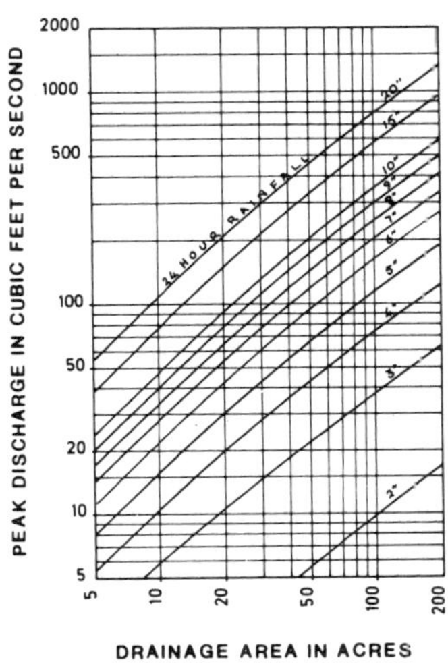

Source: "A Method for Estimating Volume and Rate of Run-off in Small Watersheds", U.S. DA SCS, April, 1973.

PEAK RATES OF DISCHARGE FOR SMALL WATERSHEDS

TYPE STORM DISTRIBUTION

1

Slopes STEEP
Curve Number 80
24 Hour Rainfall

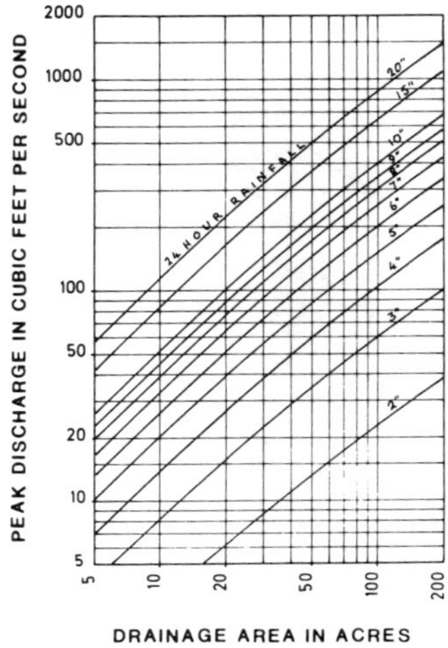

Source: "A Method for Estimating Volume and Rate of Run-off in Small Watersheds", U.S. DA SCS, April, 1973.

PEAK RATES OF DISCHARGE FOR SMALL WATERSHEDS

TYPE STORM DISTRIBUTION

1

Slopes STEEP
Curve Number 85
24 Hour Rainfall

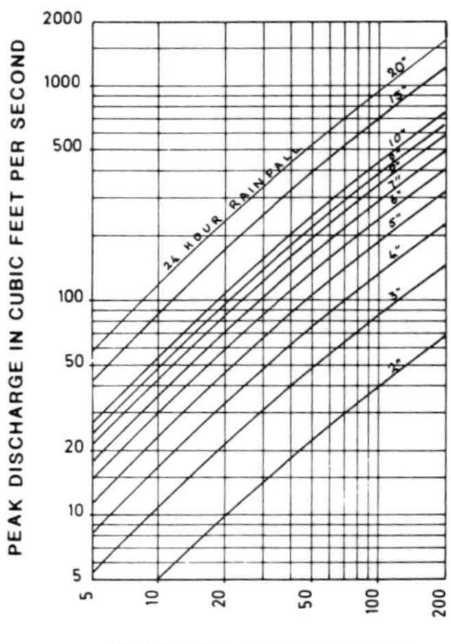

Source: "A Method for Estimating Volume and Rate of Run-off in Small Watersheds", U.S. DA SCS, April, 1973.

PEAK RATES OF DISCHARGE FOR SMALL WATERSHEDS

TYPE STORM DISTRIBUTION

1

Slopes STEEP
Curve Number 90
24 Hour Rainfall

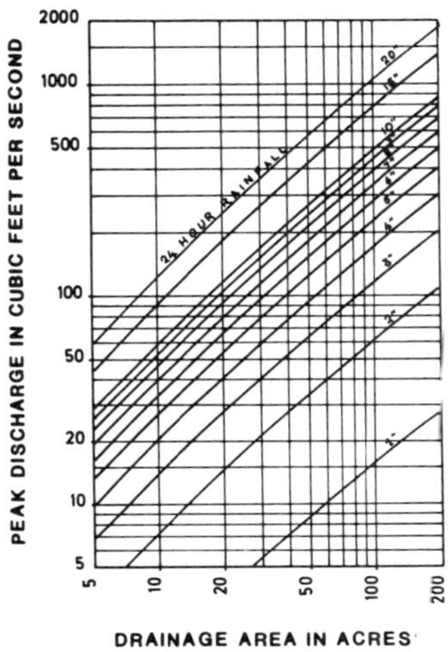

Source: "A Method for Estimating Volume and Rate of Runoff in Small Watersheds", U.S. DA SCS, April, 1973.

PEAK RATES OF DISCHARGE FOR SMALL WATERSHEDS

TYPE STORM DISTRIBUTION

2

Slopes FLAT
Curve Number 60
24 Hour Rainfall

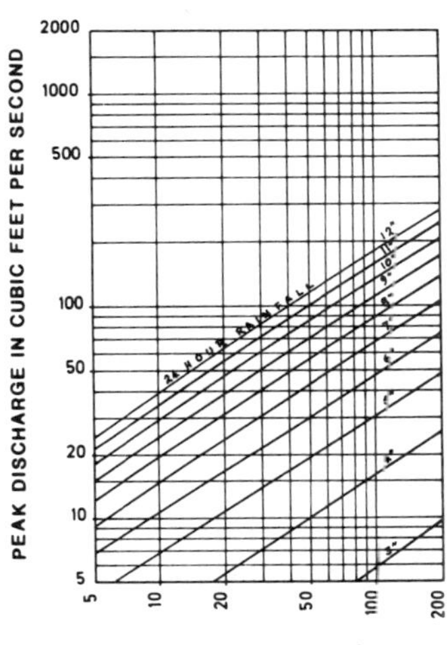

Source: "A Method for Estimating Volume and Rate of Runoff in Small Watersheds", U.S. DA SCS, April, 1973.

PEAK RATES OF DISCHARGE FOR SMALL WATERSHEDS

TYPE STORM DISTRIBUTION

2

Slopes FLAT
Curve Number 65
24 Hour Rainfall

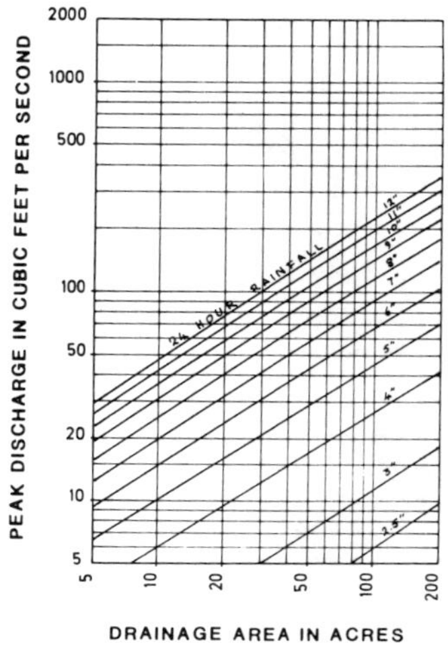

Source: "A Method for Estimating Volume and Rate of Runoff in Small Watersheds", U.S. DA SCS, April, 1973.

PEAK RATES OF DISCHARGE FOR SMALL WATERSHEDS

TYPE STORM DISTRIBUTION

2

Slopes FLAT
Curve Number 70
24 Hour Rainfall

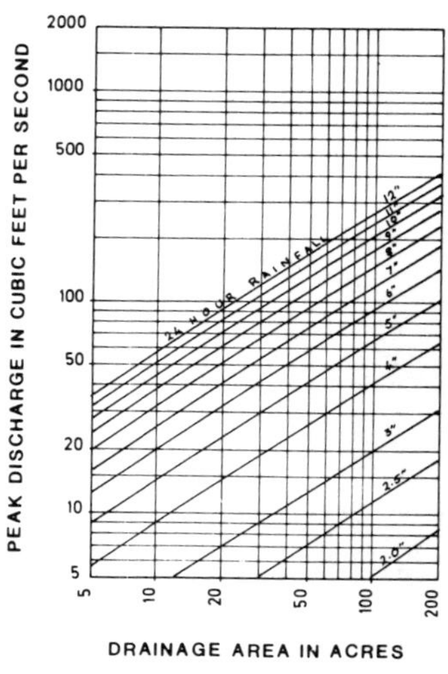

Source: "A Method for Estimating Volume and Rate of Runoff in Small Watersheds", U.S. DA SCS, April, 1973.

PEAK RATES OF DISCHARGE FOR SMALL WATERSHEDS

TYPE 2 STORM DISTRIBUTION

Slopes FLAT
Curve Number 75
24 Hour Rainfall

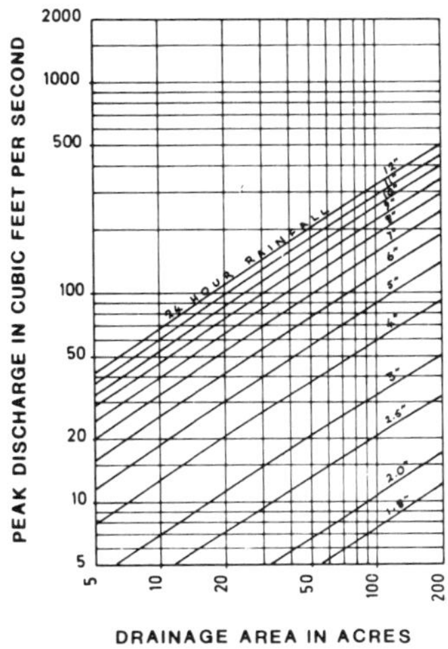

Source: "A Method for Estimating Volume and Rate of Run-off in Small Watersheds", U.S. DA SCS, April, 1973.

PEAK RATES OF DISCHARGE FOR SMALL WATERSHEDS

TYPE 2 STORM DISTRIBUTION

Slopes FLAT
Curve Number 80
24 Hour Rainfall

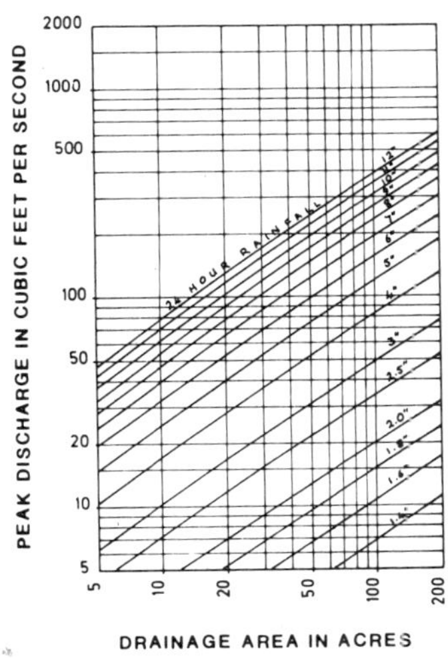

Source: "A Method for Estimating Volume and Rate of Run-off in Small Watersheds", U.S. DA SCS, April, 1973.

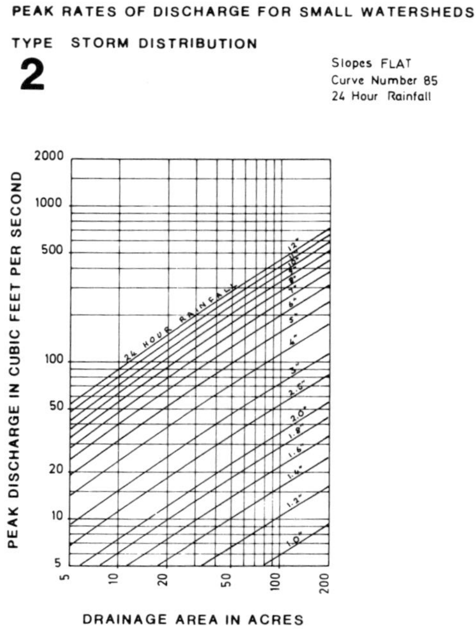

Source: "A Method for Estimating Volume and Rate of Runoff in Small Watersheds", U.S. DA SCS, April, 1973.

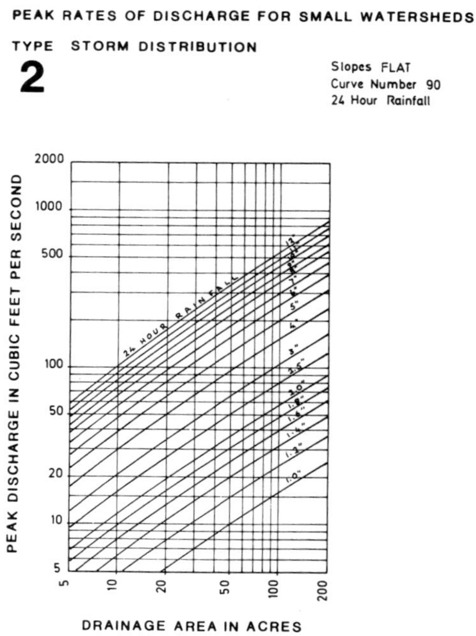

Source: "A Method for Estimating Volume and Rate of Runoff in Small Watersheds", U.S. DA SCS, April, 1973.

PEAK RATES OF DISCHARGE FOR SMALL WATERSHEDS

TYPE STORM DISTRIBUTION

2

Slopes MODERATE
Curve Number 60
24 Hour Rainfall

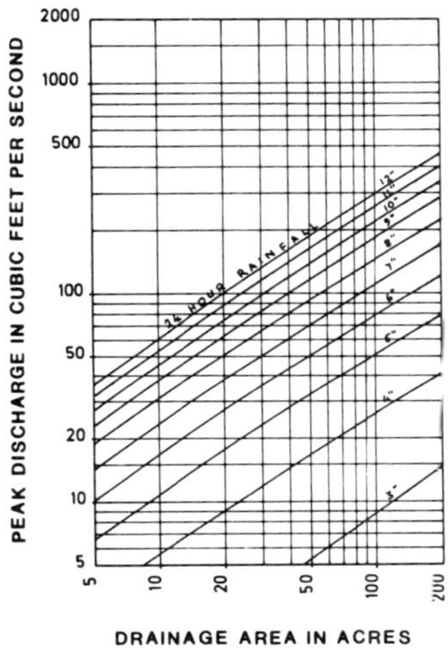

Source: "A Method for Estimating Volume and Rate of Run-off in Small Watersheds", U.S. DA SCS, April, 1973.

PEAK RATES OF DISCHARGE FOR SMALL WATERSHEDS

TYPE STORM DISTRIBUTION

2

Slopes MODERATE
Curve Number 65
24 Hour Rainfall

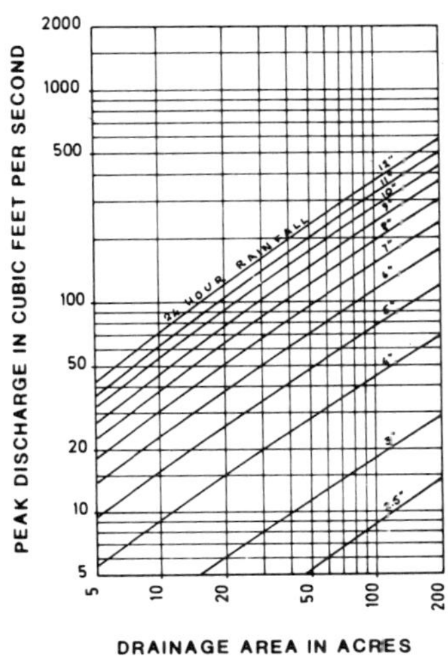

Source: "A Method for Estimating Volume and Rate of Run-off in Small Watersheds", U.S. DA SCS, April, 1973.

PEAK RATES OF DISCHARGE FOR SMALL WATERSHEDS

TYPE STORM DISTRIBUTION

2

Slopes MODERATE
Curve Number 70
24 Hour Rainfall

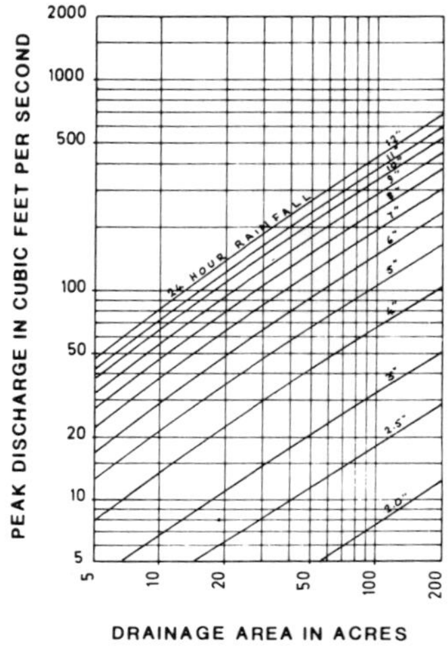

Source: "A Method for Estimating Volume and Rate of Run-off in Small Watersheds", U.S. DA SCS, April, 1973.

PEAK RATES OF DISCHARGE FOR SMALL WATERSHEDS

TYPE STORM DISTRIBUTION

2

Slopes MODERATE
Curve Number 75
24 Hour Rainfall

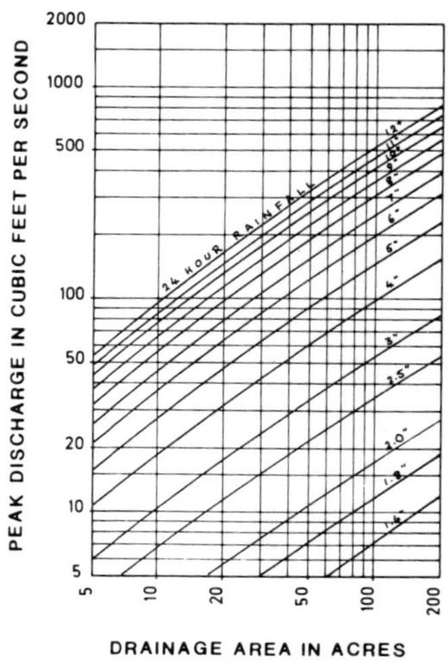

Source: "A Method for Estimating Volume and Rate of Run-off in Small Watersheds", U.S. DA SCS, April, 1973.

PEAK RATES OF DISCHARGE FOR SMALL WATERSHEDS

TYPE STORM DISTRIBUTION

2

Slopes MODERATE
Curve Number 80
24 Hour Rainfall

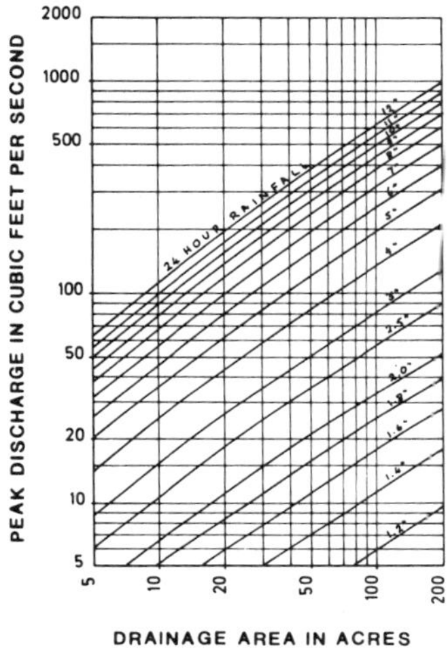

Source: "A Method for Estimating Volume and Rate of Run-off in Small Watersheds", U.S. DA SCS, April 1973.

PEAK RATES OF DISCHARGE FOR SMALL WATERSHEDS

TYPE STORM DISTRIBUTION

2

Slopes MODERATE
Curve Number 85
24 Hour Rainfall

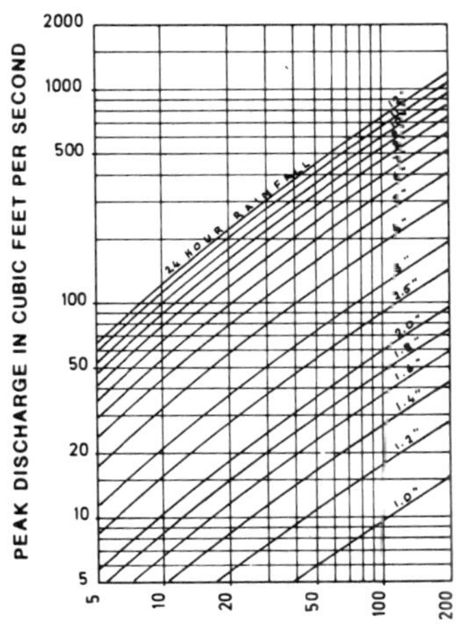

Source: "A Method for Estimating Volume and Rate of Run-off in Small Watersheds", U.S. DA SCS, April, 1973.

PEAK RATES OF DISCHARGE FOR SMALL WATERSHEDS

TYPE STORM DISTRIBUTION

2

Slopes MODERATE
Curve Number 90
24 Hour Rainfall

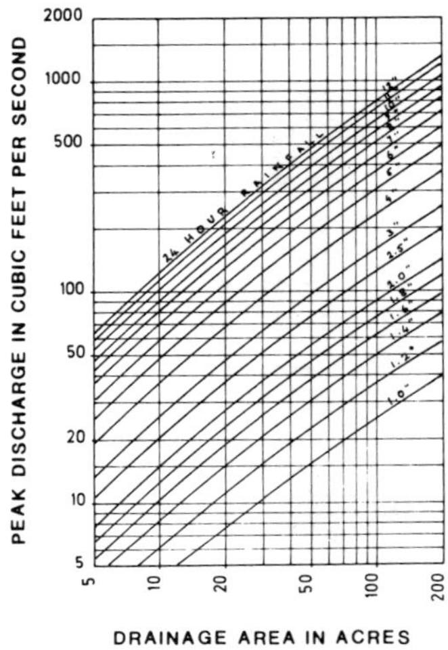

Source: "A Method for Estimating Volume and Rate of Run-off in Small Watersheds", U.S. DA SCS, April, 1973.

PEAK RATES OF DISCHARGE FOR SMALL WATERSHEDS

TYPE STORM DISTRIBUTION

2

Slopes STEEP
Curve Number 60
24 Hour Rainfall

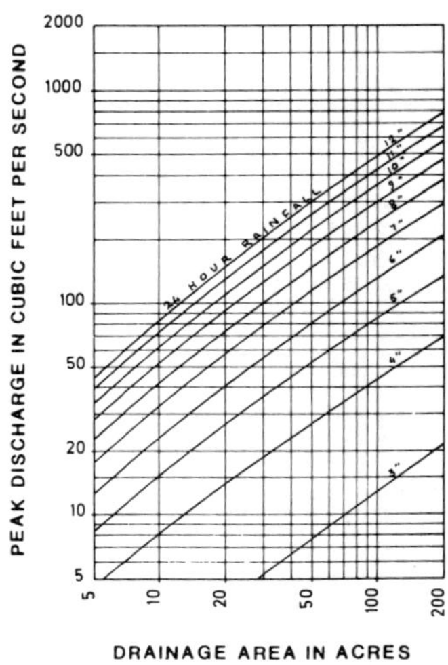

Source: "A Method for Estimating Volume and Rate of Run-off in Small Watersheds", U.S. DA SCS, April, 1973.

PEAK RATES OF DISCHARGE FOR SMALL WATERSHEDS

TYPE STORM DISTRIBUTION

2

Slopes STEEP
Curve Number 65
24 Hour Rainfall

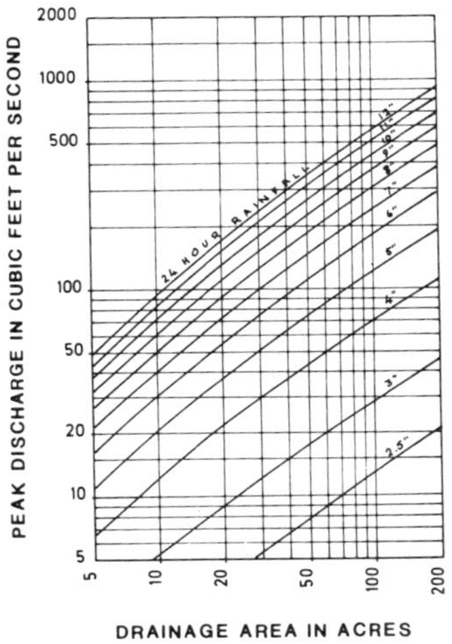

Source: "A Method for Estimating Volume and Rate of Run-off in Small Watersheds", U.S. DA SCS, April, 1973.

PEAK RATES OF DISCHARGE FOR SMALL WATERSHEDS

TYPE STORM DISTRIBUTION

2

Slopes STEEP
Curve Number 70
24 Hour Rainfall

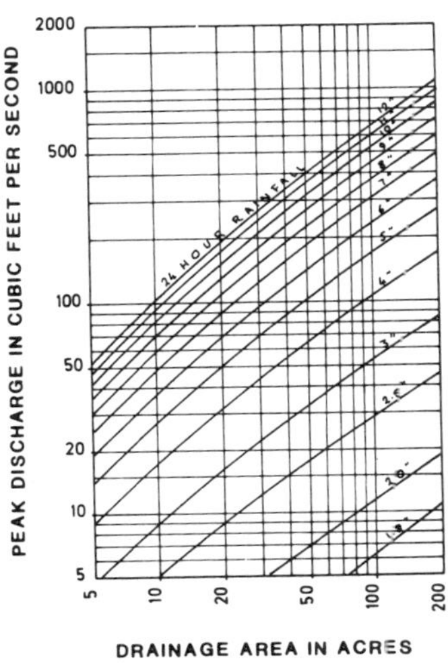

Source: "A Method for Estimating Volume and Rate of Run-off in Small Watersheds", U.S. DA SCS, April, 1973.

PEAK RATES OF DISCHARGE FOR SMALL WATERSHEDS

TYPE STORM DISTRIBUTION

2

Slopes STEEP
Curve Number 75
24 Hour Rainfall

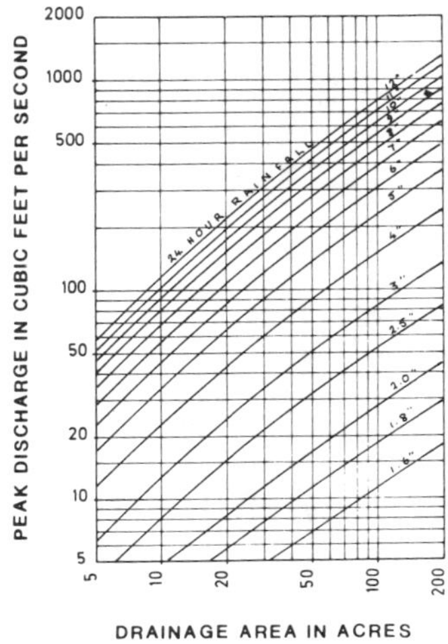

Source: "A Method for Estimating Volume and Rate of Run-off in Small Watersheds", U.S. DA SCS, April, 1973.

PEAK RATES OF DISCHARGE FOR SMALL WATERSHEDS

TYPE STORM DISTRIBUTION

2

Slopes STEEP
Curve Number 80
24 Hour Rainfall

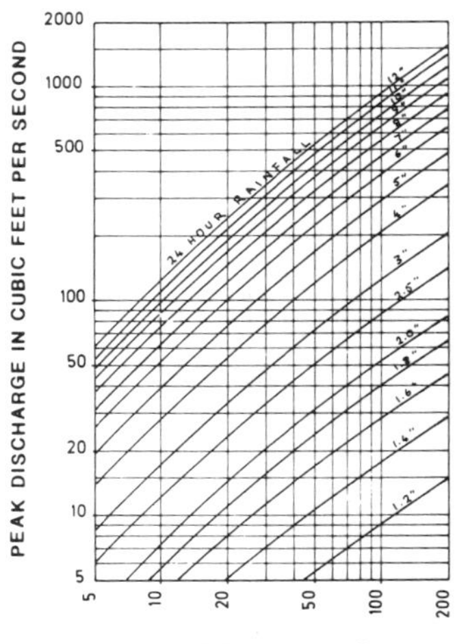

Source: "A Method for Estimating Volume and Rate of Run-off in Small Watersheds", U.S. DA SCS, April, 1973.

PEAK RATES OF DISCHARGE FOR SMALL WATERSHEDS

TYPE STORM DISTRIBUTION

2

Slopes STEEP
Curve Number 85
24 Hour Rainfall

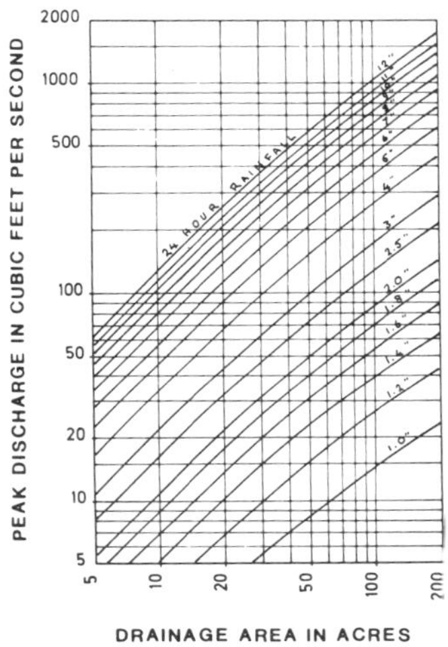

Source: "A Method for Estimating Volume and Rate of Run-off in Small Watersheds", U.S. DA SCS, April, 1973.

PEAK RATES OF DISCHARGE FOR SMALL WATERSHEDS

TYPE STORM DISTRIBUTION

2

Slopes STEEP
Curve Number 90
24 Hour Rainfall

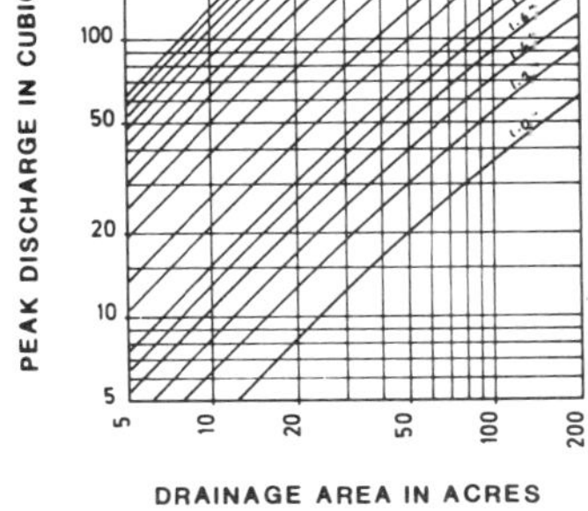

Source: "A Method for Estimating Volume and Rate of Run-off in Small Watersheds", U.S. DA SCS, April, 1973.

APPENDIX E

Mannings Nomograph

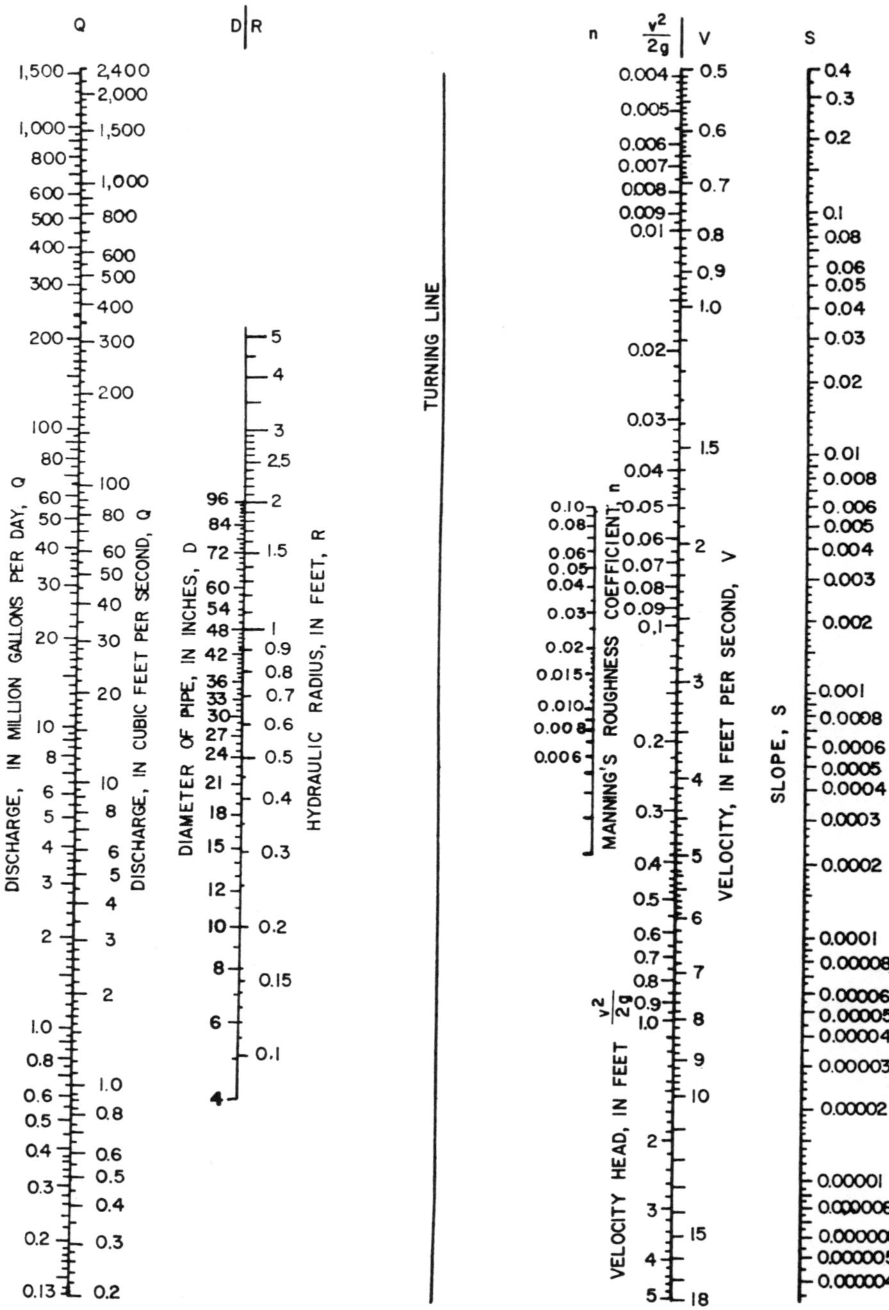

APPENDIX F

How to Calculate Injury Rates

I. COMPUTE HOURS OF EXPOSURE

	Full-time Permanent (1)	Part-time Permanent (2)	Full-time Temporary (3)	Part-time Temporary (4)	Other (5)	Other (6)	Other (7)	Other (8)
Approximate number of workers (averaged over year)	(a)	(a)	(a)	(a)	(a)	(a)	(a)	(a)
Average number of weeks worked per year per worker	48*(b)	48*(b)	(b)	(b)	(b)	(b)	(b)	(b)
Average number of hours worked per week (including overtime) per worker	(c)	(c)	(c)	(c)	(c)	(c)	(c)	(c)
Work Force hours per year (MULTIPLY (a) x (b) x (c) = (d) for each column)	(d)	(d)	(d)	(d)	(d)	(c)	(d)	(d)

*48 weeks is used as an average for permanent positions. 48 weeks is 52 weeks, less two weeks for vacation, 1 week to allow for holidays and 1 week for sick and other leave. You may change 48 to some other number if you like, but 48 is probably close enough in most cases.

TOTAL HOURS OF EXPOSURE (Add up all the numbers in the (d) boxes. Call this Z.)

[ ] Z

1. Choose a 1 year period (e.g., calendar year 1974).
2. Divide the work force during this year into work force groups by:
   1) the average number of weeks worked during the year
   2) the average number of hours worked per week, including overtime; per worker.

DO NOT HESITATE TO MAKE ESTIMATES OR USE AVERAGES. PLACE EMPLOYEES WITH UNUSUAL WORK HOURS IN THE WORK FORCE GROUP WHICH MOST NEARLY APPROXIMATES THEIR WORK HOURS.

NOTE: In counting the number in a work force group, count the number of <u>positions</u> rather than the number of employees. For example, if an employee in a Heavy Equipment Operator position quit after 3 months, and another employee was hired to fill the position for the other 9 months, these two employees would be counted as 1 full-time permanent employee.

3. Complete the above table using the employee groups you have identified.

<u>EXAMPLE</u>:
An organization had, during 1974, an average of approximately 125 filled permanent positions, working an average of 48 weeks a year at an average of 40 hours per week. The organization had no part-time employees. It had hired 50 full-time temporary employees for 12 weeks during the summer, who worked an average of 45 hours a week including overtime. In addition it had an average of 25 landfill workers who worked all year but worked a 48 hour week. The hours of exposure are, in column 1, 125 x 48 x 40 = 240,000; in column 3, 50 x 12 x 45 = 27,000; and in column 5, 25 x 48 x 48 = 57,600, for a total of 324,600, in box Z.

<u>ALTERNATE METHOD</u>: Using payroll records
You may be able to obtain the number of hours worked per year by your employees from your payroll records. Keep in mind, however, that hours of exposure is equal to hours <u>worked</u> not hours <u>paid</u> (if they are different). If your payroll records of hours do not separate hours paid from the hours that are not worked, but are paid (such as vacations, sick leave and holidays), a correction is necessary. In this case, use 92% of your hours paid to compute your hours worked.

II. LIST INJURIES BY TYPE

1. Review your injuries that occurred during the same time period for which you computed hours of exposure. Use either the OSHA Log (OSHA Form 100) or other injury records.
2. Starting from the top of the list of categories below, enter the number of injuries at your organization which fall into each category.

BE CERTAIN THAT YOU COUNT EACH INJURY ONLY ONCE. If an injury falls into more than one category, place it in the highest category on the list in which it falls. For example, an injury that resulted in both medical treatment and lost time would be put in the "lost time injury" category.

DO NOT INCLUDE INJURIES RESULTING IN FIRST AID ONLY.

Category and Brief Definition	Place Number of Injuries in These Boxes	
DEATH	[ ]	A
PERMANENT DISABILITY (Total or Partial)	[ ]	B
LOST TIME INJURY - An injury which results in time lost from regular job, not including the day of the injury or the day of return to work.	[ ]	C
TRANSFER TO ANOTHER JOB (temporarily or permanently) e.g., an injury that required return to a "light duty" job assignment.	[ ]	D
RESTRICTION OF WORK OR MOTION, OR LOSS OF CONSCIOUSNESS	[ ]	E
MEDICAL TREATMENT ONLY - Includes treatment administered by a physician or by registered professional personnel. It does not include first aid (one-time treatment or subsequent observation of minor injuries which do not ordinarily require medical care) even though provided by a physician.	[ ]	F

3. For ANSI frequency rate ADD A + B + C = X [ ] X
4. For OSHA incidence rate ADD A + B + C + D + E + F = Y [ ] Y

III. CALCULATE INJURY RATE

ANSI Frequency Rate:

Divide the number in box "X" by the number in box "Z", and multiply the quotient by 1,000,000.

[ ] = [ ] "X" / [ ] "Z" x 1,000,000

Your ANSI Frequency Rate

OSHA Incidence Rate:

Divide the number in box "Y" by the number in box "Z", and multiply the quotient by 200,000.

[ ] = [ ] "Y" / [ ] "Z" x 200,000

Your OSHA Incidence Rate

Tear off this form and use it to calculate your injury rate. If you would like to help improve the National Safety Council's data on injuries in the solid waste industry, please send a copy of your calculations to:

Mr. Ray Lascoe
National Safety Council
425 North Michigan Avenue
Chicago, Illinois 60611

SUPERVISOR'S ACCIDENT REPORT

Incident No. ________________

(To be completed immediately after accident, even when there is no injury)

Company name and address ______________________________

Plant or location address ______________________________
(if different from above)

1. Name and address of injured ______________________ SSN ______________ 2. Age ______
(or ill) person
______________________________ 3. Sex ______

4. Years of service ______ 5. Time on present job ________ 6. Title/occupation ______________

7. Department ______________________ 8. Date of accident ______________ 9. Time ________

10. Accident category (check) ☐ Motor Vehicle; ☐ Property Damage; ☐ Fire; ☐ Other ______________

11. Severity of injury or illness ☐ Non-disabling; ☐ Disabling; ☐ Medical Treatment; ☐ Fatality

12. Amount of damage $ ____________ 13. Location ______________________

14. Estimated number of days away from job ______________________

15. Nature of injury or illness? ______________________

16. Part of body affected? ______________________

17. Degree of disability? ______________________
(Temporary total; permanent partial; permanent total)

18. Causative agent most directly related to accident? (Object, substance, material, machinery, equipment conditions)

____

____

____

Was weather a factor? ______________________

19. Unsafe mechanical/physical/environmental condition at time of accident? (Be specific)

____

____

____

20. Unsafe act by injured and/or others contributing to the accident. (Be specific, *must* be answered)

____

____

____

21. Personal factors (improper attitude, lack of knowledge or skill, slow reaction, fatigue)

____

____

____

22. **Personal protective equipment required? (Protective glasses, safety shoes, safety hat, safety belt)**__________

Was injured using required equipment?__________

23. **What can be done to prevent a recurrence of this type of accident? (Modification of machine; mechanical guards; correct environment; training)**

____

____

24. **Detailed narrative description (How did accident occur; why; objects, equipment, tools used, circumstance, assigned duties. Be specific)**__________

____

____

____

(Use additional sheets, as required)

25. **Witnesses to accident**__________

____

Date prepared__________ **Signature of Foreman/Supervisor**__________

Department__________

SUPERINTENDENT'S APPRAISAL AND RECOMMENDATION

a. In your opinion what action on the part of injured (or ill) person or others contributed to this accident?

____

____

b. Your recommendation__________

____

Date__________ Signature of Superintendent__________

FOR SAFETY OFFICE USE ONLY

Temporary Total ☐	Permanent Partial ☐	Death or Permanent Total ☐
Started losing time______	Part of Body______	
Returned to work______	Per cent loss or loss of use______	
Time charge______	Time charge______	Time charge: 6,000 days

Compensation $______ Medical $______ Other $______ Total $______

Name and address of hospital__________ **Name and address of physician**__________

Issued by NATIONAL SAFETY COUNCIL • 425 North Michigan Avenue • Chicago, Illinois 60611.

A NONGOVERNMENTAL, PRIVATELY SUPPORTED, PUBLIC SERVICE ORGANIZATION

This form meets information requirements of OSHA form 101.

Form IS-IA-25M-87301 Printed in U.S.A. Stock No. 129.21

APPENDIX G

Counterdoses for Poisons or Overdoses

AMERICAN DRUGGIST

COUNTERDOSES FOR THE HOME

POISONS

- Acids · 16
- After-Shave Lotion · 8
- Camphor · 1
- Carbon Monoxide · 11
- Chlorine Bleach · 15
- Cologne, Perfume · 8
- Detergents · 15
- Disinfectant
  - *with chlorine* · 15
  - *with carbolic acid* · 3
- Food Poisoning · 6
- Furniture Polish · 14
- Gasoline, Kerosene · 14
- Household Ammonia · 13
- Insect & Rat Poisons
  - *with arsenic* · 2
  - *with sodium fluoride* · 10
  - *with phosphorus* · 12
  - *with strychnine* · 5
- Lighter Fluid · 14
- Lye · 13
- Wild Mushrooms · 6
- Oil of Wintergreen · 8
- Pine Oil · 14
- Rubbing Alcohol · 8
- Turpentine · 14

OVERDOSES

- Alcohol · 8
- Aspirin · 8
- Barbiturates · 9
- Belladonna · 5
- Bromides · 6
- Codeine · 4
- Headache & Cold Compounds · 8
- Iron Compounds · 7
- Morphine, Opium · 4
- Paregoric · 4
- 'Pep' Medicines · 2
- Sleeping Medicines · 9
- Tranquilizers · 9

DO THIS FIRST

- **CALL A PHYSICIAN—IMMEDIATELY!**
- Keep the patient warm.
- Determine if the patient has taken
  - *(1) A POISON*
  - *(2) AN OVERDOSE*
- While waiting for physician, give appropriate counterdose below.
- But do not force any liquids on the patient—if he is unconscious.
- And do not induce vomiting if patient is having convulsions, or if patient is unconscious.

To Find The Correct Counterdose

- In one of the lists printed at left, find substance causing the trouble.
- Next to that substance is a number. This refers to counterdose bearing same number in the section below.

Keep All Poisons And Medicines Out Of Reach Of Children!

1
Induce vomiting with
- Finger in throat. OR
- 1 tablespoon of syrup of ipecac, followed by a glass of water. OR
- Teaspoonful of mustard in half glass of water.

2
- Give glass of milk. OR 1 tablespoonful of activated charcoal mixed with a little water.
- Induce vomiting (see #1) but not with syrup of ipecac if charcoal was given.

3
- Induce vomiting. (See #1)
- Then give 4 tablespoons of castor oil.
- Next give glass of milk OR the white of 2 raw eggs.

4
- Induce vomiting (see #1) if patient is conscious.
- Give glass of milk, OR activated charcoal in water.
- Give 2 tablespoons of epsom salt in 2 glasses of water.
- Keep patient awake.

5
- Give glass of milk, or activated charcoal in water.
- Induce vomiting (#1) if not in convulsions. Do not use ipecac if charcoal was given.
- Keep patient quiet.

6
- Induce vomiting. (See #1)
- Next give 2 tablespoons of epsom salt in 2 glasses of water—except in cases where diarrhea is severe.

7
- Induce vomiting. (See #1)
- Give 2 teaspoons of bicarbonate of soda in a glass of warm water.
- Finally give glass of milk.

8
- Give a glass of milk.
- Next induce vomiting (#1). Give 2 teaspoons of bicarbonate of soda in a glass of warm water.

9
- Give activated charcoal in water.
- Induce vomiting (see #1) but not with ipecac.
- Give 2 tablespoons of epsom salt in 2 glasses of water.

10
- Give glass of milk or lime water.
- Then induce vomiting. (See #1)

11
- Carry victim into fresh air.
- Make patient lie down.
- Give artificial respiration if necessary.

12
- Induce vomiting. (See #1).
- Don't let vomit touch victim's skin or yours.
- Then give 4 oz mineral oil. Positively do NOT give vegetable or animal oil.
- Also give 1 tablespoon of bicarbonate of soda in a quart of warm water.

13
- Give 2 tablespoons of vinegar in 2 glasses of water.
- Then give the white of 2 raw eggs . . . or 2 ounces of vegetable oil.
- Do NOT induce vomiting!

14
- Give water or milk.
- Then give 4 tablespoons of vegetable oil.
- Do NOT induce vomiting!

15
- Give patient one or two glasses of milk.

16
- Give large quantity of water.
- Give 2 tablespoons of milk of magnesia, or a similar antacid.
- Do NOT induce vomiting!

Post This Chart On The Back Of Your Medicine Cabinet Door—Or On The Back Of Your Bathroom Door.

EMERGENCY PHONE NUMBERS

PHYSICIAN ____________________

POISON CONTROL CENTER ____________________

PHARMACIST ____________________

POLICE ____________________

HOSPITAL ____________________

AMBULANCE ____________________

FIRE ____________________

Bibliography

American City, 1947. What Happens Inside a Sanitary Landfill? Am. City, Vol. 62, No. 5, pp. 11.

American Public Works Association, 1963. Proceedings of the National Conference on Solid Waste Research. Spec. Rept. No. 29.

American Public Works Association, 1966 (2nd ed.). Municipal Refuse Disposal. APWA Public Admin. Service, Chicago.

American Society of Civil Engineers, 1959. Sanitary Landfill. ASCE Manuals of Engineering Practice, No. 39.

J. R. Andersen and J. N. Dornbush, 1967. Influence of Sanitary Landfill on Ground-Water Quality. Jour. Am. Water Works Ass., Vol. 59, No. 4, pp. 457–470.

J. R. Andersen and J. N. Dornbush, 1968. Investigation of the Influence of Waste Disposal Practices on Ground-Water Qualities. Technical Completion Report, OWRR Project A-003-SDAK, Water Res. Inst., Washington, DC.

J. R. Andersen and J. N. Dornbush, 1968. Quality Changes of Shallow Ground-Water Resulting from Refuse Disposal at a Gravel Pit. Rept. on Proj. SRI 3553, Brookings, S. Dakota.

J. R. Andersen, 1964. The Public Health Aspects of Solid Waste Disposal. Public Health Reports, Vol. 79, No. 2, pp. 93–96.

M. A. Apgar and D. Langmuir, 1971. Ground-Water Pollution Potential of a Landfill Above the Water Table. Ground-Water, Vol. 9, No. 6, pp. 76–94.

G. L. Barthauer, Z. V. Kosowski, and J. P. Ramsey, 1971. Control of Mine Drainage from Coal Mine Mineral Waters. Phase I – Hydrology and Related Experiments. Env. Prot. Agency, Water Qual. Office, Water Pollution Res. Series, Project 14010DDH.

B. A. Begg, 1967. Sanitary Landfill – A Bibliography. Drexel Inst. Techn. Civil Eng. Dept., Philadelphia, Pa.

R. E. Bergstrom, 1968a. Disposal of Wastes: Scientific and Administrative Considerations. Ill. St. Geol. Survey Env. Geol. Notes, No. 20.

R. E. Bergstrom, 1968b. Feasibility of Sub-Surface Disposal of Industrial Wastes in Illinois. Ill. St. Geol. Survey Circular 426.

R. E. Bergstrom, 1968c. Feasibility Criteria for Subsurface Waste Disposal in Illinois. Ground Water, Vol. 6, No. 5, pp. 5–9.

R. E. Bevan (editor), 1967. Notes on the Science and Practice of the Controlled Tipping of Refuse. The Inst. of Public Cleansing, London.

B. F. Bjornson, H. D. Pratt, and K. S. Littig, 1956. "Control of Domestic Rats and Mice," U.S. Department of HEW, Reprinted 1971 by Bureau of Solid Waste Management.

D. R. Brunner, D. J. Keller, C. W. Reid Jr., and J. Wheeler, 1970. Sanitary Landfill Guidelines — 1970. U.S. Dept. of HEW, Public Health Service, Bureau of Solid Waste Management.

D. R. Brunner and D. J. Keller, 1971. Sanitary Landfill Design and Operation. Env. Protection Agency, Solid Waste Management Program, Cincinnati.

R. G. Butler, G. T. Orlob, and P. H. McGauhey, 1954. Underground Movement of Bacterial and Chemical Pollutants. J. Am. Water Works Ass., Vol. 46, pp. 97–111.

California State Water Resources Control Board, Solid Waste Eng. Section, 1968. Special Study — Solid Wastes and Water Quality: A Study of Solid Waste Disposal and Their Effect on Water Quality in the San Francisco Bay — Delta Area. Task VII-1C, Cal. State Dept. of Health.

J. T. Callahan, 1972. The Role of the U.S. Geological Survey in Waste Disposal Monitoring. Ground Water, Vol. 10, No. 3, pp. 33–44.

C. K. Calvert, 1932. Contamination of Ground-Water by Impounded Garbage Waste. Jour. Am. Water Works Ass. Vol. 24, No. 2, pp. 266–270.

L. V. Carpenter and L. R. Setter, 1940. Some Notes on Sanitary Landfills. Am. Jour. Publ. Health, Vol. 30, No. 4, pp. 385–393.

K. Cartwright and M. R. McComas, 1968. Geophysical Survey in the Vicinity of Sanitary Landfills in NE Illinois, Ground-Water, Vol. 6, No. 5, pp. 23–30.

K. Cartwright and F. B. Sherman, 1969. Evaluating Sanitary Landfill Sites in Illinois. Ill. St. Geol. Survey Env. Geol. Notes, No. 27.

K. Cartwright and F. B. Sherman, 1971. Ground-Water and Engineering Geology in Siting of Sanitary Landfills. Ill. State Geol. Survey Reprint Series.

J. J. Coe, 1970. Effect of Solid Waste Disposal on Ground-Water Quality. Jour. AWWA, Vol. 62, No. 12, pp. 776–783.

Committee on Pollution, 1966. Waste Management and Control. Nat'l Academy of Sciences, Nat'l Research Council, Publ. 1400.

County of Los Angeles, Dept. of County Engineer, and Engineering Science, Inc., 1969. Development of Construction and Use Criteria for Sanitary Landfills. A Report on Grant No. D01-UI-00046, U.S. Dept. of HEW, Public Health Service, Bur. Solid Waste Management., Cincinnati.

J. W. Crosby, III, D. L. Johnstone, C. H. Drake, and R. L. Fenton, 1968. Migration of Pollutants in a Glacial Outwash Environment. Wat. Res. Research, Vol. 4, No. 5, pp. 1095–1114.

F. L. Cross and G. Noble, 1973. "Handbook on Hospital Solid Waste Management," Technomic Publishing Company.

R. L. Cummins, 1968. Effects of Land Disposal of Solid Wastes on Water Quality. Rept. SW-2ts, Dept. of HEW, Public Health Service, Cincinnati.

Drexel University, 1968. Pollution of Subsurface Water by Sanitary Landfills — Final Report. U.S. Dept. HEW, Public Health Service Grant No. 5-R01-U100516.

P. B. Du Montelle, 1970. Geologic Investigation of the Site for an Environmental Pollution Study. Ill. St. Geol. Survey Env. Geol. Notes, No. 31.

W. L. Dunn, 1968. Refuse Filling of a Swamp in a Fresh Water Lake. Civil Engineering, Jan. 1968, pp. 60–62.

Editorial, 1962. Refuse Disposal, Ground Water and Research. Public Works, Vol. 93, No. 6, pp. 7.

R. W. Eldredge, 1973. "Designing Landfills to Minimize Leachate." Presented at 1973 International Public Works Congress, September.

R. W. Eldredge, 1975. "Selection of Sanitary Landfill Equipment." Waste Age, Jan./Feb.

Engineering News Record, 1969. Solid Wastes Pile Up While Laws Crack Down and Engineers Gear Up. Eng. News Record, June 2, pp. 28–32.

Engineering-Science, Inc., 1961. Effects of Refuse Dumps on Ground-Water Quality. Cal. St. Water Pollution Control Board Pub. 24.

Engineering-Science, Inc., 1965 and 1967. In situ Investigation of Movements of Gases Produced from Decomposing Refuse. Cal. Water Quality Control Board Pub. No. 31 and 35.

R. N. Farvolden and G. M. Hughes, 1969. Sanitary Landfill Design. Ind. Water Eng., Vol. 6, No. 8, pp. 26–28.

J. C. Frye, 1967. Geological Information for Managing the Environment. Ill. State Geol. Survey Env. Geol. Notes, No. 18.

J. C. Frye, 1971. A geologist Views the Environment. Ill. State Geol. Survey Env. Geol. Notes, No. 42.

A. A. Fungaroli and R. L. Steiner, 1971. Laboratory Study of the Behavior of a Sanitary Landfill. J. Water Poll. Control Feder., Vol. 43, No. 2, pp. 252–267.

A. A. Fungaroli and R. L. Steiner, 1971. Results of a Laboratory Sanitary Landfill Study. Drexel Inst. Techn., Dept. of Civ. Eng., Philadelphia.

A. A. Fungaroli, R. L. Steiner, and I. Remson, 1968. Design of a Sanitary Landfill Laboratory Lysimeter. Drexel Inst. Techn., Dept. of Civ. Eng. Series 1, No. 9, Philadelphia.

A. A. Fungaroli, R. L. Steiner, I. Remson, and G. H. Emrich, 1968. Design of a Sanitary Landfill – Field Experiment Installation. Drexel Inst. Tech., Dept. of Civ. Eng. Series 1, No. 10, Philadelphia.

J. F. Furness, 1956. Egham Continues Wet Pit Tipping Test. The Municipal Journal, pp. 1055–1059.

J. H. Galbraith, R. E. Williams, and P. L. Siems, 1972. Migration and Leaching of Metals from Old Mine Tailings Deposits. Ground-Water, Vol. 10, No. 3, pp. 33–44.

G. A. Garland and D. C. Mosher, 1975. "Leachate Effects of Improper Land Disposal," Waste Age, March, pp. 42–48.

J. J. Geraghty, 1962. Movements of contaminants through geologic formations. Water Well Journal, Vol. XVI, No. 3, pp. 12–13, and 44–48.

J. E. Hackett, 1965. Ground-Water Contamination in an Urban Environment. Ground-Water, Vol. 3, No. 3, pp. 27–30.

D. J. Hagerty, J. L. Pavoni, and J. E. Heer, Jr., 1973. "Solid Waste Management," Van Nostrand Reinhold Environmental Engineering Series.

R. K. Ham, 1970. Large Elevation Landfilling for Refuse Disposal. Public Works, Jan. 1970, pp. 92–95.

W. M. Harrington, Jr., 1971. Sanitary Landfill Design Considerations to Protect Water Supplies. Willing Water, Vol. 15, No. 4.

S. A. Hart, 1967. Solid Wastes Management in Germany: Report of the U.S. Study Team Visit, June 25–July 8, 1967. U.S. Public Health Service Publ. No. 1812, Washington.

G. J. Hopkins and J. R. Popalisky, 1970. Influence of an Industrial Waste Landfill Operation on a Public Water Supply. Jour. WPCF, Vol. 42, No. 3, pp. 431–436.

G. M. Hughes, 1972. Hydrogeologic Considerations in the Siting and Design of Landfills. Ill. St. Geol. Survey Env. Geol. Notes No. 51.

G. M. Hughes, 1967. Selection of Refuse Disposal Sites in Northeastern Illinois. Ill. State Geol. Survey Env. Geol. Notes, No. 17.

G. M. Hughes and W. Duel, 1966. Report on the Investigation of a Landfill in Lake Co., Illinois. Ill. State Geol. Survey Open File Report.

G. M. Hughes, R. N. Farvolden, and R. A. Landon, 1967. Hydrology of Solid Waste Disposal Sites. In: Proc. Eng. Foundation Research Conference on Solid Waste Research and Development, Univ. School, Milwaukee, Wis.

G. M. Hughes, R. A. Landon, and R. N. Farvolden, 1968. Hydrogeology of Solid Waste Disposal Sites in NE Illinois. Progress Report, U.S. Dept. HEW, Public Health Service Demonstration Grant No. 50D01-UI-00006-02.

G. M. Hughes, R. A. Landon, and R. N. Farvolden, 1971. Hydrology of Solid Waste Disposal Sites in Northeastern Illinois. A Final Report on a Solid Waste Demonstration Grant Project (GO 6-EC-00006), Solid Waste Management Series SW-12d, U.S. Env. Protection Agency, Washington.

G. M. Hughes, R. A. Landon, and R. N. Farvolden, 1969. Hydrogeologic Data from Four Landfills in Northeastern Illinois. Ill. State Geol. Survey Env. Geol. Notes, No. 26.

G. M. Hughes, R. A. Landon, and R. N. Farvolden, 1971. Summary of Findings on Solid Waste Disposal Sites in Northeastern Illinois. Ill. State Geol. Survey Env. Geol. Notes, No. 45.

G. M. Hughes and J. I. Larsen, 1970. Hydrogeology and Solid Waste Disposal in the Vicinity of Belvedire, Boone Co. Ill. St. Geol. Survey Open File Report.

E. R. Kaiser, 1966. Chemical Analysis of Refuse Components. Proceedings Nat'l Incinerator Conference, ASME, New York, pp. 84–88.

K. M. Kent (retired), 1975. "A Method for Estimating Volume and Rate of Runoff in Small Watersheds," SCS-TP-149, U.S. Department of Agriculture, Soil Conservation Service.

B. T. Laksham, 1973. "Aquifer Contamination – (Causes and Effects on Potable Water)." Presented at the 25th Annual Meeting of the AWWA, Maryland.

R. A. Landon, 1968. Application of Hydrogeology to the Selection of Refuse Disposal Sites. Proc. of the 2nd Mid-Atlantic Industrial Waste Conf., Drexel Inst. Techn., Philadelphia, Pa. Also in Ground-Water, Vol. 7, No. 6.

B. E. Lane, R. R. Parizek, and G. H. Emrich, 1968. Leachate Movement in the Subsoil Beneath a Sanitary Landfill Trench Traced by Means of Suction Lysimeters. Proc. of the 2nd Mid-Atlantic Industrial Waste Conference, Drexel Inst. Techn., Philadelphia, Pa., pp. 261–277.

H. E. Le Grand, 1964a. System for Evaluation of Contamination Potential of Some Waste Disposal Sites. Jour. AWWA, Vol. 56, pp. 959–974.

H. E. Le Grand, 1964b. Management Aspects of Ground-Water Contamination. Jour. Water Poll. Control Federation, Vol. 36, pp. 1133–1145.

H. E. Le Grand, 1965. Environmental Framework of Ground-Water Contamination. Ground-Water, Vol. 3, No. 2, pp. 11–15.

H. E. Le Grand, 1968. Monitoring of Changes in Quality of Ground-Water. Ground-Water, Vol. 6, No. 3, pp. 41–18.

L. B. Leopold, 1968. "Hydrology for Urban Land Planning – A Guidebook on the Hydrologic Effects of Urban Land Use," Geological Survey Circular No. 554.

P. H. Mc Gauhey, 1968. Manmade Contamination Hazard. Ground-Water, Vol. 6, No. 3, pp. 10.

Mary McLean, 1971. "Planning for Solid Waste Management," ASPO Planning Advisory Service, Report No. 275, December.

L. L. Mink, R. E. Williams, and A. T. Wallace, 1972. Effect of Early Day Mining on Present Day Water Quality. Ground-Water, Vol. 10, No. 1, pp. 17–26.

R. C. Merz, 1969. Special Studies of a Sanitary Landfill. Final Rept. to Bur. Solid Waste Mngmt. Grant No. UI 00518-08, Univ. S. California.

R. C. Merz and R. V. Stone, 1963, 1964 and 1965. Factors Controlling Utilization of Sanitary Landfill Site. Reports to U.S. Dept. HEW, Nat'l Inst. Health, Public Health Service Projects EF-00160-03, EF-00160-04, and EF-00160-05.

R. C. Merz and R. V. Stone, 1964. Gas Production in a Sanitary Landfill. Public Works, Vol. 95, pp. 84.

R. C. Merz and R. V. Stone, 1966. Special Studies of a Sanitary Landfill. U.S. Dept. HEW, Public Health Service Grant RO1-SW00028-06. (Distributed by Nat'l Techn. Inf. Service, Springfield, Va., as PB-196148, 240 pp., 1970).

R. C. Merz and R. V. Stone, 1966. Sanitary Landfill Behavior in an Aerobic Environment. Public Works, Vol. 97, pp. 67.

R. C. Merz and R. V. Stone, 1968. Quantitative Study of the Gas Produced by Decomposing Refuse. Public Works, Vol. 99, No. 11, pp. 86.

Ministry of Housing and Local Government, 1961. Pollution of Water by Tipped Refuse. Rept. Techn. Committee on Experimental Disposal of House Refuse in Wet and Dry Pits, Her Majesty's Stationery Office, London.

National Association of Countries, 1971. "Guidelines for Local Governments on Solid Waste Management," U.S. EPA.

New York State Department of Environmental Conservation, 1971. "Sanitary Landfill," Bureau of Solid Waste Management.

R. R. Parizek and G. H. Emrich, 1967. Renovation of Pollutants from the Sanitary Landfill in Carbonate Rock Terrain. Coop. study between Dept. of Geol., Pennsylv. St. Univ. and Bur. San. Eng., Pennsylv. Dept. of Health.

R. D. Pomeroy and Westcott, 1950. Report on Probable Ground-Water Pollution from Rubbish Dumps. In: L. Angeles County Sanitation District's Report on the Collection and Disposal of Refuse, pp. 127–141.

S. R. Qasim and J. C. Burchinal, 1970a. Leaching of Pollutants from Refuse Beds. Jour. of the S.E.D. of ASCE, Vol. 90, No. SA1, pp. 49–58.

S. R. Qasim and J. C. Burchinal, 1970b. Leaching from Simulated Landfills. Jour. Water Pol. Control Fed., Vol. 42, No. 3, pp. 371–379.

Refuse Collection and Disposal for the Small Community, 1953. A joint study of the U.S. Dept. HEW and the Am. Public Health Assoc.

I. Remson, A. A. Fungaroli, and A. W. Lawrence, 1968. Water Movement in an Unsaturated Sanitary Landfill. J. San. End. Div. of ASCE, Vol. 94, No. SA2, pp. 307–317.

L. J. Ricci, 1974. "Garbage Routes to Methane," Chemical Engineering, May 27, pp. 86.

N. B. Schomaker, 1971. "Construction Techniques for Sanitary Landfills," Paper presented at the "American Association for the Advancement of Science," Philadelphia, Pa.

C. Schneider, 1953. Sanitary Fill Re-Used Safely. Am. City, Vol. 68, No. 10, pp. 83.

W. J. Schneider, 1970. Hydrologic Implications of Solid-Waste Disposal. U.S. Geological Survey Circular 601-F (Water in the Urban Environment).

R. J. Schoenberger, 1968. Chemical Characteristics of Leaching from an Incinerator – Residue Landfill. Proc. 2nd Mid-Atlantic Industrial Waste Conference, Drexel Inst. Techn., Philadelphia, Pa.

H. R. Seitz, A. T. Wallace, and R. E. Williams, 1972. Investigation of a Landfill in Granite-Loess Terrane. Ground-Water, Vol. 10, No. 4, pp. 35–41.

J. R. Sheaffer, B. von Boehm, and J. E. Hackett, 1963. Refuse Disposal Needs and Practices in NE Illinois. NE Ill. Metropolitan Area Pl. Comm. Techn. Rept. 3.

J. R. Sheaffer and A. J. Zeizel, 1966. The Water Resource in Northeastern Illinois – Planning its Use. NE Ill. Metr. Area Plan. Comm. Techn. Report 4.

T. J. Sorg and H. L. Hickman, 1968. Sanitary Landfill Facts. U.S. Dept. HEW, Public Health Service Publ. No. 1792, Cincinnati.

G. F. Sowers, 1968. "Foundation Problems in Sanitary Landfills," Journal of the Sanitary Engineering Division, Proc. of the ASCE, pp. 108.

R. L. Steiner and R. Kantz, 1968. Sanitary Landfill; A Bibliography. U.S. Dept. of HEW, Public Health Service Publ. No. 1819.

R. L. Steiner, A. A. Fungaroli, R. J. Schoenberg, and P. W. Purdom, 1971. Criteria for Sanitary Landfill Development. Public Works, Vol. 102, No. 3, pp. 77.

J. W. Stewart and R. V. Manan, 1970. Hydrologic Factors Affecting the Utilization of Land for Sanitary Landfills in Northern Hillsborough Co., Florida. Florida Bur. of Geology Map Series No. 39.

R. V. Stone, 1961. Scientific Analysis of Sanitary Landfills. Am. Public Works Assoc. Yearbook, pp. 250–261.

Symposium on Effects of Tipped Domestic Refuse on Ground-Water Quality, 1969. Water Treatment and Examination, Vol. 18, Part I.

U.S. Department of Health, Education and Welfare, 1969. "Elements of Solid Waste Management," Training Manual.

U.S. Department of HEW, 1961. Ground-Water Contamination. Proceedings of Symposium. The R. A. Taft San. Eng. Center Techn. Rept. W61-5, Cincinnati.

University of Southern California, Sanitary Engineering Research Lab., 1952a. An Analysis of Refuse Collection and Sanitary Landfill Disposal. Techn. Bull. No. 9, Series 37.

University of Southern California, Sanitary Engineering Research Lab., 1952b. Report on the Investigation of Leaching of Ash Dumps. Calif. St. Water Poll. Control Board Publ. No. 2.

University of Southern California, Sanitary Engineering Research Lab., 1952c. Waste Water Reclamation. Calif. St. Water Poll. Control Board Publ. No. 9.

University of Southern California, Sanitary Engineering Research Lab., 1954. Report on the Investigation of a Sanitary Landfill. Calif. St. Water Poll. Control Board Publ. No. 10.

University of Southern California, Sanitary Engineering Research Lab., 1955, 1956, 1958 and 1960. Reports on Continuation of an Investigation of Leaching of Rubbish Dumps. Univ. of Calif. Eng. Center Repts. 38-3, 46-3, 64-3, and 72-3, Los Angeles.

W. H. Walker, 1969. Illinois Ground-Water Pollution. J. AWWA, Vol. 61, No. 1, pp. 31.

L. Weaver, 1956. The Sanitary Landfill – Part II: Selection of Site. American City, Vol. 71, pp. 132.

L. Weaver, 1964. Refuse Disposal, Its Significance. Ground-Water, Vol. 2, No. 1, pp. 26.

L. Weaver and D. M. Keagy, 1952. The Sanitary Landfill Method of Refuse Disposal in Northern States. U.S. Dept. HEW, Public Health Service Publ. 226.

M. J. Wilcomb and H. L. Hickman, 1971. Sanitary Landfill Design, Construction and Evaluation. U.S. Env. Protection Agency, Solid Waste Mngmt. Rept. SW-88ts.

J. H. Williams, 1969. Can Ground-Water Pollution be Avoided? Ground-Water, Vol. 7, No. 2, pp. 21–23.

A. E. Zanoni, 1971. Ground Water Pollution from Sanitary Landfills and Refuse Dump Grounds – A Critical Review. Wis. Dept. of Natural Resources Research Report 69.

A. Zaporozec and D. A. Stephenson, 1972. Hydrogeologic Aspects of Solid Waste Disposal. A paper for the 1st Int'l Congress on Solid Waste Disposal and Public Cleansing, Prague, Czechoslovakia, June 1972.

A. Zaporozec, 1974. Hydrogeological Evaluation of Solid Waste Disposal in South Central Wisconsin. Technical Bulletin No. 78, Wisconsin Department of Natural Resources, Madison, Wisconsin.

INDEX

A

B

C

D

E

F

G

M

N

O

P

R

S

T

Z